The Pragmatic Turn

Toward Action-Oriented Views in Cognitive Science

Strüngmann Forum Reports

Julia Lupp, series editor

The Ernst Strüngmann Forum is made possible through the generous support of the Ernst Strüngmann Foundation, inaugurated by Dr. Andreas and Dr. Thomas Strüngmann.

This Forum was supported by the
Deutsche Forschungsgemeinschaft
(German Science Foundation)

The Pragmatic Turn

Toward Action-Oriented Views in Cognitive Science

Edited by

Andreas K. Engel, Karl J. Friston, and Danica Kragic

Program Advisory Committee:

Andreas K. Engel, Karl J. Friston, Bernhard Hommel,
Danica Kragic, Julia Lupp, Paul F. M. J. Verschure,
and Gottfried Vosgerau

The MIT Press

Cambridge, Massachusetts
London, England

This volume is the result of the 17th Ernst Strüngmann Forum,
held March 16–21, 2014, in Frankfurt am Main, Germany.

Series Editor: J. Lupp
Assistant Editor: M. Turner
Photographs: B. Fuge
Lektorat: BerlinScienceWorks

The book was set in TimesNewRoman and Arial.

Library of Congress Cataloging-in-Publication Data

Names: Engel, Andreas K., editor. | Friston, K. J. (Karl J.), editor. |
    Kragic, Danica, editor.
Title: The pragmatic turn : toward action-oriented views in cognitive science
    / Andreas K. Engel, Karl J. Friston, and Danica Kragic, eds.
Description: Cambridge, MA : MIT Press, [2016] | Series: Strüngmann forum
    reports | Includes bibliographical references and index.
Identifiers: LCCN 2015048280 | ISBN 9780262034326 (hardcover : alk. paper)
    ISBN 9780262545778 (pb)
Subjects: LCSH: Cognition. | Sensorimotor integration. | Action theory. |
    Cognitive science.
Classification: LCC BF311 .P724 2016 | DDC 153--dc23 LC record available
at http://lccn.loc.gov/2015048280

Contents

The Ernst Strüngmann Forum

Science is a highly specialized enterprise—one that enables areas of enquiry to be minutely pursued, establishes working paradigms and normative standards, and supports rigor in experimental research. Some issues, however, do not fall neatly into the purview of a single disciplinary field. Here, specialization can hinder conceptualization and limit the generation of potential problem-solving approaches. The Ernst Strüngmann Forum was created to address such topics.

Founded on the tenets of scientific independence and the inquisitive nature of the human mind, the Ernst Strüngmann Forum is dedicated to the continual expansion of knowledge. Its activities promote interdisciplinary communication on high-priority issues encountered in basic science. Through its innovative communication process, the Ernst Strüngmann Forum provides a creative environment within which experts scrutinize high-priority issues from multiple vantage points.

This process begins with the identification of themes. By nature, a theme constitutes a problem area that transcends classic disciplinary boundaries—a topic of high-priority interest that requires concentrated, multidisciplinary perusal. Proposals are received from leading scientists who are active in their field and reviewed by an independent Scientific Advisory Board. Once approved, a steering committee is convened to refine the scientific parameters of the proposal and select participants. Approximately one year later, a central gathering, or Forum, is held to which circa forty experts are invited. Expansive discourse is employed to approach the problem. Often, this necessitates reexamining long-established ideas and relinquishing conventional perspectives, yet when accomplished, new insights begin to emerge. The resultant ideas and newly gained perspectives from the entire process are communicated to the scientific community for further consideration and implementation.

Preliminary discussion for this theme began in 2012, based on an ongoing "pragmatic turn" in cognitive science: one moving away from the traditional representation-centered framework toward understanding cognition as "enactive." Although action-oriented views have been present in the individual fields of cognitive science, neuroscience, psychology, robotics, and philosophy of mind, strong links across domains were lacking. Thus, need was expressed for a collective discussion to confront different approaches and types of data, to elaborate key concepts, and to facilitate interdisciplinary interactions.

From October 25–27, 2013, the Program Advisory Committee (Andreas Engel, Karl Friston, Bernhard Hommel, Danica Kragic, Julia Lupp, Paul Verschure, and Gottfried Vosgerau) met to define the scientific framework for this Forum, which was held in Frankfurt am Main from October 26–31, 2014.

This volume communicates the synergy that emerged from a very diverse group of experts and is comprised of two types of contributions. Background information is provided on key aspects of the overall theme. These chapters, drafted before the Forum, were reviewed and subsequently revised. In addition, Chapters 4, 10, 15, and 20 report on the extensive discussions of the working groups. These chapters are not consensus documents; they transfer the essence of the multifaceted discourse, expose areas where opinions diverge, and highlight topics in need of future enquiry.

An endeavor of this kind creates its own unique group dynamics and puts demands on everyone who participates. Each invitee played an active role and embraced the process with a willingness to probe beyond that which is evident. For their efforts and commitment, I am grateful to all. A special word of thanks goes to the Program Advisory Committee, to the authors and reviewers of the background papers, as well as to the moderators of the individual working groups (Gottfried Vosgerau, Bernhard Hommel, Paul Verschure, and Tony Prescott). The rapporteurs of the working groups (Giovanni Pezzulo, James Kilner, Anil Seth, and Peter Dominey) deserve special recognition: to draft a report during the Forum and finalize it in the months thereafter is no simple matter. Most importantly, I extend my sincere appreciation to Andreas Engel, Karl Friston, and Danica Kragic: as chairpersons of this 18th Ernst Strüngmann Forum, their commitment ensured a most vibrant intellectual gathering.

A communication process of this nature relies on institutional stability and an environment that encourages free thought. The generous support of the Ernst Strüngmann Foundation, established by Dr. Andreas and Dr. Thomas Strüngmann in honor of their father, enables the Ernst Strüngmann Forum to pursue its work in the service of science. In addition, the following valuable partnerships are gratefully acknowledged: the Scientific Advisory Board, which ensures the scientific independence of the Forum; the German Science Foundation, for its supplemental financial support; and the Frankfurt Institute for Advanced Studies, which shares its intellectual setting with the Forum.

Long-held views are never easy to put aside. Yet, when this is achieved, when the edges of the unknown begin to appear and the resulting gaps in knowledge are able to be identified, the act of formulating strategies to fill such gaps becomes a most invigorating activity. On behalf of everyone involved, I hope that this volume will convey a sense of this lively exercise and further the emergence of a novel understanding of cognition.

Julia Lupp, Program Director Ernst Strüngmann Forum
Frankfurt Institute for Advanced Studies (FIAS)
Ruth-Moufang-Str. 1, 60438 Frankfurt am Main, Germany
https://esforum.de/

List of Contributors

Bar, Moshe   The Gonda Brain Research Center, Bar-Ilan University, Israel

Barsalou, Lawrence W.   Institute of Neuroscience and Psychology, University of Glasgow, Glasgow G12 8QB, U.K.

Blanke, Olaf   Laboratory of Cognitive Neuroscience, Center for Neuroprosthetics, Swiss Federal Institute of Technology (EPFL), 1202 Geneva, Switzerland

Bohg, Jeannette   Autonomous Motion Department, Max Planck Institute for Intelligent Systems, 72076 Tübingen, Germany

Butz, Martin V.   Cognitive Modeling, Department of Computer Science and Department of Psychology, Faculty of Science, University of Tübingen, 72076 Tübingen, Germany

Dominey, Peter F.   Inserm U846, Stem Cell and Brain Research Institute, Human and Robot Cognitive Systems, 69675 Bron Cedex, France

Engel, Andreas K.   Department of Neurophysiology and Pathophysiology, University Medical Center Hamburg-Eppendorf, 20246 Hamburg, Germany

Ford, Judith M.   San Francisco Veterans Administration Medical Center and University of California, San Francisco, U.S.A.

Friston, Karl J.   Wellcome Trust Centre for Neuroimaging, Institute of Neurology, UCL, London WC1N 3BG, U.K.

Frith, Chris D.   Wellcome Trust Centre for Neuroimaging at University College London, London WC1N 3BG, U.K., and Interacting Minds Centre, Aarhus University, 8000 Aarhus C, Denmark

Frith, Uta   Institute of Cognitive Neuroscience, University College London, London WC1N 3AR, U.K.

Gallagher, Shaun   Department of Philosophy, University of Memphis, Memphis, TN 38152, U.S.A. and Faculty of Law, Humanities and the Arts, University of Wollongong, Australia

Hamilton, Antonia F. de C.   Institute of Cognitive Neuroscience, University College London, London, WC1N 3AR, U.K.

Heed, Tobias   Faculty of Psychology and Human Movement Science, University of Hamburg, 20146 Hamburg, Germany

Heyes, Cecilia   All Souls College, University of Oxford, Oxford OX1 4AL, U.K.

Hill, Elisabeth   Department of Psychology, Goldsmiths College, University of London, London SE14 6NW, U.K.

Hoffmann, Matej   iCub Facility, Istituto Italiano di Tecnologia, 16163 Genoa, Italy

Hohwy, Jakob   Cognition and Philosophy Lab, Monash University, Clayton, VIC 3800, Australia

Hommel, Bernhard   Leiden University, Cognitive Psychology Unit, 2333 AK Leiden, The Netherlands

Iriki, Atsushi   Laboratory for Symbolic Cognitive Development, RIKEN Brain Science Institute, Wako-shi, Saitama 351-0198, Japan

Jacob, Pierre   CNRS Emeritus Research Director, Institut Jean Nicod, Ecole Normale Supérieure, 75005 Paris, France

Jörntell, Henrik   Neural Basis for Sensorimotor Control, Department of Experimental Medical Science, 22184 Lund, Sweden

Jost, Jürgen   Max Planck Institute for Mathematics in the Sciences, 04103 Leipzig, Germany

Kilner, James   Sobell Department of Motor Neuroscience and Movement Disorders, University College London, London WC1N 3BG, U.K.

König, Peter   Institute of Cognitive Science, University Osnabrück, 49076 Osnabrück; and Department of Neurophysiology and Pathophysiology, University Medical Center Hamburg-Eppendorf, 20246 Hamburg, Germany

Knoblich, Günther   Department of Cognitive Science, Central European University, 1051 Budapest, Hungary

Kragic, Danica   Centre for Autonomous Systems, Computational Vision and Active Perception Lab, School of Computer Science and Communication, KTH Royal Institute of Technology, 10044 Stockholm, Sweden

Kyselo, Miriam   Berlin Center for Knowledge Research, Technical University of Berlin, 10623 Berlin, Germany

Maye, Alexander   Department of Neurophysiology and Pathophysiology, University Medical Center Hamburg-Eppendorf, 20246 Hamburg, Germany

McGann, Marek   Department of Psychology, Mary Immaculate College, University of Limerick, Ireland

Menary, Richard   ARC Centre of Excellence in Cognition and its Disorders, Centre for Agency, Values, and Ethics, Department of Philosophy, Macquarie University, Sydney NSW 2109, Australia

Metzinger, Thomas   Philosophisches Seminar, Gutenberg Research College, Johannes Gutenberg-Universität, Mainz, 55099 Mainz, Germany

Morsella, Ezequiel   Department of Psychology, San Francisco State University, San Francisco, CA 94132-4168; and Department of Neurology, University of California, San Francisco, CA 94117, U.S.A.

Nagel, Saskia K.   Department of Philosophy, University of Twente, 7500 AE Enschede, The Netherlands

O'Regan, J. Kevin   Laboratoire Psychologie de la Perception – CNRS UMR 8242, Université Paris Descartes, 75006 Paris, France

Oudeyer, Pierre-Yves   Inria, 200, 33405 Talence, France

Pezzulo, Giovanni   Institute of Cognitive Sciences and Technologies, The National Research Council, 00185 Rome, Italy

Prescott, Tony J.   Department of Psychology and Sheffield Robotics, University of Sheffield, Sheffield S10 2TP, U.K.

Prinz, Wolfgang   Max Planck Institute for Human Cognitive and Brain Sciences, 04103 Leipzig, Germany

Pulvermüller, Friedemann   Brain Language Laboratory, Department of Philosophy and Humanities, Freie Universität Berlin, 14195 Berlin, Germany

Rupert, Robert D.   Department of Philosophy, University of Colorado at Boulder, UCB232, Boulder, CO 80309-0232, U.S.A.; and School of Philosophy, Psychology, and Language Sciences, University of Edinburgh, Dugald Stewart Building, Edinburgh EH8 9AD, U.K.

Sánchez-Fibla, Marti   SPECS, Universitat Pompeu Fabra, 08018 Barcelona, Spain

Schwartz, Andrew   Department of Neurobiology, University of Pittsburgh, Pittsburgh, PA 15261, U.S.A.

Seth, Anil K.   Sackler Centre for Consciousness Science and Department of Informatics, University of Sussex, Brighton BN1 9QJ, U.K.

Southgate, Victoria   Centre for Brain and Cognitive Development, School of Psychology, Birkbeck College, University of London, London WC1E 7HX, U.K.

Tramacere, Antonella   Department of Neuroscience, University of Parma, 43100 Parma, Italy

Tsotsos, John K.   Department of Electrical Engineering and Computer Science, York University, Toronto, ON M3J 1P3, Canada

Verschure, Paul F. M. J.   SPECS Laboratory of Synthetic Perceptive, Emotive and Cognitive Systems, Department of Technology, Center of Autonomous Systems and Neurorobotics, Universitat Pompeu Fabra, Barcelona, Spain and Institució Catalana de Recerca i Estudis Avançats (ICREA), Barcelona, Spain

Vigliocco, Gabriella   Division of Psychology and Language Sciences, University College London, London WC1H OAP, U.K.

Vosgerau, Gottfried   Institut für Philosophie, Heinrich-Heine-Universität Düsseldorf, 40225 Düsseldorf, Germany

1

Introduction

Where's the Action?

Andreas K. Engel, Karl J. Friston, and Danica Kragic

Abstract

Cognitive science is witnessing a pragmatic turn away from the traditional representation-centered framework of cognition toward one that focuses on understanding cognition as being "enactive." The enactive view holds that cognition does not produce models of the world but rather subserves action, as it is grounded in sensorimotor skills. The conclusions of this Ernst Strüngmann Forum suggest that strong conceptual advances are possible when cognition is framed by an action-oriented paradigm. Experimental evidence from cognitive science, neuroscience, psychology, robotics, and philosophy of mind supports this position.

This chapter provides an overview of the discourse surrounding this collaborative effort. Core topics which guided this multidisciplinary perusal are identified and challenges that emerged are highlighted. Action-oriented views from a variety of disciplines have started to cross-fertilize, thus promoting an integration of concepts and creating fertile ground for a novel theory of cognition to emerge.

Overview

Over the last two decades, a "pragmatic turn" has started to emerge in cognitive science away from the traditional representation-centered framework toward a paradigm that focuses on understanding cognition as being "enactive"; that is, cognition as a form of practice (Varela et al. 1992; Clark 1998; Noë 2004). In contrast to classical cognitivist models, an enactive view holds that cognition should not be understood as serving to make models of the world, but rather as subserving action and grounded in sensorimotor skills. Accordingly, cognitive states and their associated neural activity patterns need to be studied with respect to their functional role in action generation. Moreover, this paradigm shift stipulates new views on the functional relevance and the presumed "representational" nature of neural processes.

First evident in robotics, an action-oriented approach to cognition took longer to gain influence in cognitive psychology and neurobiology. Currently, action-oriented approaches are evolving in parallel in robotics, cognitive science, neuroscience, psychology and philosophy of mind. However, strong conceptual links across these domains have, for the most part, been lacking.

This Ernst Strüngmann Forum was convened to examine the key concepts of an emerging action-oriented view of cognition from multiple perspectives (e.g., robotics, cognitive science, neuroscience, psychology and philosophy of mind). The Forum provided the prerequisite intellectual setting in which to explore the preconditions and possible consequences of such a paradigm shift, and successfully brought together leading proponents from wide-ranging fields. Its interdisciplinary nature and open discussions enabled us to evaluate critically the different approaches and types of data. Importantly, it permitted us to search for novel and more integrated perspectives.

This book is the result of an extended dialogue between fifty colleagues and is made up of two types of contributions: articles that provide information on key aspects of action-oriented perspectives and collaborative reports of the discussion that ensued (see Pezzulo et al., Kilner et al., Seth et al., Dominey et al., this volume). As evidenced by the various chapters, action-oriented views from multiple fields have started to cross-fertilize, thus enabling conceptual integration and creating fertile ground for a novel theory of cognition to emerge. Action-oriented views are not only conceptually viable, they are supported by substantial experimental evidence. Numerous empirical findings overtly demonstrate the action-relatedness of cognitive processing or can be reinterpreted using this new framework.

In this introductory chapter, we present an overview of the "pragmatic turn" (i.e., how science has reached this point) and highlight challenges that emerged from the ensuing shift in paradigms. The core topics that guided our collaborative effort are identified to provide context, and main findings from the discussion groups are summarized.

The Pragmatic Turn

Since its formation as a discipline—intending to provide a naturalistic account of the mind and its processes—cognitive science has been dominated by a computational-representational view of cognition. The key assumptions that characterize this classical representation-centered paradigm include the following:

- Cognition is understood as computation over mental representations.
- The subject of cognition is a detached observer with a "bird's eye" view of the world.
- Intentionality is explained by the representational nature of mental states.

- The architecture of cognitive systems is conceived as being highly modular.
- Processing in subsystems is assumed to be largely context invariant.
- Computations are thought to occur in a substrate-neutral manner (functionalism).
- Models of cognition take into account only the inner states of a cognitive system (individualism).

These assumptions, which go back to the work of Fodor (1981), Newell and Simon (1972), and other protagonists of the representational theory of mind, seem to be present in most theoretical accounts of cognition, albeit with varying degrees of emphasis. Although the paradigm was highly fruitful in stimulating important research in the early decades of cognitive science, massive criticism arose, and with it claims that the classical view may be highly biased, if not misleading, in nature (Winograd and Flores 1986; Brooks 1991b; Varela et al. 1992; Dreyfus 1992; Clark 1995, 1998; O'Regan and Noë 2001; Noë 2004; Engel 2010; Engel et al. 2013).

Out of this criticism emerged the beginnings to an action-oriented paradigm (Varela et al. 1992; Clark 1998; O'Regan and Noë 2001; Noë 2004; O'Regan 2011). Initially, the paradigm shifted and was most explicitly developed in the field of robotics (Winograd and Flores 1986; Brooks 1991b; Dreyfus 1992; Pfeifer and Bongard 2006). More recently, it gained influence in cognitive psychology (Hommel et al. 2001; O'Regan and Noë 2001; Schütz-Bosbach and Prinz 2007) and neuroscience (Jeannerod 2001; Beauchamp and Martin 2007; Friston 2010; Friston, Daunizeau, et al. 2010; Pulvermüller and Fadiga 2010; Engel 2010; Engel et al. 2013).

The basic idea behind an action-oriented paradigm holds that cognition should not be understood as a capacity for deriving world models, which in turn would provide a database to support thinking, planning, and problem solving. Instead, cognitive processes are closely intertwined with action. Cognition is thus best understood as "enactive"; that is, as a form of practice itself. This enactive view (advocated by Varela et al. 1992; Clark 1998; Noë 2009; Engel 2010; O'Regan 2011; Engel et al. 2013) can be summarized as follows:

- Cognition is understood as the capacity to generate structure by action.
- The cognitive agent is immersed in its task domain.
- System states acquire meaning through their functional role in the context of action.
- The functioning of cognitive systems is thought to be inseparable from embodiment.
- A holistic view on the architecture of cognitive systems prevails, emphasizing the dynamic nature and context-sensitivity of processing.
- Models of cognition take into account the "extended" nature of cognitive systems.

The concept of action, as used here, is neither coextensive with that of behavior nor with that of movement (Mead 1938; Engel et al. 2013). Expression of action is used in a wider sense, including acts not involving any overt movements (e.g., thinking, calculating, imagining, or deciding). The description of acts or actions typically makes references to goals, whereas behavior can be described without making any reference to mental states.

An action-oriented paradigm is supported by a number of prominent and highly discussed conceptual approaches. The notion that cognition can only be understood by considering its inherent action-relatedness is a key postulate of the "enactive" approach developed by Varela, Thompson, and Rosch (1992). In their view, cognition should not be considered as producing veridical representations of the environment but rather as the capacity of generating structure by action (Varela et al. 1992). A related, strongly action-oriented view of cognition has also been advocated by Clark (1995, 1998).

Of particular relevance in this context is the sensorimotor contingency theory put forward by O'Regan and Noë (2001). This approach builds on earlier approaches to explain the fundamental role of action for perception and awareness as in, for example, Gibson's affordances (Gibson 1979). It also relates to older neurobiological concepts, such as the "reafference principle" of von Holst and Mittelstaedt (1950), who discovered that an efference copy is needed for the unambiguous interpretation of sensory signals (Wolpert and Flanagan 2001; Friston 2010; Wolpert et al. 2011). According to O'Regan and Noë (2001), an agent's sensorimotor contingencies (SMCs) are constitutive for cognitive processes. In this framework, SMCs are defined as law-like relations between movements and associated changes in sensory inputs that are produced by the agent's actions. Once acquired, an agent can use these SMCs to predict consequences of its own actions.

Recent work in cognitive robotics suggests that the learning of such predictions may mediate the acquisition of object concepts (Krüger et al. 2007; Bergström et al. 2011; Maye and Engel 2012), thus grounding knowledge of objects in repertoires of actions that can be performed on them. This theoretical perspective is also closely related to the active inference approach to action and perception (Friston and Stephan 2007; Friston 2010; Friston, Daunizeau, et al. 2010) as well as to models of predictive coding (Rao and Ballard 1999).

An action-oriented paradigm has the potential to change our view of the brain and its function profoundly. If mapped to the neuroscientific level of description, the conceptual premises of the pragmatic stance may lead to a redefinition of some of the basic explananda. Then, neuroscience would not need to explain how brains act as world-mirroring devices (Marr 1982; Churchland et al. 1994) but rather as "vehicles of world-making" (Varela et al. 1992): vehicles which support, based on individual learning history, the construction of the experienced world and the guidance of action. Data from developmental and cognitive neuroscience seem to advocate such a departure from more classical views on cognition and brain function.

Challenges and Controversies

The shift toward inherently action-oriented views of the brain and cognition has brought with it a number of challenges. In our brief discussion of these issues which follows, we highlight chapters in this volume where critical discussion is offered.

Questioning Representations

Understanding the functional role and semantics of neural states constitutes a key challenge. The insight that cognition may be fundamentally grounded in action seems to enforce a radical change in how we conceive of the functional significance of neural activity patterns. Some argue that brain states prescribe possible actions, rather than describe states of the outside world (Clark 1998). Thus, brain states might better be understood as "directives" that guide action, rather than as "representations." Such "directives" could be conceptualized as dispositions for action embodied in dynamic activity patterns (Engel et al. 2013).

Gallagher (this volume) argues that a holistic, enactive conception of cognition focuses on the rich dynamics of brain-body-environment systems; thus, it may have higher explanatory power than classical views which rely on "representation-in-the-head" models. Menary (this volume) examines this issue further, reflecting on the classical work of American pragmatism. He proposes an action-oriented view which suggests that cognitive systems do not encode representations that are then processed computationally, but rather explore and sample the environment in the service of action.

Contradicting the enactive view, Barsalou (this volume) argues that action-oriented accounts have difficulties in providing a comprehensive view of mediating processes, which are characteristic of human cognition (including conceptualization, affect or self-regulation) and may require a notion of representation. Friston (this volume) proposes the concept of probabilistic representations within the framework of predictive coding. On his account, even if reformulated in probabilistic terms, internal states in a cognitive system must still stand in for or represent external states (see also Kilner et al., this volume).

Cognitive Role of Motor Brain Structures

To what degree can the function of motor regions be understood as directly supporting cognition, as opposed to a view which assigns merely "output" functions to these circuits? For instance, attention and decision making may rely much more on motor regions than previously assumed (Rizzolatti et al. 2002; Engel et al. 2013). Imaging studies show that object concepts in semantic memory do not solely rely on sensory features but depend critically on motor properties associated with the object's use (Martin 2007; Beauchamp and Martin 2007). An intriguing finding is that motor and premotor systems, basal ganglia, and cerebellum are also active during mental simulation of events as occurs, for instance, during mental rotation of objects (Jeannerod 2001). If subjects are trained to perform functional tasks on certain objects, premotor regions become active during visual perception of these objects (Weisberg et al. 2007).

Strong support for the cognitive role of motor circuits has been provided by research on the mirror neuron system (Rizzolatti et al. 2002; Rizzolatti and Craighero 2004), which suggests that the processing of social events (e.g., observing and coordinating with the actions of other subjects) involves action-generating neural systems. Importantly, evidence shows that the mirror neuron system also includes primary motor cortex (Hatsopoulos and Suminski 2011). The observation of visual and somatosensory responses in primary motor cortex suggests that this area may also be involved in predicting future sensory consequences of actions. Similar conclusions have emerged from studies which demonstrate an involvement of motor and premotor cortex in speech perception and language comprehension (Pulvermüller and Fadiga 2010; Pulvermüller, this volume).

From the viewpoint of an integrated sensorimotor approach, does it make sense to use the classical categories of "motor" and "sensory" cortex? These cortical regions could instead be viewed as proprioceptive and exteroceptive sensorimotor areas that encode SMCs. For example, visual cortex might be considered to be the recipient of top-down predictions about the consequences of oculomotor acts.

Role of Sensorimotor Contingencies

The concept of SMCs refers to the learning and deployment of the patterns of correlation between movements and associated changes in sensory inputs that are produced by an agent's actions. Clearly, SMCs play an important role in basic sensorimotor integration, because they are necessary for an organism to distinguish self-generated sensory changes from those not related to the organism's own action (von Holst and Mittelstaedt 1950; Crapse and Sommer 2008). The importance of this principle has been well established in the context of eye movements as well as grasping or reaching movements (Crapse and Sommer 2008). A critical question is to what extent more complex cognitive functions can be achieved by learning SMCs (Maye and Engel 2012, 2013).

Similar principles of predicting sensory inputs may also play a key role in more complex cognitive processes, such as language comprehension (Pickering and Garrod 2007) or predictions about sequences of abstract stimuli (Schubotz 2007). Furthermore, prediction of sensory outcomes of actions is critical for the sense of agency; that is, the conscious experience of oneself as the initiator and executor of one's own actions (David et al. 2008). Malfunction of action-outcome contingencies have been implicated in the pathogenesis of psychiatric disorders such as schizophrenia (Frith et al. 2000b; Ford et al. 2008).

The challenging question is whether SMC knowledge is sufficient to implement complex cognitive functions, or whether higher cognitive processes based on other principles are required. Recent work in cognitive robotics suggests that SMC learning may mediate acquisition of object concepts (Bergström et al. 2011; Maye and Engel 2012; Bohg and Kragic, this volume), grounding knowledge of objects in repertoires of actions that can be performed on them (Beauchamp and Martin 2007). Maye and Engel (this volume) suggest that the concept of SMCs can be extended to accommodate more complex types of action-related contingencies. They propose that this might include object-related contingencies; that is, sets of SMCs that describe the multisensory impression an object leaves upon actions of the agent. Furthermore, intention-related contingencies might comprise the long-term correlation structure between complex action sequences and the resulting outcomes or rewards which the agent learns to predict over extended timescales. After learning, these intention-related SMCs could be used to predict whether an action will be rewarding or not, and rank alternatives. At the same time, such contingencies could provide the basis for action plans that involve several steps to reach an overall goal. Another consequence is that anticipation and anticipatory behavior as well as the sense of agency might be grounded in SMCs.

Jost (this volume) discusses the issue of SMCs with respect to principles that can serve to optimize the sensorimotor interaction of a system with its environment. One of the relevant principles is empowerment: the amount of information that an agent can inject into the environment through use of its effectors and recapture through its sensory organs (Klyubin et al. 2005). Jost emphasizes the limitations of the SMCs approach and suggests that structural priors will be required for learning in SMCs-based systems, since acquiring correlation structures is difficult for high-dimensional data sets.

Role in Development

One of the key aims of this Forum was to evaluate action-oriented concepts of cognition against a developmental background and to discuss to what extent evidence from developmental studies is able to support an action-oriented framework for cognition (see Pezzulo et al., this volume). In developmental robotics, this issue is pertinent to address issues, for example, related to how robots can achieve mastery of high-dimensional action spaces.

From a developmental perspective, Pezzulo (this volume) emphasizes that action and cognition can hardly be viewed as separate domains. Instead, pragmatic skills (e.g., the mastery of SMCs) are integral components required to develop higher cognitive capabilities. Although this basic developmental relevance is undisputed, this premise can be integrated into several accounts which differ in how they conceptualize the exact developmental role of action. Under an interactionist view, influential in the field of robotics, cognitive development is an incremental process of self-organization in which the results of a given agent-environment interaction can give rise to increasingly more complex skills and cognitive capabilities. According to a "cognitive mediation" account, sensorimotor processes involved in action control are seen as mediating, but not as being constitutive for, higher cognitive abilities. Still another account predicts that action-related mechanisms are important for the acquisition of cognitive capacities, but not for their deployment once they have been learned. Pezzulo concludes that further experimental evidence is needed to evaluate these three accounts.

Hamilton, Southgate, and Hill (this volume) discuss evidence from developmental studies in humans relevant to this issue. Studies of links between motor and cognitive systems in young children suggest that motor skills are relatively weakly linked to executive function (e.g., prediction and planning). More robust links seem to exist, however, to the development of social skills, with changes in motor skills predicting later performance in communication and social interaction.

Predictive Coding and Active Inference

A key mechanism for cognitive processing seems to be the optimization of predictions and the minimization of prediction errors (Rao and Ballard 1999). It has been suggested that new views unifying perception, cognition, and motor control may emerge from this basic principle (Friston 2010; Friston, Daunizeau, et al. 2010). Implications of this principle for the understanding of the relation between action and cognition were another key topic of the Forum (see Kilner et al., this volume). Friston (this volume) reviews the notion of active inference, in which the brain tries to infer the causes of its sensory input while sampling that input to minimize uncertainty about its inferences. This view implies that action and perception cannot be separated because both are needed to suppress prediction errors by optimizing the states and parameters in the brain's model of its exchanges with the world (Friston and Stephan 2007; Friston 2010; Friston, Daunizeau, et al. 2010).

Hohwy (this volume) explores the implications of the prediction error minimization principle for aspects of embodiment, such as the sense of body ownership and the sense of agency. He suggests that within this framework, experience of body ownership can be conceptualized as a case of perceptual inference, and that the sense of agency is not only related to error minimization in predictive forward models but also to the ability to reason counterfactually about possible actions. Menary (this volume) relates the principle of active inference to concepts of exploratory inference in classical pragmatism. He shows that the pragmatist "abductive" approach developed by Peirce (1931) already encompasses important elements of the predictive coding frameworks that emerged much later. However, he also emphasizes that "predictive processing is a subpersonal account of neural processes that fits within a larger account of the brain-body-niche nexus" (Menary, p. 228, this volume).

Action-Relatedness of Phenomenal States

In more classical accounts, conscious awareness is largely detached from action and the activation of motor circuits. An important question thus involves whether implications of the pragmatic turn also relate to current models of consciousness. The sensorimotor account (O'Regan and Noë 2001; O'Regan 2011) claims to provide a radically novel approach to consciousness and phenomenal states. Accordingly, conscious awareness is the process of exploiting the mastery of SMCs for planning, prediction, reasoning, and generating behavior (e.g., speech). For instance, being visually aware of a scene means to gear relevant sets of SMCs to "see" the scene. Thus, the sensorimotor account predicts that action plays a constitutive role in the emergence of phenomenal states (O'Regan and Noë 2001; O'Regan 2011). This view was strongly debated at the Forum (see Seth et al., this volume), with one of the points of controversy stipulating that there could be no conscious phenomenology without (potential) voluntary action.

Considering the relation between action and consciousness from the reverse perspective, Frith and Metzinger (this volume) emphasize the role of consciousness in optimizing human behaviors. In their account, conscious experience is particularly relevant for optimizing flexible social interactions as well as for the emergence of cultural phenomena, including cultural narratives.

Verschure (this volume) advocates an action-related view which also holds that consciousness is defined through repertoires of SMCs of embodied and situated agents. Verschure, however, suggests that a number of conceptual ingredients be added to the basic sensorimotor framework, including virtualization of action-effect predictions in a dedicated memory system, the assumption of intentionality priors that interpret novel states as being caused by other agents, and the notion of consciousness as an integrated sequential process. In his view, consciousness is seen as a form of memory that unifies and interprets the states of the agent to facilitate the optimization of its parallel real-time control loops that are driving action.

Joint Action and Social Cognition

Social aspects of cognitive processing have, by many accounts, been conceptualized as mainly involving "theory of mind"-type representations in the individual brain. A key question is whether this largely disembodied approach fully captures the nature of social interactions. Alternatively, a perspective aimed at grounding social cognition in joint action (including, e.g., synchronized movements) might have great potential. Ambitious questions are whether enactive approaches to social cognition might also extend to interactions between humans and robots, and what mechanisms might establish social cognition in artificial agents. These issues are addressed by Dominey et al. (this volume).

In the classical framework, which has also been termed the "spectator theory" of social cognition (Schilbach et al. 2013), the primary mode of social interaction is that of a detached observer who theorizes and produces inferences about other participants. The pragmatic turn has inspired an alternative view which holds that even complex modes of social interaction, which may be grounded in basic sensorimotor patterns, enable the dynamic coupling of agents (Di Paolo and De Jaegher 2012). A key hypothesis deriving from this view is that learning and mastery of action-effect contingencies may be critical to predict the consequences of actions from others and, thus, to enable effective coupling of agents in social contexts.

This agrees well with a model of social cognition, which predicts that shared intentionality can arise from joint action (Sebanz et al. 2006). Along similar lines, Prinz (this volume) emphasizes the role of action for alignment in social contexts (e.g., in social imitation or mirroring). According to his view, this raises the need for common coding mechanisms for action perception and action production.

Pragmatic Cognitive Science

It remains to be seen whether the conceptual shifts implied by the pragmatic turn can actually lead to the development of novel experimental paradigms and strategies. A radical action-orientedness would violate many practical constraints and theoretical premises of cognitive science as they have functioned in the past. For instance, if the representational stance is largely abandoned, a new view on the functional roles of neural states will need to be developed: rather than encoding information about pregiven objects or events in the world, neural states support the capacity of structuring situations through action. An interesting consequence of this view is that the "meaning" of neural states would eventually be determined by their functional role in the guidance of action, not by a mapping to a stimulus domain as assumed in many representationist accounts. Thus, the action-oriented view advocated here has the potential to open up a novel perspective on the grounding of neural semantics (Engel 2010; Engel et al. 2013).

The pragmatic turn may eventually bring about consequences in actual research praxis. Will the inherent conceptual shifts lead to the development of different experimental settings and paradigms, or new styles of experimentation? Clearly, research in a framework for pragmatic neuroscience requires us to avoid studying passive subjects and to use, instead, paradigms with active exploration. If neural states are individuated through their functional role in action generation, then the primary focus of experimentation should be on studying the relation of neural activity patterns to action contexts, rather than on investigating their dependence on external stimuli—a view which has dominated classical neurophysiology for decades.

Prescott and Verschure (this volume) discuss the implications of the pragmatic turn for the agenda of cognitive science with respect to a number of relevant application domains, including biomimetic approaches in robotics, enactive approaches to the development of sensory substitution devices or commercial gaming equipment, and immersive virtual reality and telepresence technologies. As they argue, these domains highlight action-oriented approaches as cases of "mode-2 science": the traditional distinction between basic and applied research will become increasingly blurred, and research will occur within transdisciplinary groupings and a stronger mix of research cultures and involve an increased diversity of actors from a broad set of stakeholders. They also emphasize that research into action-oriented cognition has real-world consequences and entails social risk. Thus such research should be performed openly and in dialogue with the wider public.

Dimensions of Discourse

To examine the key concepts of an emerging action-oriented view of cognition and the consequences of such a paradigm shift, working groups met to address:

1. The role of action in the development and acquisition of cognitive capabilities (Pezzulo et al., this volume)
2. Action-oriented models of cognitive processing (Kilner et al., this volume)
3. The relevance of action-oriented approaches to understanding consciousness and phenomenal experience (Seth et al., this volume)
4. The potential implications of a shift toward action-oriented views in cognitive science (Dominey et al., this volume)

Each group defined their own approach to their specific theme and were asked to consider, in addition, a number of overarching questions: Which methodological and theoretical principles does the pragmatic turn suggest? What are testable hypotheses that derive from those principles and which critical experiments could serve to validate these? Which empirical data presently support action-oriented models of cognitive processing? What are the limitations of action-oriented explanatory strategies? Below, we briefly summarize the main aspects of the groups' discussions.

Development, Acquisition, and Adaptation of Action-Oriented Processing

Despite decades of research, cognitive science lacks a comprehensive framework to study and explain cognitive development. The paradigm of action-based cognition implies that cognitive development itself is an active process, not a passive, automatic, and self-paced maturational process. In this context, it is important to note that "active" refers not only to sensorimotor activity but also to autonomous exploration, as present in active perception or active learning.

In their discussions, Pezzulo et al. (this volume) asked: What mechanisms are involved in acquiring action-derived cognitive processing? What synergies exist between cognition and action in development? To what extent do these synergies provide a scaffold for adaptive behavior and cognition in the mature agent? Can action be exploited to acquire higher cognitive capabilities, like mastery of abstract concepts? What are the brain mechanisms that support the learning of SMCs?

Pezzulo et al. explore how an emphasis on action affects our understanding of cognitive development and concluded that an action-based approach offers a much-needed integrative theory for cognitive development. Their report reviews multiple factors and mechanisms that influence development (e.g., sensorimotor skills; genetic, social, and cultural factors; and associated brain mechanisms), focusing on how these can be incorporated into a comprehensive action-based framework. They take the position that a research agenda for action-based cognitive development must consider how all factors are integrated, how they interact over time, and what action-oriented aspects of development explain higher cognitive abilities.

In addition, Pezzulo et al. present key challenges to such a research agenda (e.g., problems inherent in explaining higher-level cognitive abilities or in the construction of novel experimental methodologies). Emergent from their discussions is a picture of a novel field that is beginning to take form—an action-based approach to cognitive development. Still in its infancy, this field holds great promise to improve scientific understanding of cognitive development and is likely to have further, important implications for education and technology.

Action-Oriented Models of Cognitive and Functional Processing

Kilner et al. (this volume) consider action-oriented processing from a model-oriented point of view. Key questions addressed by this group included: Which cognitive functions can be grounded in sensorimotor processes? How can this be modeled and formalized exploiting, for example, predictive coding, active inference, or information theory? What is the role of action in understanding social interactions? Which neuroscientific evidence or constraints specifically speak to an action-oriented account of cognition? Does an action-oriented approach furnish novel hypotheses on neural mechanisms of cognitive processes?

In their report, Kilner et al. discuss possible relationships between action and cognition, in abstract or conceptual terms, and scrutinize models of their interrelationships as well as their role in mediating cognitively enriched behaviors. Examples of conceptual models inspired by an action-oriented paradigm are briefly surveyed, with a particular focus on ideomotor theory. Subsequently, Kilner et al. introduce formal versions of these theories, drawing on formulations in systems biology, information theory, and dynamical systems theory. An attempt is made to integrate these perspectives under the enactivist version of the Bayesian brain, namely active inference. Implications of this formalism and more generally of action-oriented views of cognition are addressed.

Kilner et al. consider issues that need to be addressed before action-oriented models can be tested. These relate, in particular, to experimental approaches to study the activation of simultaneously active neural circuits in the brain when responding to naturalistic stimuli. Necessary advances include the design of software to make and annotate naturalistic stimuli, the use of virtual reality to allow more naturalistic interaction while maintaining experimental control, the use of mobile measures of neural signals, as well as novel analytic tools and data-constrained modeling based on this data.

Action-Oriented Understanding of Consciousness and the Structure of Experience

Given the emphasis on the role of action in shaping (or constituting) perception, cognition, and consciousness, Seth et al. (this volume) examine how an action-oriented approach might alter our understanding of consciousness and the structure of experience, combining viewpoints from philosophers, neuroscientists, psychologists, and clinicians. This is an exciting area of enquiry, since most of the resurgent activity in consciousness science has focused on the neural, cognitive, and behavioral correlates of perception, independently of action. Throughout their wide-ranging discussion, Seth et al. scrutinize how actions shape consciousness and what determines consciousness of actions. They consider the specific context of self-experience, from its bodily aspects to its social expression.

Their report focuses on specific theoretical frameworks that emphasize the role of action in cognition. Four candidate frameworks are discussed which put specific emphasis on action: (a) the Bayesian brain, equipped with mechanisms of active inference (see also Friston, this volume); (b) sensorimotor contingency theory (O'Regan and Noë 2001; O'Regan 2011), (c) distributed adaptive control (see also Verschure, this volume), and (d) enactive autonomy and autopoiesis. All four frameworks converge on the notion that action shapes and structures conscious experiences in ways that extend beyond the trivial case of selecting sensory samples. Action emphasizes the openness of consciousness to extrapersonal influences. More controversial is the suggestion that emerges, in particular from SMC theory and enactive autonomy approaches; namely, that actions (possibly social actions) are constitutive of conscious experiences.

Seth et al. identify a number of potential challenges for action-oriented theories of consciousness that stem, for example, from work in patients with disorders of motor control. If action is important in shaping (or even constitutive in) conscious contents, then motor control disorders (e.g., amyotrophic lateral sclerosis, locked-in syndrome) should dramatically affect consciousness. Establishing changes in consciousness in such patients may not, however, be straightforward, and a number of relevant strategies are discussed.

Implications of Action-Oriented Paradigm Shifts in Cognitive Science

In their discussions, Dominey et al. (this volume) reviewed the status and implications of the action-oriented paradigm shift, posing questions such as: What are epistemological implications of the pragmatic turn (e.g., for our view of reality and our ways of acquiring knowledge)? What are the societal implications of action-oriented approaches (e.g., for educational programs, structuring of social processes)? What are the implications of a pragmatic turn for research programs and experimental strategies in cognitive science? What are implications for the modeling of cognitive processes and the implementation of artificial sentient systems? What are the potential implications for a better understanding of cognitive dysfunctions and the pathogenesis of neuropsychiatric disorders?

An action-oriented perspective changes the concept of an individual from a passive observer to an actively engaged agent who interacts in a closed loop with the world. Crucially, this interaction involves engaging with others and, within a landscape of cognition and action, cognition exists to serve action. Surveying this landscape, Dominey et al. address the current and potential influence that an action-oriented perspective could have on the study of cognition (including perception, social cognition, social interaction, sensorimotor entrainment, and language acquisition) as well as on neuroscience.

In addition, Dominey et al. discuss the impacts on science. They find that an action-oriented perspective has already changed the way perception, social cognition and interaction, as well as their underlying neurophysiological mechanisms, are viewed, and they note its potential to alter approaches to engineering. Further impacts include the application of enactive control principles to couple action and perception in robotics and the construction of more holistic systems design in engineering. Practical applications range from using an action-oriented approach in education to the design of therapeutic approaches in developmental and psychopathological disorders to the future development of neural prostheses. Dominey et al. conclude with a discussion of the possible societal implications that could result from the pragmatic turn.

Conclusion

The pragmatic turn has permitted a novel action-oriented framework for cognition to emerge—one that is receiving increased support from researchers trying to cope with problems not adequately solved by orthodox cognitive science. At this point in time, the pragmatic turn entails more of an agenda for the future rather than a paradigm already in place. According at least to its more radical proponents, the ultimate goal is eventually to transform the whole theory of cognition into a theory of action. Notably, this is not a behaviorist move, since the dynamics of the cognitive system lie at the very heart of the enterprise, and clear reference is made to internal states of the cognitive system. Conceptually, this view is seamlessly compatible with embodiment and "extended mind" approaches.

Concurrent with the conceptual implications of the pragmatic turn stands an increasing body of experimental evidence in support of an action-oriented framework of cognition. Will these conceptual shifts eventually lead to a different style of experimentation, to different settings and to new "laboratory habits"? An increasing number of researchers are already implementing approaches inspired by concepts of pragmatic cognitive science: from the use of natural stimuli to complex sensorimotor paradigms, massively parallel recording techniques, and less restrained subjects. Above all, the pragmatic turn inherently implies that the return of the active cognizer to the lab is a matter of practice, rather than of theory.

Acknowledgments

This work has been supported by the EU (FP7-ICT-270212, ERC-2010-AdG-269716, H2020-641321).

Development, Acquisition, and Adaptation of Action-Oriented Processing

2

The Contribution of Pragmatic Skills to Cognition and Its Development

Common Perspectives and Disagreements

Giovanni Pezzulo

Abstract

From both an evolutionary and a developmental perspective, sensorimotor skills precede "higher" cognitive abilities. According to traditional cognitive theories, abilities such as thinking, planning, and mind reading are not action dependent. Action-based theories, however, do not view action and cognition as separate domains; they argue that pragmatic skills (or a "mastery of sensorimotor contingencies") are integral components of higher cognitive abilities, essential for their development. For example, pragmatic skills might afford (active) perception, (active) learning, and the acquisition of conceptual knowledge as well as "intellectual" skills (e.g., thinking or calculating). Despite accumulating empirical support for action-based theories, it is unclear to which extent pragmatic skills contribute to cognition and its development, with contrasting proposals in the field. This chapter reviews three (not mutually exclusive) perspectives: (a) the coordinated self-organization of behavior and cognition; (b) the role of "cognitive mediators" across sensorimotor and higher cognitive domains; and (c) the action-based construction of abstract and amodal cognitive domains. Common perspectives and disagreements between these views are discussed, and open issues, opportunities for theoretical debates, and empirical tests are highlighted, all of which might contribute to a research agenda for the emerging "pragmatic" view of cognition.

Introduction

How agents—living organisms or robots—develop "higher" cognitive abilities based on existing sensorimotor skills is a topic of debate in many disciplines (e.g., cognitive science, psychology, neuroscience, philosophy, and robotics).

Contrasting proposals have emerged to describe sensorimotor and higher cognitive domains as separate, integrated, or coextensive.

Action-based theories do not regard action and cognition as separate domains. They argue that action is inherent to cognition and its development. Accordingly, the primary role of cognition is to support (and enhance) action-control abilities rather than to produce "encyclopedic" knowledge that is detached from action and perception systems. All cognitive operations usually mentioned in psychology textbooks (e.g., perception, memory, reasoning) are organized around—and ultimately functional to—the demands of action control and goal achievement:

- Perception and learning consist of picking up regularities in the information flow as structured by my actions (O'Regan and Noë 2001; Pfeifer and Scheier 1999).
- Attention is understood as selection-for-action or focusing on the regularities that are useful for the current action demands (Allport 1987; Ballard 1991).
- Conceptualization and memory consist of "encoding patterns of possible physical interaction and reenacting them in the service of the current interaction or to select the next one" (Glenberg 1997) and thus even memory is action dependent because "it is only when patterns of sensorimotor experience have been structured that I can memorize them" (Verschure et al. 2003).
- Cognitive development is "scaffolded by action development" (Byrge et al. 2014; Piaget 1954; von Hofsten 2004).
- Decision making is not independent of action processes; rather, action performance should be considered as a proper part of a decision process, not merely as a means to report the decision (Lepora and Pezzulo 2015).

Despite recent progress, it is unclear if, and how, this new "pragmatic" view extends to the domains of higher cognition.

A common starting point of action-based theories is that higher cognition is not modularized—as in the famous "cognitive sandwich metaphor" which segregates perception, cognition, and action domains—but is instead deeply integrated with, or even dependent on, action-perception systems. Support for this view comes from evolutionary arguments, which hold that the architecture of cognition can be traced back to the sensorimotor architecture of our earlier evolutionary ancestors. The brain's architecture developed to meet the needs of interactive behavior not cognitive abilities, such as playing chess or doing complex exercises in logic and mathematics; the demands of situated action control might also have somehow bootstrapped and shaped higher cognition (Cisek and Kalaska 2010; Pezzulo and Castelfranchi 2009). As a consequence, cognition is better described as a set of adaptive skills that exist *in continuity with* action-control mechanisms; they do not form a separate, modularized domain.

A similar argument might hold at a developmental timescale. A recurrent theme within action-based theories is that during development, the acquisition of pragmatic skills guides the acquisition of cognitive abilities. In other words, the driving force of development is the acquisition of practical skills (e.g., learning to control one's own actions and to achieve more complex goals; learning to predict longer-range consequences of actions), not the direct acquisition of cognitive skills: the former provides a scaffold for the latter (see Pezzulo et al., this volume). It has been demonstrated that action structures the perceptual domain (Gibson 1979; O'Regan and Noë 2001), the memory domain (Verschure et al. 2003), and the ability to exert proactive behavior, among others (von Hofsten 2004).

The theory of *sensorimotor contingencies* (SMCs) (O'Regan and Noë 2001) offers one possible explanation. Broadly speaking, learning sensorimotor skills means learning to exploit systematic relations between actions and (changes in) the world and leads to a *mastery* of SMCs. In action-based theories, which have both cognitive and enactivist flavors, such systematic relations are key to both action control and cognition (Clark 1998; Maturana and Varela 1980). Under cognitive views, systematic relations must first be internally represented (e.g., as internal generative or forward models) before they can be successively reused during overt sensorimotor interactions or covert tasks, such as imagery. According to an enactivist view, systematic relations are directly enacted to engage in sensorimotor interactions without being internally represented.

Sidestepping this dispute, skilled action control has *epistemic* effects. As clearly recognized by ecological psychologists, sensations help to determine actions but, in the process, they also create new sensations, thus creating a continuous action-perception loop (Gibson 1979). This implies that agents not only change the world by acting (a pragmatic effect), they also unveil and inject information into the world through their actions (an epistemic effect) and, in doing so, they create structure in the sensorimotor flux. The information and structure created by acting can be exploited to steer *active* forms of perception, cognition, and learning. Consider, for example, *optic flow*: only through action are the stimuli required to recognize, for example, the shape, distance, or movements of objects created. A tenet of SMC theory is that action is integral to perception and that sensory experience is an active mode of exploration (O'Regan and Noë 2001), and thus a sole property of active agents. Support for the importance of action for perception comes from studies which show that the development of the latter is impaired in animals that only experience the world passively (Held and Hein 1963). Cognitive robotic experiments have further assessed the importance of sensorimotor engagement for various "active" (i.e., action-mediated) strategies, such as active vision or active learning, where perception and learning depend on the robot's ability to select its next stimuli by acting (Pfeifer and Scheier 1999; Verschure et al. 2003).

Recognizing that actions have epistemic effects is necessary to link the domains of skills and sensorimotor control to knowledge and cognition, which are separated in traditional cognitive theories. In principle, one might argue that action is not only inherent to perception (*no action, no perception*), it is also essential for all domains of cognition, including learning, memory, concept formation, and beyond (*no action, no learning, no concepts, no cognition*). Accordingly, the domain of "action" can also be extended to include mental operations or "intellectual skills," such as *thinking or calculating* (Engel et al. 2013; Rosenbaum et al. 2001). However, mechanistic theories of the relations between sensorimotor and higher cognitive skills are at best largely incomplete, especially in terms of developmental aspects. How might "pragmatic skills" support higher cognitive abilities and/or bootstrap them during development?

Pragmatic Skills and Cognitive Development

The idea that sensorimotor development promotes cognition was pioneered by Piaget (1954). More recently, several working examples have emerged in embodied cognitive science (Barsalou 2008; Byrge et al. 2014; Thelen et al. 2001) and robotics (Nolfi and Floreano 2001; Pfeifer and Scheier 1999; Verschure et al. 2003). However, contrasting proposals exist, which I classify as follows:

1. The *emergentist* perspective emphasizes the self-organization of increasingly more complex (inter)action patterns during development.
2. *Cognitive mediation* stresses that certain abilities, developed initially for the demands of situated action, are mediators of higher cognitive abilities. For example, prediction (or other information-processing mechanisms) can be reused and adapted from the domain of action control to novel, more cognitive domains.
3. The *abstract-and-amodal* perspective emphasizes that action-based processes help develop cognitive abilities, but that once established they become autonomous from perceptual and motor systems, and are thus abstract and amodal domains.

These perspectives are not mutually exclusive and do not clearly cover the full spectrum of possible views. However, they do exemplify the potentialities of action-based approaches and point to current controversies that are ripe for debate.

The Coordinated Self-Organization of Behavior and Cognition

According to an emergentist or interactionist view, cognitive development is an incremental process of self-organization in which the learned products of a given agent-environment interaction (e.g., a grasping skill) can be exploited to acquire increasingly more complex skills (e.g., a reaching-and-grasping skill). For example, robotic experiments show that robots endowed with quite generic *fitness functions* are able to generate an incremental repertoire of behaviors:

they first acquire low-level behaviors (e.g., locomote, avoid obstacles) and are able successively to acquire higher-level behaviors (e.g., push objects) by integrating and recombining lower-level behaviors (Martius et al. 2013; Nolfi 2009; Verschure and Pfeifer 1993). By acquiring such behaviors, the robots automatically acquire some "cognitive" abilities, at least for some simple forms of cognition. An example of how this is possible is offered by an evolutionary robotics experiment: robots constituted of simple feedforward neural networks learned to "recognize" or "discriminate" objects of different shapes (a wall vs. a cylinder) by producing different behavioral patterns as they interacted with the objects (e.g., linear vs. back-and-forth trajectories) but without using internal states that "represent" the categories (Morlino et al. 2015; Nolfi 2009). In this example, the ability to categorize emerged through situated interaction without corresponding "internal states" or specialized cognitive processes (e.g., the comparison with exemplars of the category or the accumulation of evidence in favor of the perceptual alternatives), as would be more typical in cognitive modeling. This example shows that in naturalistic tasks, "categorization" does not necessarily correspond to an explicit cognitive operation but can be an intrinsic aspect of a successful interaction with objects. This might be different during psychology experiments where subjects are explicitly asked to report a category name. Indeed, further studies suggest that more sophisticated and "explicit" forms of categorization can emerge if the task demands are appropriate. For example, in another evolutionary robotics study, where several robots were able to communicate (with auditory beeps), a more discrete category-recognition ability emerged in which the robot communicated a discrete state (e.g., I am or I am not close to a cylinder) (Nolfi 2009). Other experiments have assessed whether adaptive agents can store the results of previous interactions in memory (e.g., sequences of navigation actions) and use them successively to improve their performance, thus linking pragmatic skills to memory function and planning (Verschure et al. 2003, 2014).

The emergentist idea can, in principle, be extended to most or all cognitive domains. During development, as a child grows and learns new skills, it creates (or unveils) increasingly more structure in the input through its actions (e.g., imagine an infant who learns to crawl and then to walk); in turn, it can learn this structure in the form of novel SMCs to be mastered. By acquiring new SMCs, children influence their developmental trajectory, determining a circular causality between development and cognition. Byrge et al. (2014) summarized this nicely: *changes in the action system due to growth and development change the inputs of the brain and have cascading effects on cognition and behavior in a circular process.* For a child, learning to crawl or to walk opens up entirely new opportunities for visual experiences and manipulations (as the child has many more objects within its reach), as well as for social and linguistic exchanges; these, in turn, support a range of changes in object memory, object discrimination, and view-invariant object recognition (Byrge et al. 2014). Accordingly, changes in the action system support cognitive achievements:

learning more skills implies experiencing more new patterns, and thus new opportunities to learn even more skills or to increase one's memory, decision, and social abilities. In principle, even the most advanced cognitive and social skills should be understood as emerging from brain-body-environment dynamics rather than from some form of "internalized" cognitive process, and for this they are better characterized and studied using the tools of dynamical systems (Kelso 1995; Richardson et al. 2008).

Open Issues and Current Controversies

One important consequence of the emergentist view is that development is not a process that unfolds autonomously or proceeds in predetermined stages. Instead, development depends on continuous interactions between the brain, the body, and behavior. Indeed, it is only through this continuous interaction that the necessary information is acquired to support skill learning and cognitive achievements. Furthermore, a circular causality is introduced by developmental changes in body morphology and size (e.g., growth): brain networks support cognition and action (e.g., connectivity), and the resulting behavior shapes inputs to the brain (Byrge et al. 2014). These ideas have a clear appeal but introduce a number of difficulties for current experimental approaches, given the presence of various factors that influence one another and operate over long timescales (e.g., development), and the fact that agents should be tested in conditions where they can freely explore and select their stimuli. Some of these problems can be mitigated using a "synthetic" approach: robots which develop similar abilities over time, as in developmental robotics (Lungarella et al. 2003), can be analyzed as to how they solve cognitive problems equivalent to those faced by living organisms.

How should the driving forces that guide exploration be understood? Most theories assume an initial sensorimotor exploration phase, which is increasingly considered to be systematic and goal-driven rather than a "random" process (Gottlieb et al. 2013; von Hofsten 2004). After a while, for example, a thirsty child can control its hand to reach for a glass, thus achieving its goal of drinking from the glass. This new skill then provides the basis to learn increasingly more complex skills (e.g., throwing a glass or stacking them on top of each other). This process, however, is severely underconstrained. In most engineered settings, the control problem to be learned is known in advance and therefore fixed. Here, however, it is open-ended: it changes over time due to the combined effect of changes in body morphology (due to growth) and the organism's action and sensation patterns as the organism learns new skills (e.g., crawling vs. walking). From a machine learning perspective, the presence of so many degrees of freedom creates a difficult "autonomous exploration" problem that cannot be easily tackled by existing approaches. Insights from developmental robotics and machine learning can help elucidate the importance of using "developmental tricks" (e.g., freezing some degrees of freedom before the system is fully able to exploit them) or of focusing exploration to the most informative regions or where learning progress can be observed (Baldassarre and Mirolli 2013; Oudeyer et al. 2005; Schmidhuber 1991a). Here there are ample opportunities for new experiments.

Within this perspective, there is a huge territory that has not yet been completely charted: the social scaffold of development, or in other words, how other people influence the SMCs that an agent (e.g., a child) experiences and learns. Children develop important skills and learn about important concepts through social interaction, not just through individualistic exploration. Initial social interactions are simple (one child, one caregiver) but they become increasingly more complex (e.g., societal). In terms of information theory, the actions of other agents (e.g., caregivers) structure a child's input and thus influence its learning. In some cases, caregivers use specialized *sensorimotor communication* strategies to structure a coactor's input space (e.g., maximize the information gain of children), as exemplified by child-directed speech (motherese) or gestures (motionese) (Pezzulo and Dindo 2011; Pezzulo et al. 2013b). Furthermore, coactors continuously create "social affordances," which greatly expand action and learning possibilities in the same way that tools do, where words might also be understood as "social tools" (Borghi and Binkofski 2014). Social dynamics are studied in many laboratories, but there is a trade-off between designing ecologically valid scenarios and obtaining controlled data. It is also unclear whether a social, interactive (vs. individualistic) acquisition modality of certain concepts influences their "conceptual content": concepts are based on social SMCs and emerge from patterns of interaction between two agents and two brains (including *linguistic* interactions) rather than from patterns of interaction between one agent and the world.

Finally, this perspective recognizes the role of external stimuli and affordances in shaping perception-action loops but pays less attention to the role of internally generated processes, such as *goals*. Focusing on goal-directed behavior has important consequences: a person can, for example, structure input in different ways depending on the goal in mind; this means that a person can elicit different conceptual content. For instance, I can categorize the chair in front of me as an "obstacle" or "resting place," depending on my goal (walking through the room vs. resting), meaning that affordances can be goal dependent. Furthermore, I can act intentionally to *create* or unveil new affordances that fulfill my goals: a boxer can move closer to or farther from an opponent to create a "left jab" affordance (i.e., the area from which he can hit the opponent from the left side) or a "defense" affordance (Araújo et al. 2006). Thus, even within the boundaries of an emergentist or ecological approach to conceptual processing, an individual might need to move from a stimulus- and action-centered perspective to a *goal-centered* view of cognition.

The Role of "Cognitive Mediators" (or "Neuronal Mediators")

This perspective emphasizes that sensorimotor processes involved in action control can "mediate" higher cognitive or social abilities. However, contrasting proposals exist as to what constitutes exactly the role of a (cognitive or neuronal) "mediator." In one, the role of a cognitive mediator is played by common neuronal substrates across sensorimotor and cognitive domains. In another, (pre)motor brain areas have been proposed to play several roles well beyond action control to support imagery, planning, and action understanding (Jeannerod 2006; Rizzolatti and Craighero 2004). A related proposal is that the "semantic" systems of the brain, which support most cognitive operations, are deeply based on sensorimotor circuits. Embodied theories of cognition argue that the human's conceptual system is grounded in perceptual experience and the sensorimotor patterns elicited during previous experiences and agent-environment interactions; thus the reenactment of sensorimotor experience supports (or is a mediator of) conceptual processing (Barsalou 1999, 2008). Here, importantly, knowledge is retained in modal and multimodal brain areas, in the same format as used for sensation and action; for example, verbal memory in the articulatory control system and object concepts (and the corresponding lexicon) in the same action-perception system used to interact with the objects (Martin 2007; Pulvermüller 2005). The resulting view holds that action-perception circuits might be the neural basis for higher cognitive operations, including attention, the processing of semantic knowledge, and communication (Pulvermüller et al. 2014).

Within embodied theories of cognition, *body processes* and *resources* mediate cognitive abilities and their acquisition. Useful examples are offered by the SNARC (spatial-numerical association of response codes) effect, which suggests that number coding is highly spatially organized and dependent on the agent's perspective, as well as by experiments that reveal a close link between the way subjects move their eyes and the way they solve complex problems (Grant and Spivey 2003). One explanation as to why bodily processes might be integral to cognitive operations is that they mediate the acquisition of conceptual and problem-solving abilities. Learning a category does not consist in the passive sampling of stimuli (e.g., of exemplars of a category); it has strong situated and embodied components. For example, Barsalou (1999, 2008) proposed that categorical representations might emerge when attention is focused repeatedly on the same kind of thing in the world, where the agent's embodiment, perspective, and current activity all constrain this process. Thus, developmental processes might, in principle, explain why conceptual and even linguistic domains seem to be organized around body- and action-relevant dimensions, as in the case of "actions toward the body" versus "action away from the body" (Glenberg and Kaschak 2002).

A related but different view is the notion of "neural reuse" or "recycling" across simpler to more complex domains. Here the idea involves adapting an existing information-processing mechanism to novel, more cognitive contexts. Several examples have been proposed (not all action-centered), including the reuse of (a) hippocampal resources from spatial navigation to episodic memory and mental time travel (Buzsáki and Moser 2013; Pezzulo, van der Meer et al. 2014; Buzsáki et al. 2015), (b) cerebellar internal models from movement control to problem solving (Ito 1993), and (c) parietal systems to coordinate transformation from individual to joint actions (Iriki and Taoka 2012; see also Anderson 2010; Dehaene and Cohen 2007). Furthermore, a related view holds that changes in the brain induced by sensorimotor experience (e.g., tool-use learning) create a *neural niche*—or a newly available resource in the form of extra brain tissue—that can expand an agent's abilities in unprecedented cognitive domains (Iriki and Taoka 2012).

Perhaps the most developed hypothesis of "cognitive mediators" is a *prediction-centric* view of cognition, where predictive mechanisms (e.g., generative or forward models) originally developed for motor control have been "exapted" during evolution to promote and mediate increasingly sophisticated cognitive abilities (e.g., planning, action understanding, and problem solving), all of which now form a *motor cognition* domain (Jeannerod 2006). Hence, the most important route toward "cognitive" operations is the *covert* reuse of predictive mechanisms: "action simulation" or "what-if" loops are steered, both in the future and the past, without associated overt actions (Clark and Grush 1999; Grush 2004).

Predictive mechanisms, used in both overt and covert forms, have been linked to several important cognitive abilities. Engel et al. (2013) have proposed that prediction is central to the acquisition of SMCs. For example, learning a SMC specific for a given object corresponds to determining the conditional probability of making a sensory observation given the past movements and observations. This knowledge, in turn, can be used to categorize and "ground" objects. A sponge, for instance, can be recognized in terms of the (anticipated) softness when it is squeezed (either through imagining or memory) (Pezzulo 2011; Roy 2005). Robots can ground navigation concepts using an internal simulation of possible trajectories. Hoffmann (2007) reports that the distance from obstacles is grounded and estimated by running simulations until an agent encounters the obstacle: *dead ends* are recognized through simulated obstacle avoidance, and *passages* are understood in terms of successfully terminated trajectory simulations. Quinton et al. (2013) demonstrated that pictures of animals can be recognized by learning to predict how sensory features change as a function of eye movements. In an *active inference* generalization of this view, the object or event categorization process becomes even more active because a *hypothesis testing* mechanism allows the system to run "experiments" that disambiguate among competing hypotheses. Here, a possible "experiment" consists of directing a saccade to the most informative and discriminatory parts of the environment, rather than just passively collecting "samples" of perceptual evidence. This would discriminate among multiple competing hypotheses and

allow an object or event to be recognized faster (Friston et al. 2010a), thus supporting *counterfactual* forms of reasoning (Seth 2014).

Gorniak and Roy (2007) have proposed that the meaning of words and sentences can also be grounded in sets of anticipations learned via sensorimotor
interaction. In the "semiotic schemas" framework (Roy 2005), words for perceptual features are grounded into sensory information; for example, "red" is
grounded in some (expected) values of the robot's sensors. Object words are
grounded as a result of (actual and potential) actions; for example, the meaning
of the word "cup" is grounded in the sensorimotor patterns expected by interacting with the object. Finally, in the social cognition domain, action understanding has been explained in terms of simulated action and the reuse of one's
own motor repertoire. Here, an observer agent can simulate performing several
actions (e.g., kick and push) and compare their predicted sensory effects with
the observed results of a performer agent's action. Using this mechanism, the
best-matching hypothesis disambiguates what the performer actor is currently
doing (Kilner et al. 2007; Wolpert et al. 2003).

Open Issues and Current Controversies

The most obvious implication of the predictive view is that the maturation of
predictive abilities shapes the developmental trajectory of children. This hypothesis has motivated studies with infants and children (von Hofsten 2004)
as well as others designed to investigate whether motor expertise and superior prediction abilities (e.g., in athletes) yield cognitive benefits (Aglioti et al.
2008; Pezzulo et al. 2013a). All these studies have elucidated important analogies between the neurocomputational architecture of prediction and advanced
social and cognitive skills. They have also exposed new areas of enquiry, in
particular, into whether a "simulative" mechanism might be insufficient for
social understanding and mind reading and might instead act in concert with
nonsimulative mechanisms (Frith and Frith 2008). It is unclear what role (if
any) these nonsimulative mechanisms may play in executive function (see
Hamilton et al., this volume).

Another problem is whether the "action simulation" approach can explain
declarative forms of knowledge. Whereas in traditional cognitive theories, procedural and declarative kinds of knowledge are considered to be completely
separated, the idea of simulation offers at least one possible mechanism to
reuse (in simulation) sensorimotor skills to access that part of knowledge encoded in procedural format in the internal models and make it declarative.
Consider the following example: with what finger do you press the "L" button on your keyboard? The act of imagining your use of a keyboard brings
forth a series of predictions and (depending on task demands) elicits various
types of information: the finger that is actually used to press the "L" button,
the weight and color of the keyboard, and so on. Here, the knowledge elicited
through reenactment is not confined to action control but can be used for other

cognitive operations (e.g., for reasoning or linguistic communication) in open-ended ways (Pezzulo 2011). An intriguing proposal is that the reason why this knowledge is normally not available in declarative format during overt sensorimotor engagement is that action control is too fast; this knowledge might become (consciously) available when the demands are less strict, such as during off-line imagery (Jeannerod 2006). Accordingly, the distinction between procedural and declarative knowledge would not be one of format, but rather depend on usage and temporal constraints. Thus the empirical question becomes: Which part of procedural knowledge is or is not accessible?

A related dispute is whether, and how, predictive and internal simulation abilities can support reasoning in sensorimotor as well as more "abstract" domains. It has been proposed that prediction mechanisms can be temporarily detached from the overt sensorimotor loop to be fully "internalized," supporting thought processes. For example, mechanics can assemble or disassemble an engine in their mind before doing it in practice; a climber can simulate climbing a wall before actually doing it—a form of *embodied problem solving* (Koziol et al. 2014). This ability plausibly involves the coordinated re-enactment of exteroceptive, proprioceptive, and interoceptive information as well as, in some cases, some overt body movements (e.g., eye and arm movements that one would have executed in the real situation, as well as pantomime movements; see Figure 2.1). Despite this suggestive evidence, it is not clear in which domain the idea of *thinking as internalized action* holds (Cotterill 1998; Hesslow 2002; Pezzulo and Castelfranchi 2009). Indeed, it is important to recognize that most prediction-based views focus on *forward models*; these are not "generic" predictors, because they only generate predictions that are conditionally dependent on one's own actions, and are constrained by one's own embodiment, sensorimotor system, and experience (Pezzulo et al. 2013a). Reusing forward models for higher cognition means that all cognitive operations remain in some way tied to the same constraints as sensorimotor operations. Is this, however, true for all domains of cognition, or do some domains require amodal symbolic manipulations that are detached from perception and action systems?

The Construction of Abstract and Amodal Domains of Cognition

Within this perspective, action-control mechanisms are viewed as being capable of directly mediating cognitive processing. They are seen as playing a role during the *acquisition* of higher cognitive skills but they are not as important for their deployment: once they have been acquired, cognitive skills become part of an abstract (or amodal or symbolic) cognitive domain that is separate from perception and action systems. In this abstract-and-amodal view, there is no need to use action-perception loops or "action simulations" for higher cognitive operations. Action-centered mechanisms are considered—at best—to be "facilitators," not "mediators," of higher cognition.

Figure 2.1 Previewing climbing routes as an example of an *embodied problem solving*. Each figure captures climbers in the process of previewing a (novel) climbing route before a competition. Generally, climbers mimic, imagine, and plan their future climbing movements (both overtly and covertly). In a competition, routes are unknown to the athletes: "route setters" ensure that they include nontrivial sequences of movements. Hence it is important for an athlete to plan in advance how to approach the route. This form of problem solving is embodied in the sense that it requires consideration of the athlete's embodied knowledge (e.g., limb length, finger strength, potential opportunities provided by the various kinds of climbing holds, possible or impossible kinematics). The best climbers are able to anticipate a lot of information (e.g., proprioceptive information, body posture at critical points, how much force to use) and make important decisions (e.g., where to rest) before they begin to climb. Of course, these decisions are subject to revision once they are underway. Pezzulo et al. (2010) reported an advantage of expert climbers in a memory task (i.e., remembering sequences of holds in a route), but only when the climbing route had the right affordances (i.e., when it was "climbable," not when the sequences formed a non-climbable route). Photos reproduced with permission from Luca Parisse (risk4sport.com).

The empirical evidence for or against a separate, amodal domain of higher cognition is at the heart of a debate between proponents and opponents of an embodied view of cognition (Barsalou 2008; Caramazza et al. 2014). From a developmental perspective, a key issue involves which (brain and computational) mechanism might produce "disembodiment" or a complete "detachment" of cognitive processing from sensorimotor experience. In one of the earlier computational implementations of Piaget's ideas, Drescher (1991) proposed a "constructivist" schema mechanism that incrementally constructs new components on top of sensorimotor experience. Here, an artificial agent is initially endowed with simple sensorimotor schemas (essentially,

context-action-prediction triplets) that permit it to interact with a simple artificial environment. However, the goal of the schema architecture is to build
increasingly more complex schemas that include nonperceptual elements
and conceptual knowledge. A specific example can help to clarify this point.
Initially, the first elements of schema triplets (context) include only sensor
measurements but successively the agent interactively enlarges its "ontology"
by learning new so-called synthetic items (i.e., common causes from a set of
related interactions involving action-outcome effects). This schema learning
mechanism essentially discovers nonperceptual states that make schema predictions more accurate, thus increasing the agent's knowledge (e.g., the concept of a "cup" along with a schema that encodes a new regularity: "if a cup
is in front of me and I grasp it, it will be in my hand"). These new schemas, in
turn, can be used as a new starting point to develop (at least in principle) an
open-ended repertoire of skills, including more abstract actions (e.g., counting
actions that permit counting the cups, or naming actions that permit linguistical
referencing). Although this mechanism might seem similar to the predictive
view, the synthetic objects created through this learning process need not have
sensorimotor components; in other words, they are amodal, not modal or multimodal, states. Once acquired, the new schemas create new abstract, symbolic
domains. In the examples above, this would consist of a numerical domain and
a linguistic domain that are segregated from action and perception systems,
which is at odds with the aforementioned embodied views of cognition.

Open Issues and Current Controversies

Are domains of abstract cognition (e.g., numerical or linguistic cognition) multimodal (in keeping with embodied theories of cognition) or are they amodal
(as described above)? A third possibility is that both conditions exist. It has
been argued (Pezzulo and Castelfranchi 2009) that it is possible to compose
music in both "Mozartian" (i.e., by hearing or rehearsing auditory information)
and "Bachian" (i.e., by using only the symbolic music notation—a new code
which has its own rules) ways. This "modality" debate is central in contemporary cognitive science, and numerous controversies exist (Barsalou 1999,
2008; Caramazza et al. 2014; Martin 2007; for discussion of language, action semantics, and the modal versus amodal debate, see Pulvermüller, this
volume).

This debate has implications for the design of cognitive models and autonomous agent architectures. To date, we have not yet resolved how to devise an architecture that can incrementally build up new skills and knowledge
and extend its abilities from simpler to increasingly more demanding cognitive domains. Should such an architecture have separate modules for sensorimotor skills and higher cognition (Anderson 1983), or should the latter
be based on (roughly) the resources that were used in the former (Pezzulo et
al. 2011; Pezzulo, Verschure et al. 2014)? The goal of achieving *open-ended*

development was present in the early "schema mechanism" architecture discussed above. If we adapt (roughly) the same concepts to the modern language of generative models and hierarchical architectures for active inference (Friston et al. 2010a; Friston et al. 2015; Pezzulo et al. 2015), synthetic items and schemas would be latent states and models at high hierarchical levels, which encode the "hidden" (i.e., not directly perceivable) causes of perceptions and actions (see Friston, this volume). These hidden states are not necessarily tied to a specific modality, but can generate predictions in multiple modalities—proprioceptive, exteroceptive, and interoceptive—which in turn directly engage action, perception, and emotion processes (Adams et al. 2013a; Clark 2013b; Pezzulo 2014; Seth 2015; Stoianov et al. 2016). During development, new hidden nodes and models are learned that encode regularities (and permit prediction and control) at longer timescales and in different domains (e.g., social domains), thus expanding the scope and potential for control of the agent, and—at least in principle—supporting cognitively demanding cognitive operations such as counting or reading.

In practice, the potential for this architectural scheme (or others) to mimic cognitive development and extend to the domains of higher cognition remains to be fully explored (for further discussion, see Pezzulo 2011). Furthermore, it is still unclear what resources (e.g., representational) would be required for such an architecture. It has been argued that children face very challenging "structure learning" problems during their development, for which structured and symbolic representations might help (Tenenbaum et al. 2011). However, even in the cognitive modeling literature, it is unclear whether the development of abstract cognitive abilities is based on modal (e.g., interoceptive) processes or amodal and symbolic states, and whether these are innate or can be learned (König and Krüger 2006).

Conclusions

The novel "pragmatic" view of cognition offers a way to change how we conceptualize living organisms, their brains, and their behavior. Since the field is young, controversies exist and important elements of action-based theories await in-depth investigation.

Most action-based theories agree that the acquisition of cognitive abilities is guided, during development, by the acquisition of pragmatic skills or a mastery of SMCs. However, there are contrasting proposals as to how exactly pragmatic skills contribute to cognitive development. These proposals can be grouped into three (not exclusive) perspectives:

1.   How pragmatic skills contribute to development is tied to the potential of enabling new SMCs in an incremental process of self-organization of both brain networks and behavior. This view tends to assume that

cognitive abilities (e.g., categorization or problem solving) are not discrete operations but rather emerge during agent-environment interactions, without the necessity of representation or "internalization."

2. The key to cognitive development is the presence of "cognitive mediators" (e.g., a core set of prediction abilities that are reused across action control and higher cognition). A popular example is the idea that cognition can be seen as an *internalized* form of action, where the same mechanisms mediate both. This view tends to "cognitivize" action control in the same way it makes cognition action-based; it emphasizes that intentional action has a complex structure and is mediated by sophisticated mechanisms like internal models.

3. Action-control mechanisms contribute to the development of abstract and amodal domains of cognition. Once established, these domains, however, do not depend on action-perception brain systems for deployment. Thus action-control mechanisms are "facilitators" of cognitive development.

This taxonomy points to numerous open issues: How should we understand and study the circular causality between development and cognition? What (if any) are the most important "cognitive mediators"? Do amodal domains of cognition exist and, if so, how are they developed? In other words, are we "Mozartian" or "Bachian" or both?

From a more epistemological perspective, other questions include: Can prediction-based mechanisms "detach" from the overt sensorimotor loop, and would this count as a "representational" role? A related set of questions involves the neuronal implementation of the proposed architectural schemes: Does the brain use symbols to mediate thought processes? Is there evidence of a causal role of sensorimotor representations in higher cognition? Is there evidence of truly amodal brain representations and, if so, how can we recognize them? Are these innate or do they form (and if so, how) during cognitive development?

We must continue to assess the merits of existing action-based theories by designing novel experiments that test emergentist, modal, and amodal views of higher cognition, and by realizing robots that embody these views. Developing more advanced theoretical proposals must be an important objective within the agenda for a "pragmatic" cognition.

3

The Development of Action Cognition

Antonia F. de C. Hamilton, Victoria Southgate,
and Elisabeth Hill

Abstract

Humans learn motor skills over an extended period of time, in parallel with many other
cognitive changes. The ways in which action cognition develops and links to social and
executive cognition are under investigation. Recent literature is reviewed which finds
evidence that infants advance from chaotic movement to adult-like patterns in the first
two or three years of life, and that their motor performance continues to improve and
develop into the teenage years. Studies of links between motor and cognitive systems
suggest that motor skill is weakly linked to executive function and more robustly predicts social skill. Few, if any, models account directly for these patterns of results, so the
different categories of models available are described.

Introduction

Humans are born with very limited motor skills and yet, over an extended
period, develop into independent individuals capable of the precise control of
skilled actions. Throughout this process, an increasing ability to control actions may also contribute to the development of other cognitive faculties such
as language, executive control, and social interaction. Thus, action cognition
concerns two topics: how the motor control system actually works, and how
motor control relates to other cognitive processes. We begin by reviewing adult
models of the motor system because it is useful to understand the end point of a
developmental process. Thereafter we discuss the developmental changes that
occur on the way to that end point, both within the motor system and in links
between motor and other cognitive systems. We conclude with a consideration
of some of the different theoretical frameworks that have been put forward to
account for the *development* of action cognition.

Control of Human Movement

The task of reaching out to pick up a toothbrush, applying toothpaste, and brushing one's teeth may seem trivial to the adult who does this daily without much thought. However, learning this skill is not simple. Children develop some motor skills rapidly after birth, but many take years before expertise is fully achieved. To control a highly nonlinear and redundant system of muscles and bones in an efficient fashion, their motor systems must contend with signaling delays and sensorimotor noise (Franklin and Wolpert 2011). Despite this complexity, a preschool child can easily surpass the visuomotor skill of the best robots available today. The study of action cognition is the study of the information-processing systems that underpin motor abilities. While it draws heavily on basic computational motor control, action cognition[1] includes more abstract processes, such as motor planning and motor sequencing, which are sometimes studied in relation to executive control. It is important to explore the relationship between action cognition and other cognitive systems, in particular executive function and social cognition.

A large number of different models have been proposed to understand sensorimotor control in typical adults, and from these, two major categories of model emerge. Computational models describe human movement in terms of optimal feedback control (Todorov and Jordan 2002) and forward/inverse models (Kawato and Wolpert 1998), considering in detail the type of engineering required to control the human motor system. An alternative approach simplifies control to the idea of an equilibrium point and suggests that the spring-like properties of the musculature can be adjusted to move the equilibrium of the arm (Feldman et al. 1998). A similar principle is found in the more recent active inference model (Friston et al. 2011). However, a key difference between these two classes of models concerns whether prediction is separate to or fully integrated with control (Pickering and Clark 2014). We will draw primarily from the former class of models, because they have been tested in more detail in developmental populations.

To summarize current knowledge about the development of motor systems and action cognition, we will refer to recent studies in this area. The vast majority of published papers on motor development focus on clinically relevant behaviors (e.g., walking, writing) without regard to the underlying cognition. Here our discussion focuses on performance of specific motor tasks that link closely to particular computational components of motor control. We distinguish between multisensory integration, visuomotor mapping, forward models, motor planning, and action comprehension, asking how each develops from infancy to adulthood.

---

[1] The term "cognition" is used as a synonym for "information processing." It does not imply a particular symbolic form of representation or a contrast to "affective" information but rather refers simply to any neural patterns of information between sensory input and muscle activation.

Multisensory Integration

The human brain has many sources of sensory information which allow it to determine the current state of the world (e.g., visual, tactile, proprioceptive, and auditory input channels). A single physical event often impacts on many channels at once, and thus detecting congruency between different input channels and integrating inputs is helpful in building an accurate model of the state of the world. In adults, these different sensory information sources are integrated in a Bayes-optimal fashion (Ernst and Banks 2002). However, it is not yet clear how infants and children learn to integrate different senses. At a very young age, infants are sensitive to contingencies between different sensory modalities. An early study demonstrated that five-month-old children prefer to view a video of their own leg movements than a video of time-delayed leg movements (Bahrick and Watson 1985). However, this study did not distinguish which modalities (visual, tactile, proprioceptive, motor) are focused on by the infants.

Studies of visual and tactile integration suggest that this pairing is important from a very early age. Newborns (12–103 hrs old) prefer to view a face that is touched in sync with a touch to the infant's own face, than a face which is touched out of sync (Filippetti et al. 2013). This suggests they are able to detect synchrony between a face touch and a visual event. Similarly, seven-month-old infants prefer viewing a leg touched in sync with touches to their own leg (Zmyj et al. 2011), and the strength of this effect increases from seven to ten months. The integration of postural, visual, and tactile information also improves over the six- to ten-month age range, as shown by changes to somatosensory-evoked potentials in infants (Rigato et al. 2014). In older children (six-year-olds), visual-tactile integration can be similar to adults but is not mandatory, whereas adults cannot avoid integrating cues (Jovanovic and Drewing 2014). Visual-haptic cues for size discrimination are also not integrated in children before eight years of age (Gori et al. 2008). Similar results have been found for integration of different visual cues to depth (Nardini et al. 2010).

Visual-tactile-proprioceptive integration has also been examined using the rubber hand illusion (Botvinick and Cohen 1998). In older children (four- to seven-year-olds), the illusion is present and its magnitude remains constant over this age range (Cowie et al. 2013). However, the same children showed larger errors than adults in pointing to the true location of their hand, suggesting that visual-proprioceptive integration has not yet matured in this group. Further evidence of late maturation of visual-proprioceptive integration was found in a study of 7- to 13-year-old children in a hand localization task (King et al. 2010). Younger children in the sample were more reliant on visual information, which resulted in larger proprioceptive errors. A further study showed that noisy proprioceptive information could account for worse motor performance in six-year-olds compared to 12-year-olds (King et al. 2012). Together, these studies suggest very early sensitivity to sensorimotor congruency, together with very protracted development of the ability to integrate the senses.

Visuomotor Mapping

To obtain accurate control of hand actions, an infant needs more than multisensory integration; it must link motor commands and sensory input. This process is central to action cognition and has been studied from a variety of perspectives. A substantial number of studies that recorded from single neurons in monkeys found an occipito-parietal premotor pathway with a core role in transforming visual signals to motor actions (Cisek and Kalaska 2010). Within this pathway, mirror neurons are active when participants perform and observe actions (Rizzolatti and Sinigaglia 2010). These neurons might provide a basic social mechanism for understanding other people (Gallese et al. 2004), but alternative interpretations are also available (Hamilton 2013a; Hickok and Hauser 2010). In cognitive terms, the link between visual and motor systems has been explored in the associative sequence learning model (Heyes 2001). Central to all these models is the idea that visual information (about objects in the world and the hand) must be mapped to motor information about the actions that the hand is performing. Specifically, the infant must learn to link the retinal image of a skin-colored moving shape to the motor outputs it sends to its own hand and arm muscles. This must involve transforming the retinal information into other, intermediate representations (e.g., visual primitives, kinematics, action goals, motor primitives). Such coordinate transforms have been studied in detail for spatial tasks (Andersen et al. 1997) but have been less often considered in studies of action cognition. In particular, it is not yet known what types of intermediate representation are required for action cognition or how these can best be studied. Nevertheless, it is clear that infants can learn and use visuomotor mappings. For example, the more opportunity infants have to acquire a visuomotor mapping for leg actions (via live video feed of their own legs), the more active their motor system becomes when they observe leg actions (de Klerk et al. 2014). Thus, forming early links between visual images and motor systems is critical to the developing motor system and may also contribute to social cognition.

There is evidence that building robust visual motor mappings is important for infants, even from the earliest days of life. Neonates (10–24 days old) will move their arm to keep it within a beam of light where they can see it (van der Meer 1997). Around four months of age, infants begin to make reaching movements toward objects, but their hand trajectories do not follow a smooth, adult-like path until at least three years of age, after which it continues to improve (Konczak and Dichgans 1997). The development of grasping is also prolonged. For example, an adult will typically use a large grip aperture when reaching for an orange but a small grip aperture when reaching for a grape. Infants reaching for objects of different sizes always use the same grip aperture at six months, but begin to scale their grip to the object by 13 months (von Hofsten and Ronnqvist 1988). One recent study suggests that four- to eight-month-old infants grasp as if their eyes were shut, relying on haptic cues; they only develop visual control of grasp by 24 months (Karl and Whishaw 2014). A study of four- to 12-year-olds reaching and grasping for objects showed clear improvements in trajectory and smoothness over this age range, with only the oldest children showing adult-like patterns of grip aperture scaling when reaching in the dark (Kuhtz-Buschbeck et al. 1998).

Another way to examine visuomotor mappings is to change these mappings, by asking a participant to make movements but giving false feedback about the location of the hand. Adults can adapt when a rotation of 45° is applied to the visual feedback given as the participant makes center-out movements, and then show aftereffects in the opposite direction when the false feedback is removed (Krakauer et al. 1999). The same method has been used to examine visuomotor transformation in four- to ten-year-old children. Results show that these children adapt to the new feedback like adults; however, the younger children showed smaller aftereffects when the false feedback was removed (Contreras-Vidal et al. 2005). This implies that younger children may have a broader tuning function in their visuomotor mapping than older children and adults. Overall, these studies show similarities to the studies of multisensory integration: early disorganized movements take on a recognizable pattern over the first year of life, and refinement of these movements continues for over a decade as the child gradually acquires adult levels of performance.

Prediction and Planning

A major challenge in motor control is the inherent delays in the visuomotor system. Sending a signal from motor cortex to the muscle takes around 20–30 msec (Matthews 1991), with another 25 msec required to translate that signal into a change in muscle force (Ito et al. 2004). If a visual input is required, delay in retinal and early visual systems must also be considered, giving rise to a delay in involuntary visual responses of around 110–150 msec (Day and Lyon 2000). Forward models or predictors can be used to circumvent these delays; that is, a copy of the outgoing motor command (efference copy) is used to predict what the sensory consequences of an action should be, and the predicted consequences are compared with the actual consequences (Davidson and Wolpert 2003; Miall and Wolpert 1996). One of the clearest examples of the use of forward models in the motor system can be seen in the programming of grip force. If you need to pick a raspberry from a bush without crushing it, it is important to grip inward and pull the fruit away from the bush with just the right force and timing. Studies in adults have revealed that grip force (the force inward between the finger and thumb) is closely correlated to load force (the upward force against gravity) when an object is lifted (Johansson and Cole 1992). This can best be explained by the use of a forward model in which the motor command to increase load force is also used to generate a prediction of the required grip force, so that grip and load can be controlled in parallel (Davidson and Wolpert 2003).

Studies of the development of grip force and load force over childhood demonstrate a very protracted developmental trajectory. The correlation between grip force and load force increases gradually over two to eight years of age (Forssberg et al. 1992) and continues to improve up to 14 years of age (Bleyenheuft and Thonnard 2010). Grip force dexterity in the more complex task of compressing a spring also improves over the four- to 16-year age range (Dayanidhi et al. 2013). A different way to measure predictive processes is to examine stability to unloading. In these tasks, a participant holds a heavy object in his/her left hand, thus requiring activation of muscles in the left arm to hold the object stable. In different trials, either an experimenter lifts the object from the participant's left hand (other-lift) or the participant lifts the object with their right hand (self-lift). In a self-lift, a participant is normally able to relax the left arm at just the same time as the object lifts, thus holding the left hand stable. In contrast, the timing of the other-lift cannot be predicted and so muscle activation in the left hand remains high for longer, with the left hand moving upward as the object is lifted. Performance on this task improves substantially from 4–16 years of age, but even 16-year-olds do not demonstrate the same level of performance as adults (Barlaam et al. 2012; Schmitz et al. 2002).

While predictive processes in motor control are helpful on a very short timescale (hundreds of milliseconds), planning processes can help performance on a longer timescale. One aspect of longer-term planning is seen in chaining tasks, where the kinematics of an action differs according to the next action performed. For example, a grasp followed by a throw has different kinematics during grasping compared to a grasp followed by a placing action (Becchio et al. 2012; Johnson-Frey et al. 2004). This is true for school-age children (Cattaneo et al. 2007; Fabbri-Destro et al. 2009) as well as ten-month-old infants (Claxton et al. 2003); however, detailed developmental studies have not been performed. More is known about planning based on end-state comfort. For example, when lifting a bar to place it in a particular location, adults will often begin the action with an awkward posture so as to end their action in a comfortable posture (Rosenbaum et al. 1990). The effect of end-state comfort provides a measure of action planning, and performance improves from three to ten years of age (Jongbloed-Pereboom et al. 2013; Stöckel et al. 2012; Weigelt and Schack 2010). Overall, these studies illustrate the very gradual development of predictive and planning abilities in the motor system, with changes in performance continuing up to 16 years of age.

Comprehension of Other People's Actions

An important component of action cognition is social (i.e., the ability to understand what another person is doing now and intends to do next when viewing their actions). Many studies have examined how this process develops in infancy and how it relates to other skills. From a young age, infants are able to interpret others' actions as movements directed toward goals and they use a variety of cues to identify a goal-directed action (Hernik and Southgate 2012). The dominant theory in this area suggests that if mirror neurons are central to action understanding (Gallese et al. 2004), then an infant's ability to understand an observed action should be dependent on their ability to perform that action. Data in support of this show links between performance and comprehension of actions. For example, three-month-old infants are not yet able to reach and grasp objects themselves and do not appear to understand actions as goal directed (Sommerville et al. 2005). However, if three-month-olds are provided with experience of grasping for objects by wearing Velcro gloves, which help them to pick up objects in their vicinity, they subsequently evidence an understanding that an observed reach and grasp action is goal directed (Skerry et al. 2013; Sommerville et al. 2005). Numerous other studies also demonstrate a relationship between developing action skill and various measures of action understanding (Cannon and Woodward 2012; Kanakogi and Itakura 2011).

One difficulty with these studies is that it is not always clear what it means for an infant (or an adult) to understand an action. Is it enough to predict what is next, or is a more elaborate representation of intention required? There is evidence that motor and mirror systems have a role in the former (Southgate et al. 2009, 2010, 2014). However, it seems that infants may also recruit their motor system during the prediction of others' actions that are outside of their own motor repertoire (Southgate and Begus 2013; Southgate et al. 2008). Thus, while infants' own motor skill does appear to influence their action understanding, the mechanisms mediating this relationship are unclear. There are also several reasons to believe that, at least in adults, intention understanding requires more than just motor prediction (Csibra 2007; Spunt et al. 2010). The relationship between motor and social cognition will be discussed in more detail in the next section.

Summary

Action cognition encompasses a variety of skills and computational components which must work together to allow coordinated and efficient action. Developmental studies suggest that infants rapidly learn motor skills in the first year of life, moving from helplessness to a state with some basic control systems in place. However, the acquisition of full motor skill remains very protracted. Even everyday skills, such as grasping objects, draw on a complex system for visuomotor transformation and predictive control, and performance continues to develop and improve into adolescence. Developmental trajectories for motor skills are likely to be nonlinear—with progress in an area followed by stagnation, followed by more progress—and different motor skills do not develop in synchrony, even in children.

How Does Action Cognition Relate to Other Types of Development?

Learning a new motor skill can change how a child engages with the world as well as how the world engages with a child. For example, an infant who can grasp an object might now perceive the potential of a cup for grasping in a way that a younger child might not. The grasping infant may also receive different social inputs from adults, who might place objects within reach (or remove them) and talk about the objects differently. Thus, learning a new motor skill has the potential to impact both a child's cognitive development and social development. Here we review work in this area to trace how different cognitive skills might be linked.

Intellectual and Executive Development

There are many reasons to believe that motor and intellectual skills are connected. In longitudinal studies, motor skill has been linked to later motor, social, physical, and mental health outcomes as well as to academic achievement (Bart et al. 2007; Ekornås et al. 2010; Emck et al. 2011). Several studies have focused particularly on executive function—a broad term used to describe skills, including working memory, inhibition, and cognitive flexibility (Diamond 2013), related to measures of intelligence. Some aspects of executive control can be seen even in infancy (Johnson 2012) but the development of these skills continues into adulthood (Best and Miller 2010). There is mixed evidence for a relationship between executive function skills and motor skills. Five-year-olds with motor difficulties show differences in executive function measured one year later (Michel et al. 2011). In a study of 100 typical seven-year-olds, only some correlations between executive function and motor performance were reliable (Roebers and Kauer 2009). Reliable but small correlations were also reported in a study of motor skill and intelligence in 250 children (Jenni et al. 2013). However, other studies report positive associations. For example, throwing and catching skills correlated in particular with IQ (Rigoli et al. 2012a), an effect that might be mediated by working memory (Rigoli et al. 2012b). Other work suggests that links between motor and cognitive skills might be mediated by visual performance (Davis et al. 2011).

Some theories claim strong links between motor skill, cognitive skill, and the development of the cerebellum (Diamond 2000; Wang et al. 2014). For example, in children with cerebellar tumors, there is a correlation between motor and cognitive skills (Davis et al. 2010), and cerebellar function has been linked to autism (Wang et al. 2014); direct evidence in typical children, however, is limited. Developmental coordination disorder (DCD) can also be examined as a test case. A study by van Swieten et al. (2010) demonstrated developmentally inappropriate *motor* planning in six- to 13-year-old children with DCD, but appropriate *executive* planning (using a Tower of London task) in seven- to 11-year-olds in this group. Pratt et al. (2014) identified, however, significant difficulties with both types of planning in a different group of children with DCD, compared by age and IQ to typically developing children. Leonard and Hill (2014) show that children with DCD performed worse than typically developing controls on *nonverbal* measures of working memory, inhibition, planning, and fluency, but not on tests of switching or verbal equivalents of the same tasks. Overall, these studies give mixed support to the claim that motor cognition and executive function are linked. There may be weak correlations between these cognitive systems, but the association is not a tight one.

Social Development

Much better evidence is available to suggest that motor skill contributes to social development. During infancy, motor development can affect how infants interact with individuals around them. For example, as infants improve in manipulating objects, they also show altered patterns of attention to others in the environment (Libertus and Needham 2010). The onset of crawling and walking is linked to greater joint attention and social referencing, perhaps because of the altered type and number of interactions the young child is then able to have with its caregivers (Campos et al. 2000; Karasik et al. 2011; Leonard and Hill 2014). Specifically, the ability to move around and explore the environment as well as manipulating objects and sharing them with others provides more opportunities to engage in joint attention and changes the types of vocalizations and expressions the infant receives from the caregiver. Evidence for relationships between motor development, language, and social communication skills can be seen from the outset in typical development through the tight coupling of motor and language milestones throughout infancy (Iverson 2010). There is also a feedback loop between produced and heard speech: children with autism produce less speech and, in return, receive less speech input from their caregivers (Warlaumont et al. 2014). Thus, motor skills can directly influence the child's social environment and opportunities to develop social skills.

Many studies of links between action cognition and other types of cognition have examined children with developmental disorders, in particular autism spectrum conditions diagnosed on the basis of poor social cognition. As many as 80% of children with autism have substantial motor difficulties (Green et al. 2009), and interest in the links between autism and motor cognition is increasing (Fournier et al. 2010; Gowen and Hamilton 2012). In particular, there is evidence for dyspraxia (poor performance of skilled hand actions) in autism beyond other possible motor impairments (MacNeil and Mostofsky 2012; Mostofsky et al. 2006). Infant siblings at risk of developing autism demonstrated differences in standardized motor tasks and face-processing tasks (Leonard et al. 2013). In children at high risk of developing autism, greater autism symptoms were also seen in those with poorer motor skills (Bhat et al. 2012; Leonard and Hill 2014).

Despite clear links between overall levels of motor and social skills, it is harder to identify specific differences in cognitive systems, and here results are more variable. Children with autism show poor performance in specific tasks involving motor planning in some studies (Hughes 1996) but not in others (Hamilton et al. 2007; van Swieten et al. 2010). Some studies report difficulties in chaining actions together in sequences (Cattaneo et al. 2007), but others do not (Pascolo and Cattarinussi 2012). Detailed testing of visuomotor adaptation in children with autism did not find group differences (Gidley Larson et al. 2008). Similar variability is found in studies of how children with autism understand other people's actions—a social component of motor cognition. Some studies report difficulties in answering questions about why a person performed an action (Boria et al. 2009) or in predicting what will come next in a movie (Zalla et al. 2010). Other studies, however, find no differences in the ability to make sense of hand gestures (Hamilton et al. 2007). Studies of imitation show intact performance on emulation tasks (copying the goal of an action) but poor performance on mimicry tasks (copying precise kinematic features) (Edwards 2014; Hamilton 2008). Some of these differences may be explicable in terms of links to executive function or top-down control (Wang and Hamilton 2012). Overall, there is no single aspect of motor cognition that can be directly linked to poor social cognition. More research is needed to understand how motor and social developmental processes interact.

Summary

Overall, data show reliable but small links between motor cognition and executive function, and larger more robust links between motor cognition and social cognition. In particular, changes in motor skill seem to drive changes in the child's social environment and predict later performance in situations involving communication and interaction. However, it is less clear what specific cognitive processes drive these effects. To assign motor-social links to a single brain system (such as the mirror neuron system) is probably premature (Hamilton 2013b). Instead, it will be important to consider how different systems *interact* in development, and how the acquisition of one skill gives the child more opportunities to learn other skills, in a complex interplay between the child and the social environment.

Theories for Understanding the Development of Action Cognition

There are many different theoretical frameworks under which we could try to make sense of the development of action cognition and its relationship to social cognition. Here we provide a brief overview of the different options, before ending with suggestions for future directions.

Cognitive Theories

The traditional way to understand information processing in the human brain is to develop cognitive or computational models that can reproduce that processing (Marr 1982). To understand action cognition, computational models provide a powerful way to test and explore the problems which the human brain must solve to move in the world (Franklin and Wolpert 2011). Similar computational mechanisms could be applied to social cognition: the motor control mechanisms that allow a person to predict and control a tennis racket might also allow a person to predict and control the actions of another person (Wolpert et al. 2003). Such models can incorporate gradual motor learning but do not say anything specific about development.

A related approach to action and cognition can be found in the mirror neuron framework (Rizzolatti and Sinigaglia 2010), which postulates how motor performance and action understanding could be linked to the same neural systems. The mirror neuron model has been set within a developmental context (Gallese et al. 2009), with strong claims that the failure to develop mirror systems in autism might account for difficulties in social cognition (for a critique, see Hamilton 2013b). This account also places a strong emphasis on prenatal and innate mechanisms of action and cognition, and does not leave much space for development after birth. Thus, neither of these models has much to say about the rapid improvements in action cognition during the first year of life.

One way to expand the cognitive approach, so as to consider developmental change, is to study developmental disorders. Developmental causal modeling (Morton 2004) and the ACORN framework (Moore and George 2011) provide tools for specifying and testing cognitive models of child development and developmental disorders. Using these tools, a developmental process can be formally specified in terms of the biological, neural, and cognitive changes that take place at different developmental time points, as well as the ways in which these influence each other. Such a formal model is more amenable to testing and clinical use than more weakly specified theories. For example, a developmental causal model of autism suggests that a primary difficulty in theory of mind can account for many of the observed difficulties in social cognition (Frith et al. 1991), and this has been tested in detail (U. Frith 2012). A key question for cognitive approaches to development is to identify the different modular systems and to determine if and how they might interact. For example, is the development of motor systems essential to theory of mind (Gallese et al. 2009), or not (U. Frith 2012)? It would be possible to place these questions and the relevant data on action cognition into a more formal modeling framework of causation to test out theories of developmental change, but this has yet to be attempted.

Interactionist Theories

An alternative approach that is gaining ground is to focus the study of child development on the process of development itself (Karmiloff-Smith 2012). Rather than starting from the adult end state and considering the child as an adult with some bits missing, this approach considers fully how new capacities can emerge out of the interaction between the infant and the social-motor environment. In the motor domain, dynamical systems models have been used to describe how motor skills emerge in infants from the interaction of the child and the environment (Thelen and Smith 1996). In social cognition, embedded and embodied accounts (Reddy 2008) view social skills as emerging from the interaction between infant and caregiver, rather than being internalized by the infant. A key prediction in these models is that developmental changes emerge out of the relationship between the child and the environment. For example, if a child who can walk obtains different physical and social inputs to a child who cannot walk, this will initiate the development of particular social and motor skills. The emphasis here is on a longitudinal, two-way relationship between the child and the social-physical world.

Interactionist theories are part of the push toward thinking of cognition not in isolation, but grounded in reality, embodied in the world, and created by the interaction between child and world. This push is similar to the new emphasis on "second person neuroscience" (Schilbach et al. 2013), where the emphasis is strongly on the interrelation between the developing child and that child's social-motor environment. This is a promising approach which is coherent with reports of close links between the child's social-motor experience and their further development (Leonard and Hill 2014)). However, the major limitation of this approach is its complexity. If the decomposition of behavior according to cognitive processes is abandoned, it is not clear how development should be decomposed. Yet without any decomposition, the problem of understanding a process and formulating testable models is difficult. Overall, interactionist models are intriguing but it remains very hard to find tractable experimental approaches to test their validity.

Future Directions

In this review of the development of action cognition, data suggest that motor development is a very protracted process—one that is linked to other areas of cognition, in particular to the development of social skills. Two categories of key unanswered questions include:

1. What is the best framework for understanding the development of motor control? Can we break down motor control into specific cognitive processes and track the development of each? Or is this only feasible through a holistic, interactionist account?

2. What processes link motor and social cognition? Are there specific cognitive mechanisms which might be shared between motor and social cognition and, if so, what are they? Alternatively, are the associations we observe in data between motor and social skills driven instead by changes in the child's opportunities to learn, or other facets of the environment?

To address these questions, more data is needed on how motor cognition actually develops. In young children, it would be particularly helpful to consider the substantial individual differences that are apparent at certain ages (e.g., walking at nine vs. 18 months) but resolved at later ages (almost all four-year-olds can walk in a similar way). It is also critical to consider the interplay between the child and the environment. Longitudinal studies which track the child's skill and social surroundings over time would be particularly valuable in this regard. Finally, the study of children with a range of developmental disorders (not just autism) are needed to understand why motor cognition sometimes goes wrong and what the implications of this are for both typical and atypical development.

First column (top to bottom): Gottfried Vosgerau, Cecilia Heyes, Pierre-Yves Oudeyer, Peter König, Antonia Hamilton, Antonia Hamilton and Pierre-Yves Oudeyer, Robert Rupert
Second column: Giovanni Pezzulo, Atsushi Iriki, Peter König, Antonella Tramacere, Gottfried Vosgerau, Antonella Tramacere, Uta Frith
Third column: Uta Frith, Robert Rupert, Henrik Jörntell, Saskia Nagel, Cecilia Heyes, Giovanni Pezzulo, Atsushi Iriki

4

Acting Up

An Approach to the Study of Cognitive Development

Giovanni Pezzulo, Gottfried Vosgerau, Uta Frith,
Antonia F. de C. Hamilton, Cecilia Heyes, Atsushi Iriki,
Henrik Jörntell, Peter König, Saskia K. Nagel,
Pierre-Yves Oudeyer, Robert D. Rupert,
and Antonella Tramacere

Im Anfang war die Tat. [In the beginning was the deed.]
—Goethe: Faust

Abstract

Despite decades of research, we lack a comprehensive framework to study and explain cognitive development. The emerging "paradigm" of action-based cognition implies that cognitive development is an *active* rather than a passive, automatic, and self-paced maturational process. Importantly, "active" refers to both sensorimotor activity (in the narrow sense) as well as to autonomous exploration (e.g., as found in active perception or active learning). How does this emphasis on action affect our understanding of cognitive development? Can an action-based approach provide a much-needed integrative theory of cognitive development?

This chapter reviews key factors that influence development (including sensorimotor skills as well as genetic, social, and cultural factors) and their associated brain mechanisms. Discussion focuses on how these factors can be incorporated into a comprehensive action-based framework. Challenges are highlighted for future research (e.g., problems associated with explaining higher-level cognitive abilities and devising novel experimental methodologies). Although still in its infancy, an *action-based approach to cognitive development* holds promise to improve scientific understanding of cognitive development and to impact education and technology.

Introduction

During their first years of life, children greatly increase their action repertoire to acquire sophisticated cognitive and interactional abilities (e.g., conceptual, inferential, and linguistic). How this is accomplished has been the focus of much research. To date, however, we lack a comprehensive scientific framework capable of explaining cognitive development and the relations (if any) between action and cognitive development.

In traditional cognitive science, "action" and "cognition" are often studied in isolation, with the role of action reduced primarily to the execution of a motor response. New, promising action-based theories emphasize, however, that it is only through action that a living organism can "know" or "cognize" its environment; thus action must be considered integral to cognitive processing (Engel et al. 2013; Pezzulo et al. 2011; Pezzulo 2011; Thelen and Smith 1996). How does an emphasis on action affect our understanding of cognitive development?

An action-oriented perspective views cognitive development as an *active* process rather than one that depends solely on automatic, self-paced maturational processes. More specifically, if children are to discover (and learn to predict) increasingly complex and profound regularities in their bodies and the external world, they must engage *actively* with their physical-social environment. This engagement, in turn, forms the basis for increasingly sophisticated action and cognitive abilities.

The dependence of perception and cognitive processing on action has been captured nicely through the sensorimotor contingency (SMC) theory (O'Regan and Noë 2001). Accordingly, the cognitive processing of a child (or, more generally, a living organism) does not originate from a *stimulus* but rather from an *action* (and usually an *intention*). By acting, a child causes regularities in sensorimotor patterns, which are then successively experienced. This permits a child to master the regularities in perception-action patterns, or SMCs. Although SMC theory has been used primarily to study perception, the mastery of SMCs could be considered crucial for the development of both action and cognition. This mastery guides an individual's own actions toward goals (intentional action) and permits a person to predict the consequences of the action (anticipation). It also permits an individual to produce sensory stimuli that are maximally informative; that is, to recognize an agent or object (perception and discrimination) and to learn its characteristics over time (learning and memory). Since actions are usually performed in social domains, actions also afford social communication and signaling, which create new opportunities for social and cultural learning. In sum, SMC theory emphasizes the tight link between perception and action systems and the importance of active, exploratory activities on cognitive development.

It is important to emphasize that the word "active" refers not only to sensorimotor action but also to exploration (as in active perception or active learning). Furthermore, although an action-oriented view of cognitive development gives prominence to active exploration and engagement with the environment, it does not dismiss other factors that do not appear to be linked to action, such as the importance of genetic regulation or sociocultural factors, which are likely to produce "biases" to action-based processes (e.g., by focusing our attention on important novel events or on actions performed by our conspecifics). Viewing cognitive development from an action-oriented perspective holds the promise of providing a comprehensive framework—one capable of contextualizing all factors important to cognitive development. Similarly, an action-oriented approach targets the whole range of phenomena used typically to explain development theories (e.g., the acquisition of conceptual, inferential and linguistic abilities); it is not confined to a portion that seems more closely related to sensorimotor action. We propose that research should focus on how all these (and other) abilities depend on, or at least link to, action—directly or indirectly. The research program that we envision and discuss is in some aspects new, but it connects well with existing theories of cognitive development.

Factors and Mechanisms Influencing Developmental Processes

The importance of action for cognitive development has long been recognized by "behaviorists" (Thorndike 1932; Skinner 1938) and "constructivists" (e.g., Piaget 1952) as well as recently by others (Thelen et al. 2001; von Hofsten 2004). However, the specific mechanisms and factors that underlie action-based cognitive development await detailed identification.

Since an action-oriented approach emphasizes the importance of sensorimotor learning, it is natural to assume that the most important candidates for cognitive development would be those that underlie intentional action control. Contemporary theories of action control highlight the importance of internal models (e.g., forward or inverse models) and associated "efference copies" or "corollary discharges" of motor control signals as basic mechanisms of sensory processing, prediction, and motor control (Adams, Shipp et al. 2013; Blakemore et al. 2001; Crapse and Sommer 2008; Grush 2004; Shadmehr et al. 2010). These mechanisms may also be crucial for cognitive development and it has been suggested that a copy of the efferent motor command rerouted to the sensory pathway is necessary for organisms to differentiate between reafference and exafference, so that the organism can distinguish the source of sensory inputs; that is, whether an input results from the organism's own (spontaneous) movement or emanates from the environment (von Holst and Mittelstaedt 1950). During the early stages of life, this process immediately creates a feedback loop through which both sensory and motor information are connected, thus forming the basis for sensorimotor learning and potentially providing a way to distinguish self-movements from other movements. It has also been suggested that predictive (forward) models implied in motor control might be key for the acquisition of cognitive and social abilities: they might permit cognitive agents to understand objects in the environment as well as the actions of other agents, in terms of anticipated sensorimotor patterns, thus permitting the acquisition of various interactive abilities such as coordination and action prediction (Jeannerod 2006; Pezzulo and Dindo 2011; Pickering and Garrod 2013b; Sebanz and Knoblich 2009). Forward models might also support exploratory strategies (e.g., hypothesis testing) for perceptual processing, belief revision, and skill learning (Friston, Adams et al. 2012; Gottlieb et al. 2013).

Despite their importance, mechanisms which directly link to sensorimotor learning do not offer a sufficient explanation to construct a comprehensive action-based theory of cognitive development. Traditional theories of development have proposed several potential causal factors: genes, social and cultural factors, emotions, experience, etc. An action-based approach to cognitive development needs to integrate most or all of these causal factors and recast them on the basis of their contribution to action-based processes. Below we review some of the most important factors to be considered in such an ambitious synthesis.

Common Paths of Development and the Action-Experience-Sociality Triad

Models that emphasize the role of action in cognition, or claim that cognition is action, tend to emphasize the role of experience (and especially social interaction) rather than that of genetic factors in development. Why is this? One explanation may be found in the historical treatment of action and experience (e.g., pragmatism and behaviorism). Alternatively, action and experience may be seen as the joint product of a more encompassing view of organisms as agents: makers rather than takers of their fate.[1]

However, even the link between empiricism-about-action and sociality might contain a strong element of necessity, which might reveal common paths of development despite the vast differences in individual experiences. The development of cognition needs guidance from somewhere, and if that guidance is not genetically driven, it may come from the social world. The constraints provided by the physical environment—the brute properties of objects—are insufficient, although they could be complemented by social aspects (e.g., caregivers, or co-actors more generally, often support the learning process by providing a pedagogical context) (Csibra and Gergely 2011; Pezzulo, this volume). Furthermore, members of each culture want their children to grow up performing actions in distinctive ways. The cultural specificity of action serves to

---

[1] A focus on organisms as "makers of their fate" has further impact on the understanding of autonomy and responsibility, thus opening various ethical and social questions (Nagel 2010).

enable the cumulative cultural inheritance of skills (Tomasello 2014), thus providing a source of shibboleths, or "ethnic markers," to identify who is eligible for reciprocal altruism (Cohen 2012; Hamlin et al. 2013; Riečanskỳ et al. 2014).

Hamilton et al. (this volume) point out that some motor skills show a highly protracted developmental trajectory (e.g., grip aperture, scaling, grip force, end-state comfort). In such cases, later stages of skill acquisition might represent adjustments to cultural norms rather than to the raw physical requirements of the motor tasks. Identifying the role of cultural influences on cognitive development and their potential adaptive value (e.g., as in the proposal of "ethnic markers") is an important direction for future research.

Identifying Biases and Constraints in Action-Oriented Processing and Development

A related, important issue involves the identification of various "biases" that guide and shape development (e.g., by constraining the space of sensorimotor learning and the acquisition of SMCs). There is a danger of assuming that if we correctly understand only the earliest and most basic processes and learning mechanisms (e.g., associative mechanisms), then the rest of development will magically emerge. A more comprehensive view needs to consider that experience-dependent learning is constrained and that not all SMCs are learned with equal ease. For example, it is very easy for monkeys to learn to fear a snake by watching on video the response of a model that is afraid of a snake; however, it is almost impossible to learn fear of a flower, even if the video is identical (substituting the flower for the snake) (Mineka and Cook 1993).

One example of anatomical and physiological constraints to experience-dependent learning is the brain's division into ventral and dorsal streams: the ventral stream is dedicated to the "what" (i.e., object) and the dorsal stream to the "where" (place). This division constrains what can be acquired during development (Milner and Goodale 2008). Another example of bias in development is the presence of reflexes that guide the initial exploration and shaping of the SMCs to be learned (Verschure et al. 2003).

It is important to consider what the anatomical requirements of associative learning systems are that putatively support the acquisition of SMCs. One such requirement is that the relevant sensory and motor domains must be connected to each other, thus providing the basis for associative linkage. This argument is important, for example, in the domain of language. Macaques do not have a strong dorsal connection between auditory and motor systems in their brains, whereas humans have such a data highway for auditory-motor association: the arcuate fascicle. Associative learning of speech may critically depend on the availability of ample connectivity in this domain-specific auditory-motor system (Pulvermüller and Fadiga 2010).

We still lack a systematic taxonomy and understanding of the various biases that constrain action-oriented processes and guide cognitive development.

These biases might be very diverse, emphasizing once again the necessity of an integrative research program. To exemplify the diversity of the possible biases, we review two classes: genetic factors and the mechanisms that drive autonomous learning and exploration.

Genetic Influences on the Learning of SMCs and Cognitive Development

Almost every account of development recognizes that there are contributions from genetics, learning, and social interaction. Thus, it should not seem odd to highlight the importance of genetic processes within an action-based approach. Nevertheless, most action-oriented approaches emphasize the contributions of learning over those of genetic inheritance and suggest that the genetically inherited components are quantitative biases rather than whole, dedicated cognitive processes. They also stress the action-related aspect of these genetically inherited biases.

The idea that genetics can exert important influences on development should not, however, be sidestepped. Genetics has often been identified with fixed, inflexible, and phylogenetic traits and plasticity with nongenetic processes. As a consequence, human action and the development of cognition were held to be too flexible to be contextualized in a molecular perspective. We now have the tools to revise this picture. There is every reason to assume that innate mechanisms, honed by millions of years of evolution, are not fixed and rigid behavior programs but rather pre-prepared "startup kits" which lead not only to flexible behavior but may also reveal individual differences (Carey and Gelman 2014; U. Frith 2012).

To understand better how evolution and genetics can play a role in development, it is important to note that nature ultimately influences, through natural selection, the *phenotype* (bodily patterns of both morphology and behavior) and not the micromorphology (including neural circuitry and its operating mechanisms), which realizes behavior. Accordingly, genetics and neural mechanisms are instrumental to support adaptive phenotypes. In turn, actions (behavioral phenotypes) are constrained by bodily structure, neural control mechanisms, and learning processes (including social learning), all of which interact dynamically, thus making a complete genetic specification unlikely (Edelman 1987). Certain important constraining factors to development are more likely to be encoded genetically. The human face preference in neonates is a good example (Johnson et al. 1991): From birth, human infants "track" the movement of a face-like stimulus (a triangle of dark blobs on a light background, with two blobs at the top) longer than a control stimulus (a similar triangle but with two blobs at the bottom). This bias makes the infant highly receptive to information from other people. In addition, this bias is action oriented in that it involves "tracking" (i.e., moving the head to keep the stimulus in view, which itself is a moving target). As this example clearly illustrates, the action oriented research program should not dismiss the importance of genetic factors but rather clearly identify their synergistic impact on action-based processes.

Mechanisms That Drive Autonomous Learning and Active Exploration

Another class of constraints involves mechanisms that support autonomous learning and active exploration. To acquire mastery of SMCs, children face several challenges: First, the sensorimotor spaces to be modeled are high-dimensional and nonlinear. Second, different from most machine learning algorithms (used, e.g., by web companies to classify texts or images), infants do not have access to preconstructed databases containing millions of learning examples. Instead, they have to learn incrementally through physical interaction and by performing sensorimotor "experiments," which carry costs in time and energy. Which learning and exploration methods should children use to solve the formidable challenge of SMC learning? This question can be approached based on contributions from both computational neuroscience and robotics research.

We need to understand which types of learning mechanisms can derive regularities from these data. Recent advances in theoretical neuroscience suggest that different (e.g., perceptual and structural) learning mechanisms might interact and operate on different timescales (Friston, Daunizeau et al. 2010). Furthermore, computational models of cognitive development emphasize that the developmental process needs some guidance from so-called "inductive biases" that pre-shape the space of what can be learned (Tenenbaum et al. 2006). Understanding how different learning processes and biases might interact synergistically requires the development of integrated architectures, possibly embedded in physical robots that operate in realistic sensorimotor contexts—all important open problems at the forefront of computational modeling and robotic research (Friston 2010; Verschure et al. 2014).

We also need to understand which mechanisms *guide* exploration to collect informative and useful data. Action-based theories emphasize how important it is for children to explore their environment actively and test their hypotheses—an idea which links well with the view of the "child as scientist" (Gopnik and Schulz 2004). Indeed, learning theory and models in developmental robotics show that collecting data through randomly chosen experiments is bound to fail and that constraints are needed (Oudeyer et al. 2013). It is also important to consider the fact that not all SMCs in the world are identifiable and learnable by organisms, due to limitations in time, energy, computational or inferential resources. Some contingencies may also become learnable only after certain prerequisites have been acquired. Thus, exploration strategies should include mechanisms which focus sensorimotor experimentation on those subspaces/activities that are currently "learnable," given the cognitive agent's prior knowledge and skills. These map onto what Vygotsky (1978) called the "zone of proximal development."

Current research in developmental robotics has helped elucidate several families of (interacting) mechanisms that guide and constrain exploration. This includes the biomechanical and physiological properties of the growing body, with developing neural synergies and perceptuo-motor systems. Another family of such guiding mechanisms is social shaping of the learning environment, which drives the attention and activity of learners through a diversity of social strategies (e.g., imitation, emulation, teaching strategies). Finally, *motivational* mechanisms are key in driving the organism to select particular actions and particular sensorimotor experiments. It is important not to limit research to *extrinsic* motivational systems, where the organism is driven to search for things like food or social bonding, but to include *intrinsic* motivational systems, where the organism's brain assigns value to information gain or competence gain, leading to spontaneous exploration (Baranes et al. 2014; Mirolli and Baldassarre 2013). Such intrinsic motivational mechanisms can be viewed as proximal mechanisms which favor curiosity-driven and novelty-seeking behavior as well as, ultimately, the acquisition of good predictive models for adaptive action. In developmental robotics, the development of an integrated approach to the modeling of these families of "guidance mechanisms" for exploration and their interactions is a challenge that has not yet been met.

Open Issues: Studying Development as a Continuous yet Nonhomogeneous Process

How do all the factors identified thus far operate and interact over time? The view, illustrated in Figure 4.1, is that *development is a continuous yet nonhomogeneous process* that might proceed at a different pace at different periods, where different mechanisms, genetic, associative and cultural (or their combinations) might play a more prominent role.

During the early (fetal and neonatal) stages, movement triggered by endogenous input (including intrinsic motivation) orients an individual's initial movements, thus creating the primitive establishment of SMCs. During these phases, the ways in which an organism interacts with its environment are constrained, and we can talk of a maturation of SMCs or their prerequisites, such as an initial development of basic motor abilities. In the postnatal period, maturation and a more sophisticated learning of SMCs (e.g., based on associative mechanism) overlap and interact, and infants enter into new phases of cognitive development which last quite a long time (Hilgard 1991; von Hofsten 2004). Different factors (from molecular regulation to social interactions) become extremely important; in addition, brain plasticity plays a crucial role in correlating cognitive development to the external context in a way that is not predetermined. During later phases, SMC learning becomes increasingly powerful: generalization becomes possible, capitalizing on the fact that children have increasingly more sophisticated experiences. For example, when children learn to stand still, they also learn to recognize and manipulate

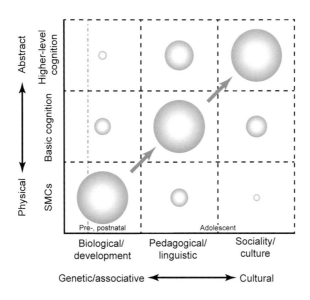

Figure 4.1   Development of physical and cognitive abilities is shaped by various genetic, associative, and cultural factors whose relative importance changes over time. The size of the dots represents the relative importance of the various factors during developmental time: from pre- and postnatal periods, when the primitive SMCs are established (bottom left), to later phases (e.g., adolescence), when social and cultural factors become dominant in the development of higher cognitive skills (top right).

a greater number of objects, and they interact in richer ways with their care-givers. Successively, children are exposed to increasingly richer contexts (social, linguistic, and pedagogical), which reach a high level of complexity in adolescence.

Essential "phase transitions" may exist in the developmental process and might be induced, for example, through the growth of the brain and the increasing complexity of social interactions supported by language development (following the maxim that "more is different"). Some of these phase transitions might reflect qualitatively different and discontinuous characteristics of higher abstraction and concept formation, perhaps powered by linguistic "start-up kits," where concept formation and the simultaneously developing language skills shape one another in a virtuous circle. From this perspective, the property that emerges from the phase transition would be qualitatively different yet physically based on existing systems. The massive and rapid expansion of the human brain during the recent evolution of hominids would contribute to the emergence of so-called human-specific higher cognitive functions (Iriki and Taoka 2012). Further research is needed to provide empirical support for these initial hypotheses as well as to link phase transitions and the classical concept of "stages" in cognitive development (Piaget 1952).

The motivation to learn new skills has roots in social cognition and imitation of valued adults or peers. Learning context, in no small measure, guides
the onset of learning, the form the practice takes, and the pleasure the learner
is experiencing in mastery. This example emphasizes the gap in current research: we need to identify a link between sensorimotor and motivational, intentional, and cultural goal processes in development. Some aspects of this
have been studied within the tradition of associative learning theory (Balleine
and Dickinson 1998; Klossek et al. 2008). However, this research, and other
theoretical and empirical resources from the same tradition, could be produc-
tively integrated with research that is more obviously part of an action-oriented
and culturally/socially embedded approach to cognitive development.

The Development of Higher Cognition

Action-oriented theories seem particularly well suited to explain the development of simple cognitive abilities which have externally perceivable or ma-
nipulatable objects as referents (e.g., the categorization of objects) (Chao and
Martin 2000). A crucial challenge for action-oriented approaches is to explain
the development of higher cognitive abilities at large, including our linguistic
abilities, capacity to use abstract concepts, and cognitive operations such as
planning, reasoning, and the engagement in social exchanges.

The action-based perspective of cognitive development diverges from
other more classical approaches in suggesting that linguistic and conceptual
knowledge acquired during development is grounded in perceptual and motor
systems rather than being a symbolic modular system or an "encyclopedia"
of concepts unrelated to action and perception systems (Barsalou 1999). This
applies to domains traditionally considered to be symbolic (e.g., language) as
well as to abstract concepts (e.g., "truth" or "democracy"). How is this possible? Below we highlight several required steps toward an action-oriented
account of abstract concepts and their development, which ultimately point to
the necessity of a broad notion of action-oriented processing.

Toward an Operational Definition of "Concept"
That Is Not Too Restrictive

A first important step toward the specification of action-oriented theories of
concepts is methodological: we must define a minimal constraint for the use
of the notion of "concept." In several disciplines (e.g., philosophy or psychology), possessing a "concept" means more than just possessing discrimination
abilities. Discriminations do not necessarily require cognitive abilities: a red
detector is able to discriminate reliably between "red situations" and "non-red
situations." In contrast, we usually explain more complex abilities with the
possession of concepts. Consider, for example, the concept "red." Possession

of the concept requires more than discriminatory abilities; one must be able to group together objects with other colors but not, for example, shapes (for an ability-based account of concepts, see Newen and Bartels 2007). The possession of the concept "tool" might be related to the action possibilities it affords (Chao and Martin 2000; Maravita and Iriki 2004), a characteristic that is especially relevant from an action-oriented point of view. Thus, one desideratum for future empirical research would be to identify the specific action possibilities a specific concept offers and to test for the presence of such actions instead of only testing discriminatory behavior. This seems to be most relevant for "abstract" concepts. The possession of the concept of *democracy*, for example, does not only lead to the ability to discriminate between democracies and non-democracies. It relates to a number of different practical skills: from voting and taking part in fair discussions to representing others in the community or accepting opinions different from one's own.

**Abstract Concepts and Their Relation to Language,
Interoceptive Systems, and Sociality**

Explaining abstract concepts is a key benchmark for action-based theories of cognitive development, because they seem prima facie to be not particularly related to action (at least in the restricted sense of sensorimotor action). It is generally assumed that abstract concepts are linguistically coded (e.g., Paivio 2007), in line with physiological evidence of greater engagement of the left perisylvian language network for abstract rather than concrete words (Binder et al. 2009). Moreover, it has been argued that abstract words are learned by extracting distributional or syntactic information from sentences in which these words are used (Gleitman et al. 2005). Accordingly, linguistic (especially syntactic) development is taken to be a prerequisite for the learning of abstract concepts.

Contrary to the hypothesis of a separate cognitive domain for abstract and linguistic processing, there have been various attempts to characterize the acquisition of abstract concepts as intrinsically action and interaction related. Below we summarize proposals which highlight the importance of affective or interoceptive processing, mental operations, and social dynamics.

Based on recent behavioral and imaging work, which used tighter matching of items than previous studies, some argue that abstract concepts entail affective processing to a greater extent than concrete concepts (Kousta et al. 2011) and, for this reason, that their encoding comprises neural networks engaged in processing emotional stimuli (Vigliocco, Kousta et al. 2014). There is a statistical preponderance of affective associations underlying abstract word meanings. According to Kousta et al. (2011), our internal affective experience, linked to interoceptive responses, would provide at least grounding at an early stage to abstract concepts: words that denote emotional states, moods, or feelings could provide examples of how a word may refer to an entity that is not externally observable but resides within the organism. Consistent with this possibility, abstract words that denote emotional states are the first abstract words to emerge during language development (e.g., Wellman et al. 1995). Such an early bootstrapping of abstract words from emotion could rely on interoception as well as on action, since several emotions are expressed in actions and these actions can be linked to words (Moseley et al. 2012).

Affective and interoceptive states are not the only "internal" referent for abstract concepts. Barsalou (1999) proposed that concepts such as "truth" might be grounded in internal cognitive operations (e.g., the "matching" of an expected and a perceived situation). How such internal operations could be linked to SMCs remains a challenge.

Another set of action-based theories of abstract concepts highlights the *social* nature of the concept and word-learning process. In the case of concrete entities, Borghi and Binkofski (2014) argue that the presence of a given object (the referent of the concept or word) constrains our sensorimotor experience, whereas in the case of abstract entities, concepts are grounded in interpersonal sensorimotor and situational experience. This different "modality of acquisition" might distinguish how concrete and abstract concepts are learned and represented. Related to this, Pulvermüller (this volume) suggests that the acquisition of SMCs can be relevant for both concrete and abstract semantics, but that these would differ significantly due to the different correlation structure of referent objects of concrete and abstract terms (e.g., "eye" versus "beauty") (see Figure 9.1 in Pulvermüller, this volume). All of these proposals for an action-based foundation of conceptual processing remain to be assessed in future studies.

A Specific Example of an Abstract Concept: The Case of Morality

Emotional and social action-based processes have been implicated in the development of moral sensitivity. Over the past ten years, there has been an upsurge of empirical research into moral cognition and the evolutionary function of the human moral sense (or faculty). One proposal is that the evolutionary basis of moral cognition is the pressure generated by the need for cooperation among humans. Baumard et al. (2013), for example, argue that human moral sense evolved within a market of cooperating partners: an agent's moral behavior serves to increase his or her reputation as a cooperative partner.

We can readily imagine that moral sensitivity is based on action contingencies and affective processes that relate to reward and punishment, when outcomes of moral actions have to be judged. A central concept that has been suggested to underpin the development of morality is *doing or not doing harm*. Even eight-month-old infants prefer agents who act positively toward prosocial individuals as well as agents who act negatively (punish) toward antisocial individuals (Hamlin et al. 2011).

A highly relevant concept to the judgment of moral behavior is whether the behavior was intended or accidental. *Intentionality* is a (possibly uniquely human) capacity with a known neural signature that underpins the ability known as "mentalizing" (Frith and Frith 2012). This ability, among other things, allows us to classify actions as deliberate or accidental. Studies using fMRI in adults have shown that the mentalizing network, which includes the right temporal-parietal junction and medial prefrontal cortex, is active when moral judgments are being made (Koster-Hale et al. 2013).

The adult conception of moral judgment tends to give primary importance to an agent's intention when action is evaluated; however, the importance of the outcome of an immoral act should not be overlooked. Preschool children give a lot of importance to the outcome. Thus, they judge agents who cause accidental harm far more harshly than agents who intended but failed to harm another person. Preschool children are perfectly able to understand the content of others' intentions, so why don't they make use of this information in their moral evaluation? One possible explanation comes from an experiment by Buon et al. (2013), who showed children nonverbal cartoons depicting three conditions: (a) an agent intentionally causes harm to his victim; (b) an agent accidentally causes harm to his victim; (c) an agent is present when another gets hurt, purely by coincidence. Their results showed that five- and three-year-olds' moral evaluations were more sensitive to the agent's "causal" role than to the agent's "intentional" role. Interestingly, the same was true for adults when they had to make moral judgments under cognitive load (Buon et al. 2013).

Studies by Koster-Hale et al. (2013) provide further evidence that, in adults, it is cognitively easier to blame agents of attempted harm (who intended, but failed to cause harm) than it is to exculpate agents who caused accidental harm. As expected from the fMRI results in this study, ventromedial prefrontal cortex patients failed to condemn attempted harm because they failed to experience any emotional aversion when they retrieved the content of the agent's malevolent intention (Young et al. 2010). However, individuals with autism spectrum disorder, who have difficulty attributing intentions, failed to make use of the information about the content of the agent's false belief to exculpate the agent who caused accidental harm without any malicious intention (Moran et al. 2011).

This example demonstrates the high-level concepts that are part and parcel of cognitive development and which are amenable to being explained in terms of action-based theories. Like many other "abstract" concepts, moral concepts may include both "individualist" elements (e.g., interoceptive and affective codes) and "social" or "collective" elements; that is, moral judgments can involve actions performed by others and their outcomes. This suggests the necessity of a broad action-oriented view—one that goes beyond the restricted domain of an individual's own sensorimotor control mechanisms. It also suggests that the current focus of action-based theories on simple forms of sensorimotor control should be expanded toward richer theories that include executive control. Indeed, the adult accomplishment here and in many other capacities is strongly influenced by effortful inhibitory mechanisms that serve to facilitate interaction and cooperation, and it thus should be included in action-oriented analyses.

Sociocultural Factors in Concept Learning: The Case of Supernatural Beliefs

Our discussion of the case of morality has highlighted the complexity of the relationship between concepts or beliefs and sensorimotor representations. Another domain where this is apparent concerns abstract supernatural beliefs. The Ifaluk of Micronesia, for example, believe that if a person is feeling sad, this will cause their relatives on a different island to become ill (Lillard 1998; Lutz 1985). In this example, it is likely that a belief has no physical or sensorimotor basis but is rather acquired and maintained via social-cultural mechanisms. Nonetheless, this belief may well have sensorimotor consequences, as demonstrated by the following: a person who believes their relative's illness is caused by events on another island might not provide the same medicines to a sick person as someone who believes that illness originates physiologically. Linking abstract beliefs to consequences, however, does not always work. Sperber (1975:33) reports that the Dorze in Ethiopia regard the leopard as a Christian animal—one that does not eat meat on Fridays. This belief could lead villagers to decide that their livestock are safe from predation on Fridays, yet they secure their chickens on Fridays, just as they do every other day. This example shows how an abstract cultural belief does not relate to the actions people actually perform, even though the belief still remains embedded in that culture. Together, both cases provide examples of knowledge or beliefs that are hard to connect to sensorimotor experience alone. They point instead to a critical *sociocultural* mode of acquiring knowledge.

If we address the question of how abstract concepts (e.g., democracy or morality) are learned by thinking only about individual minds, we run the risk of underestimating the complexity of some abstract concepts or of postulating very weak "grounding" relations to SMCs to bridge the gap. These risks can be averted if we recognize that such abstract concepts are produced by large groups of minds, over many generations, as a result of cultural evolution. In the course of ontogeny, individuals do not need to create these concepts, they only need to adopt them. This is far from being a trivial task: other agents supply not only the conceptual content, they assist individuals in adopting them and support the adoption process through scaffolding and explicit instruction. Where is the influence, then, of SMC learning? At the very least, SMC learning influences the construction of mechanisms that enable the child to learn from others (e.g., caregivers), via observation and instruction (Heyes 2012), and it gives the child a "database" of empirical regularities to be explained by abstract concepts.

Sensorimotor Learning Can Provide a "Database" for Learning and Development

Action-oriented approaches emphasize the importance of sensorimotor learning. There is a consensus that this type of learning is important in the early stages of cognitive development, but there is considerable disagreement about the role that sensorimotor learning plays in the development of "higher" cognition (e.g., in the development of mentalizing). One view is that action understanding and #mind-reading abilities are based directly on mechanisms that support action control (Rizzolatti and Craighero 2004). Another view, still consistent with a "weaker" version of the action-oriented approach, holds that the linkage between action and mentalizing abilities is more indirect. For example, sensorimotor learning enables infants to encode and predict regularities in the behavior of others: when a person begins to move a spoon of porridge from a bowl, the spoon is likely to end up in the person's mouth. Subsequently, children learn an explicit "theory of mind"[2] from adults and peers in their culture. An explicit theory of mind may be acquired through conversation and instruction, rather than through sensorimotor learning (Heyes and Frith 2014), but it allows the child to interpret the SMCs coded in their database. For example, the spoon is likely to end up in the agent's mouth because the child *wants* to eat porridge and *knows* that moving the spoon to her mouth will allow her to do this. Accordingly, sensorimotor learning provides a "database" which is then interpreted by cognitive processes acquired by a different route.

Abstract Cognitive Operations

An objective of action-based approaches is to determine whether and how cognitive operations such as reasoning, planning, and understanding others' intentions might reuse (or capitalize upon) the same action-and-prediction strategy that underlies the mastery of SMCs. Various researchers have proposed that the fundamental machinery supporting intentional action—including most prominently *internal forward models*—might be reused covertly "in simulation" of overt behavior (Jeannerod 2006). The same sensorimotor loops might be reenacted during planning and action understanding. Conceptual systems have been proposed to consist of "situated simulators" that permit reenactment and recombination of previously productive situations experienced; using such

---

[2]   It might be useful to distinguish between an explicit and implicit theory of mind (also known as mentalizing). The development of the latter could be based on an innate start-up kit, which is proposed to be missing in autism (U. Frith 2012). Learning based on this start-up kit appears to be extremely fast: using eye-tracking techniques, Kovács et al. (2010) showed that seven-month-old infants are able to track an agent's belief. This mechanism for automatic tracking of beliefs (mentalizing) is still available to adults in parallel with explicit theory of mind. Although able autistic individuals (Asperger Syndrome) can acquire explicit mentalizing, they seem unable to acquire automatic mentalizing (Senju et al. 2009).

"simulators," a modal conceptual system might support the entire range of complex cognitive operations traditionally associated with the manipulation of amodal symbols (Barsalou 2009; Pezzulo et al. 2011).

In parallel, a "weaker" version of action-based theories has emerged: complex cognitive operations use mechanisms analogous to those providing a mastery of SMC, but they involve other kinds of contingencies which are not strictly sensorimotor but are so-called "second order," in the sense that they partially or totally abstract from sensorimotor loops (for a discussion, see Pezzulo, this volume). Chess playing offers a specific example of operations that do not entail strictly SMCs but other, possibly higher-order contingencies, especially if appropriately subdivided into smaller steps. Thus, chess playing could be viewed as a movement-like situation: by planning ahead for the consequences of an action, the motor aspect would map to the specific movement of a piece whereas the sensory consequences would be the outcome of the action (i.e., how good or bad is the position of the piece) or the expected responses of the opponent. Accordingly, the specific movements executed by the hand to move a piece would be less relevant than the "movement" in a more abstract action space that is shaped by the rule of chess—hence the idea of higher-order contingencies.[3]

Open Problems

Explaining the development of abstract concepts and higher cognitive abilities is a key challenge for action-oriented views. We began this article by considering the centrality of sensorimotor skills and SMCs in cognitive development and have discussed several important factors (the importance of proprioceptive and affective states, language, and social dynamics) that should be integrated within a more comprehensive account of cognitive development. As the example of moral reasoning illustrates, all factors contribute to the advanced cognitive skills that make us humans, and—at least in principle—all can be reformulated and integrated within an action-based approach, providing that the approach is not too restrictive.

Our analysis, though, exposes important questions: What is preserved from the original concept of *sensorimotor contingencies* in domains of abstract

---

[3]   This "second-order" approach to SMCs poses a potential problem for "strong" action-oriented or "enactive" views (for terminological clarifications, see Dominey et al., this volume). Imagine that to encode an SMC for future use, the cognitive system stores the conditional probability of the appearance of a certain sensory state given a certain motor command. The encoding of such a probability represents something about (or carries information about) the sensorimotor processing, without necessarily being in a sensorimotor code. Consider a discursive list of all of the paintings in the Louvre: the entries on the list are about images, but the list, as an encoding, need not itself be imagistic. An action-oriented theorist, who is only interested in grounding according to Harnad (1990), might gladly embrace such models. However, someone who thinks cognition is entirely action-constituted (or sensorimotor-constituted) might not be satisfied with this result.

cognition: the sensorimotor format, the importance of seeing knowledge in terms of contingencies, both or neither? Do abstract concepts really "abstract" from sensorimotor experiences, in the sense of forming a schematized and potentially amodal internal representation, possibly linguistically mediated? Or should abstract concepts be conceptualized as "situated simulators" (Barsalou 2009) that reenact sensorimotor experiences, or even more drastically as collections of exemplars that share a family resemblance (e.g., for democracy, exemplars of situations where I vote, I discuss, etc.)?

More generally, once an abstract concept has been acquired, does it retain a "sensorimotor signature"? Consider two opposing, relatively extreme responses. The first fully reduces abstract concepts to encodings of SMCs: an abstract concept is nothing more than a set of such contingencies, properly grouped or associated together. The concept of a university, for example, can be formed of various associations between experiences of ways one might act (e.g., walking through a building) and the experiences that follow (e.g., the visual perceptions of classrooms and offices), to oversimplify the point. The second view takes SMCs to play a role only during the acquisition of abstract concepts: sensorimotor experience is like a ladder that gets kicked away once an abstract concept has been mastered. A subject might need some kind of sensorimotor experience to master the concept of a university (e.g., a subject must see some type of large institutional building or hear a lecture in one of them). Nevertheless, once a subject has had a sufficient amount of experience, the concept is mastered and the sensorimotor experiences that facilitated acquisition play no substantive role in the subject's later deployment of that concept. The subject can reason about and solve problems related to, for example, universities, without necessarily relying on any sensorimotor routines or engaging in any distinctively sensorimotor processing—thus using "amodal" concepts (Pezzulo, this volume; Weber and Vosgerau 2012).

These two views, however, do not exhaust the space of possibilities (Gentsch et al. 2016). Although an abstract concept may be used to solve problems, for example, in keeping with logical or semantic rules (e.g., universities enroll students, students read books, and so on), abstracted from sensorimotor processing, cognitive processing involving such units might reflect certain aspects of the way in which they were related to sensorimotor processing during acquisition. Sensorimotor aspects of concepts that generally appear to be amodal (as they are applied to a variety of circumstances) might be revealed through well-designed experimental manipulations in the manner of Chen and Bargh (1999). In their experiment, subjects were asked to recognize words with positive valence. They were able to do this more quickly when asked to indicate word recognition by pulling a lever toward themselves. Thus, it might be that after acquisition the neural unit which serves as the vehicle of the relatively amodal representation *good stuff* remains causally associated with the motor processing involved in the bodily action of pulling something toward oneself

(Rupert 1998, 2001, 2009). Such concepts may be neither fully reducible to SMCs nor entirely abstract.

Brain Mechanisms Supporting a Mastery of SMCs and Cognitive Development

Thus far we have described various factors and mechanisms that underlie cognitive development and elucidated the centrality of SMCs in this process. We have also discussed how an action-oriented approach might explain the development of higher cognition. How, then, can this action-oriented view of cognitive development be realized in the neuraxis?

Action-Perception Loops in the Brain Permit Acquiring a Mastery of SMCs

Does the brain architecture in humans (as well as our early evolutionary ancestors) support the type of action-perception loops required to learn SMCs? To address this very general question, we begin with a concrete example.

Research on the sensorimotor systems that underpin vibrissal active touch in rodents has identified an underlying neural architecture of nested sensorimotor loops that extend from the hindbrain to the cortex (Kleinfeld et al. 1999). In this system, which to a large extent reflects the organization of other mammalian sensorimotor systems, it is not helpful to distinguish components as being specifically sensory or specifically motor, despite standard nomenclature (e.g., primary somatosensory cortex or primary motor cortex). Due to the existence of tight feedback loops that connect sensing (deflection of the vibrissae) to action (movement of the body and of the vibrissae), it is more beneficial to decompose the system in terms of anatomical levels (hindbrain, midbrain, forebrain) and their interactions in loops that are closed through the world (Kleinfeld et al. 2006), rather than in terms of a feedforward architecture from sensing to action. This approach has been usefully extended to the design of artificial vibrissal sensing systems for biomimetic robots (Prescott et al. 2009), suggesting that action-perception loops—of the kind necessary to learn SMCs—constitute a useful organizational principle of the brain.

Development of SMCs across Several Levels of the Neuraxis

Can action-perception brain loops support the gradual acquisition of simple-to-complex SMCs necessary for action control and, most importantly, for the development of cognitive abilities? It has often been argued that a primary requirement for a living organism is the ability to manage adaptively with situated choices, not complex cognitive operations, and that the brain circuits originally developed to address the former might have been successively reused and extended for the latter (Cisek and Kalaska 2010; Pezzulo and Castelfranchi 2009).

We propose that SMCs develop across several levels that eventually could encompass intelligent or abstract cognitive functions.[4] (For brevity we do not focus on prenatal development, but it is likely to be important in this process and should not be ignored in a holistic take on these issues.)

Spinal sensorimotor loops are the first SMCs to develop in the somatosensory system. The next step could be the emergence of transcortical loops (somatosensory feedback which reaches the neocortex and activates the corticospinal tract). This may help extract more complex patterns of physical interactions and enable the extraction of semantic meaning at the perceptual level (feature detection), first at primary sensory cortical neurons and thereafter through proper connections with primary output cortical areas. At this level, information processed in primary cortical neurons is, to a large extent, still constrained by physical rules of the external world because of their proximity to the sensory receptors and muscles at the level of the neural connectivity. This process could accompany bodily physical maturation to the adult level, to complete the acquisition of SMCs necessary for physical interactions, and also to complete perception-action loops between the cognitive agent and the environment.

At the next stage there would be the formation of corticocortical circuits with higher association areas. Here, the information processed becomes gradually detached from physical sensorimotor rules and can thus be regarded as abstracted. There would be a constraint, however, in that the areal patterns of corticocortical connections and intracortical information principles should be largely conserved across the cerebrum, including sensorimotor associations. This stage of development proceeds under the mismatch of bodily (bodily completion) and neural (late completion, especially higher association areas where development continues until the late twenties in humans) development, namely adolescence. Social complexity during puberty would largely contribute to acquisitions of mechanisms for information processing at this stage, most of which are not instrumentally measurable.

Whereas the acquisition of SMCs is largely a matter of finding correlations between motor acts and patterns of sensory feedback, a similar mechanism can, in principle, be extended to complex cognitive and social domains, based either on SMCs or on "second-order" contingencies. In the social context, an individual can probe the responses of other persons by acting in different ways, in analogy with the sensory feedback. Different social interactions will generate varying sets of responses, which might form the basis for learning a (predictive) model of the interaction between the self and other people. At a more

---

[4] Illustrative videos can be found in the webpage of the Center on the Developing Child, Harvard University (http://developingchild.harvard.edu/resources/multimedia/videos/three_core_concepts/brain_architecture/), accessed on January 19, 2015.

advanced stage, this mechanism could permit acquisition of societal (a set of persons) models and how we predict that their interrelations will change under different (probing) conditions. In social settings this mechanism is boosted by other, not strictly sensorimotor, ways of acquiring models, especially through verbal communication. Linguistically mediated acquisition plausibly capitalizes on existing models—sharing some commonalities with them—but also opens new horizons, ultimately leading to abstract concepts, in ways that are incompletely known. Emotional components to cognitive thinking could be acquired through similar associative learning (e.g., when a child gets emotionally rewarded when exploring mathematical problems). This could confer a degree of individualization if the pattern of social rewards differs from individual to individual.

Several indications suggest that not only environmental but also genetic factors influence brain development. For example, there is a robust cerebral asymmetry in the newborn (Glasel et al. 2011), and it is believed that the early maturation of the language areas in the left hemisphere of the brain provide the basis for a start-up kit for language learning (Leroy et al. 2011).[5]

Brain Mechanisms Supporting Low- to High-Cognitive Abilities: A Computational Perspective

The traditional distinction between the domains of low- and high-level cognition (König et al. 2013) might not necessarily correspond to separate cortical operations. Instead, it might support a unified description in terms of hierarchies of loops that progressively "abstract" from sensorimotor domains.

Although we lack a comprehensive computational and mechanistic framework to describe this "abstraction" process, recent advancements in computational and systems neuroscience might offer insight into how this might be possible. A series of experiments using the distributed adaptive control robot architecture illustrates how a robot can progressively acquire increasingly complex behaviors (e.g., spatial navigation abilities) by starting from a small repertoire of initial reflexes. These reflexes support the acquisition of object and affordance representations which, in turn, form the basis to acquire hierarchies of procedural plans that permit the robot to optimize behavior over longer timescales (Verschure et al. 2003, 2014). From a different but related perspective, namely *active inference*, the brain is viewed as a statistical inference machine that encodes SMCs and other regularities using hierarchical models. Importantly, the functioning of the models at different hierarchical layers is not fundamentally different; however, they operate at different timescales (faster for the lower models, slower for the higher models) and thus can drive perception

---

[5] It is commonly believed that there may also be innate reflexes, but one factor of uncertainty to be tested in future studies is to what extent the biomechanical and anatomical properties of our bodies participate in dictating the emergence of these reflexes.

and action planning on different time horizons (e.g., proximal actions at the lower levels vs. distal plans at the higher levels) (Friston 2008, 2010).

Still another related approach asks whether the neuronal coding required for lower (e.g., sensory processing) and higher cognition might emerge by using (repeatedly) common optimization principles during learning or development. "Normative" approaches based on optimization principles have long been used to study sensory processing. For example, Olshausen and Field (1996) demonstrated that the optimization of sparse (and well-discriminable) representations over a universe of natural stimuli leads to response properties akin to simple cells in primary visual cortex. Furthermore, optimally stable/slow responses offer an explanation of complex cells in primary visual cortex (Berkes and Wiskott 2005; Körding et al. 2004). Subsequent studies extended this approach to higher levels in the visual hierarchy and other modalities (Berkes and Wiskott 2007; Dähne et al. 2014; Klein et al. 2003; Wyss et al. 2006).

In the original studies of Barlow (1961), the "sparsity" of representations derived from considerations of energy efficiency. More recent work, however, has shown that they can also be seen as approximations of *optimal predictability* in light of the subject's own action repertoire (König and Krüger 2006). This latter formulation links more directly to action-based theories, because an agent essentially builds a neural code that permits encoding and exploiting SMCs efficiently rather than just capturing the statistics of sensory events. This suggests that the same principles might be reused from sensory to more complex cognitive domains. Indeed, if the principle of *optimal predictability in light of one's own action repertoire* is replicated at higher hierarchical levels, sparse activation patterns are produced, corresponding to local and analogue activation patterns such as feature maps in low-level areas, up to binary, structured and disambiguated activation patterns (possibly implementing properties of symbols and syntax) at higher levels (König and Krüger 2006). At lower levels this mechanism produces sensory-related invariances (e.g., perceptual features), whereas at higher levels it produces action-mediated invariant representations of objects and affordances, in the sense that the properties defining an object (as an invariant) are not perceptual (like its color) but action-related, or the fact that a person can only execute certain actions on it (König et al. 2013). Accordingly, the activation of high-level neurons would not express passive object properties but "directives" (Engel et al. 2013) or inclinations of interactions with the objects, also possibly supporting so-called *predictive analogies* (Indurkhya 1992) with objects or events in another domain where it is still possible to apply the same set of actions.

To summarize, these arguments suggest that the repeated application of the *optimal predictability* principle would produce a natural transition from processing steps largely governed by external properties of external objects toward an active view of object interaction, thus possibly linking the acquisition of simple-to-complex cognitive abilities.

Open Questions

These examples illustrate that an action-oriented view can tackle important problems in a novel way, by considering the neural requirements for connecting sensory and motor streams as well as the nature of the learning problems to be solved by the brain. Although the computationally motivated hypotheses reviewed here require further empirical support, they illustrate how homogeneous processes of sensorimotor learning, hierarchical modeling, and optimal predictability might—at least in principle—explain the development of higher cognition in continuity with lower-level sensorimotor operations. Some domains of higher cognition demonstrate the predictions of the mechanisms sketched here, or similar ones, and have been investigated empirically: language (Pulvermüller 2005) and tool use (Maravita and Iriki 2004). The limitations of the action-oriented framework and the best way to test these empirically remain to be explored through future research.

Learning from Experiments on the Development of New SMCs

From an action-oriented perspective, interaction is claimed to shape (or at least contributes to the shaping of) fundamental categories (e.g., space and time) as well as basic phenomenological experience. However, to what extent this is possible is currently unknown. Ongoing research is addressing this problem using "sensory substitution" systems, which allow "new" (for our species) SMCs to be experienced.

Is It Possible to Acquire New SMCs and What Does It Imply?

The SMC theory holds that the quality of perception is constituted by mastery of SMCs; that is, the statistical relation of sensory changes and one's own actions (O'Regan and Noë 2001). As a corollary, introducing a new lawful relation should result in a new type of perceptual quality. An empirical test of this prediction was performed by using a newly developed sensory augmentation device (Kaspar et al. 2014; Nagel et al. 2005): a belt which translates the reading of a magnetic compass to a vibratory signal at the waist, always pointing north. This established a new SMC and permitted the effects of learning to "master" it to be scrutinized. After extensive training, subjects reported profound changes in perception: an enlargement of peripersonal space, increased use of egocentric reference frame, and enhanced security in navigation. Thus, a basic assumption of the theory of SMCs passed this empirical test.

None of the subjects, however, reported the development of a perception of the magnetic field as such, which would count as a truly novel experience for our species. At first glance, this does not seem noteworthy, as the orientation of the local magnetic field is of no behavioral relevance, but the consequences for spatial navigation are noteworthy. However, a sense of space preexists in normal subjects (cf. blind subjects; Kärcher et al. 2012) and is supported by vision, audition, and touch. Thus, perceptual changes induced by the sensory augmentation device appear as strong modifications of a spatial sense, not as a genuinely new and unique modality. Kärcher et al. (2012) conjectured that incomplete learning of a new modality resulted because the experimental subjects were well beyond the age of twenty. Only at a young age is the full set of mechanisms of neuronal plasticity available to support learning of SMCs. This defines a specific modality in support of a *transparent* perceptual access (t-SMC). "Transparent" refers to the common feature of sense modalities that present objects to us (and not percepts), as opposed to opaque processes, such as imagination, in which not the object but a mental image is presented to us (Martin 2002). When an individual is older (i.e., after the critical period of development), not all mechanisms of neuronal plasticity are available (Hubel and Wiesel 1970). This means that learning new SMCs will happen on top of preexisting SMCs. Perception, then, is not fully transparent, but defined by the preexisting modalities with which perception is associated—hence the term *associative SMC* (a-SMC). Thus, as demonstrated by Kärcher et al. (2012), adult subjects would be incapable of learning a truly magnetic sense; they would learn only a modification (albeit profound) of an already existing space perception.

Other Possible Effects of Learning New SMCs

The insights gained using an action-oriented approach might allow us to explain aspects of other phenomena, such as synesthesia and phantom limbs. In the paradigmatic case of color-grapheme synesthesia, viewing graphemes induces the perception of colors. For example, the letter "a" printed in black and white is perceived as red. The processing of graphemes involves the visual word form area, located directly anterior of human visual areas involved in color perception. Given present knowledge, it is highly plausible that this area supported different functions (e.g., object and face processing) before reading was learned (Dehaene and Cohen 2011). Thus the process of learning to read and process graphemes involves large-scale plasticity and possibly a "recycling" of the cortical network. A reduced capacity of neuronal plasticity (a-SMC) would lead to processing of graphemes but give rise to perception building on top of the already established visual-color SMC, thus leading to the perception of synesthetic colors. Hence, the phenomenon of synesthesia can be understood as maladaptation in the process of learning caused by a reduced set of plasticity mechanisms (a-SMC). Similarly, the phenomenon of phantom limbs can be analyzed. When a limb is lost in an accident, the respective cortical region is devoid of its original input. Due to neuronal plasticity, this input is substituted by other signals originating (or targeting) neighboring cortical areas. At a very young age, this plasticity is complete and gives rise to "normal" perception (t-SMC). However, after the age of six years, the fraction of patients experiencing phantom limbs rises quickly. Touching proximal parts of the body gives rise to the perception of a phantom limb (a-SMC).

All three phenomena—perceptual changes in adult subjects trained in a sensory augmentation device, synesthesia, and phantom limbs—could be understood as consequences of a reduced capacity of neuronal plasticity not giving rise to a fully transparent SMC as in early age, but to adapted processing still associated with the quality of perception of previously established SMCs. This suggests that there might be at least two phases in the development of action-derived cognition. In the first, a foundation is established; this might include the pre-wiring of subcortical systems and the acquisition of basic action-effect associations that can be used to drive basic actions in intentional ways, as well as a critical period for acquiring t-SMCs. During the second phase, brain plasticity still allows SMC-based learning but in ways that are constrained by the existing machinery; this would lead to developing a-SMCs and not t-SMCs. Despite these initial findings, several aspects of the acquisition of novel SMCs remain to be investigated (for a controversial discussion on early experiments aimed at manipulating perception by providing an "upright vision," see Kohler 1951).

Do We Need a New and More Interactive Method to Study Cognitive Development?

It is intrinsically difficult to study experimentally a multifaceted phenomenon that involves the brain, body, and environment over extended periods of time. An action-based view of cognitive development poses additional hurdles. Focusing on "active" agents—where "active" includes also "free" spontaneous (intrinsically motivated) exploration—is problematic for classical experimental paradigms, as these measure responses to controlled stimuli and are thus too restrictive from an action-based perspective.

Novel empirical methods are needed to address action and interaction. Many classical paradigms in cognitive psychology and neuroscience attribute a passive role to participants: predetermined stimuli are given and participants respond by choosing between predefined alternatives. Currently, several groups are working on methods to study cognition and cognitive development that will be more compatible with an action-oriented perspective (for a review, see Byrge et al. 2014). Below we review two important methodological aspects that need to be taken into account.

Decomposing the Development of Action Cognition

The classical "cognitive" approach and various versions of *enactivism* that are part of the action-based framework (for terminological clarifications, see Dominey et al., this volume) imply different ways of "decomposing" the issue of how action cognition develops. This, in turn, affects the methods used to study the development of action cognition. These two approaches differ with respect to who they regard as having the authority to define units of analysis. The classical approach assumes that scientists are entitled to do this, and that the priority is to find components which allow good prediction and control. Often, the variables of interest are the behavior of an individual or its brain state. Some types of enactivism, informed by phenomenology, assume instead that the units of analysis must honor the experience of the subject or agent (Gallagher 2005; Maturana and Varela 1980). For example, in a tool-use study, a subject feels that the tool is part of its own body. In this case, the unit of analysis could be "the subject plus the tool" rather than just "the subject," because from an enactivist perspective the subject and the tool form a system that should not be disentangled.

Moreover, the enactivist approach might consider contextual task elements in a different way than they are treated in traditional cognitive approaches. For example, in a categorization study, a cognitive psychologist usually groups various stimuli into different (fixed) categories for the purpose of statistical analysis. The enactivist might claim that the only, or most important, way to understand and categorize stimuli effectively is with respect to the ways in which subjects interact with objects (construed broadly). Stimuli might be grouped together according to kinds of subject-environment interaction. Stimuli that might have been treated, in the past, as the same (different) should be treated as different (same) depending on what is found in pre-experimental trials that measure ways in which subjects interact physically with potential stimuli. This would also mean that at each specific session or depending on a specific context, action cognition and its development can be decomposed differently.[6]

One important issue for future research is to assess the relative contribution of these two approaches to the study of cognitive development and to search for possible ways to integrate them. Some clues to unify the two contrasting views could come from the use of common computational approaches and "languages," including for example dynamical systems, complex systems, and hierarchical probabilistic methods, which are being successfully used in the study of cognitive development (Byrge et al. 2014; Pezzulo et al. 2011; Tenenbaum et al. 2011; Thelen et al. 2001) but require further elaborations through future studies. For example, some of the methodologies developed

---

[6] Consider an analogy with so-called Western and Eastern ways of thinking. Enactivists often refer to Buddhism and Mindfulness through meditation and holistic ways of thinking. As a simplified illustration of this contrast, the West is interested in objects (what) as units and in trying to formulate principles to relate and structure those units, whereas the East is interested in forms of relations (how) as units and trying to formulate structures of nodes to establish procedures to unify those units. "What"-oriented ways of understandings tend to appear static and reductionistic, whereas "how"-oriented understandings appear dynamic and holistic. This simplified illustration is offered just to exemplify the fact that both stances try to understand the same subject, albeit from different perspectives.

within the dynamical systems tradition can be used to define the most valid units of analysis inductively, in a way that might align well with enactivism. In this perspective, the most useful units of analysis to describe a given phenomenon (e.g., how parent-children dyads coordinate their actions) would result from an analysis of which "order parameters" regulate the dynamical interaction of the dyad. This is analogous to the way interpersonal distance and relative velocity have been used as "order parameters" to study the dynamical interactions of dyads (e.g., attacker-defender dyads) in sports (Araújo et al. 2006; Kelso 1995).

The Role of Robotics in the Study of Cognitive Development

Developmental processes span extended periods of time and many levels of complexity in behavior. To study this, it may be beneficial to complement traditional experimental methodologies, which are usually directed at isolated levels of behavior, with robotic approaches, which are naturally interactive and have the potential to consider several important determinants of cognitive development simultaneously (e.g., bodily actions and the situated and social aspects of a given cognitive task to be developed). The synthetic approach of robotics should be (and in some cases already is) synergistically fused with the experimental approach.

Developmental robotics seems to be particularly suited for this purpose. It is a small field that is organized primarily along two strands: (a) taking inspiration from human and animal development, it aims to build machines capable of *open-ended development* in the real world (an engineering goal); (b) it uses robots as tools to expand understanding of human development. Robotic models can be used in several ways to help us understand human cognition and its development (for an analysis of these various ways and concrete examples, see Oudeyer 2010). For example, specific data can be modeled from developmental sciences, often through collaboration between roboticist and developmental psychologists, neuroscientists, or linguists (for speech and language development mechanisms, see Broz et al. 2014; Moulin-Frier et al. 2014; Yurovsky et al. 2013; for language formation, see Steels and Belpaeme 2005). In addition, novel hypothesis can be concretely formulated, leading to novel experiments with humans (e.g., regarding the brain and behavioral mechanisms of intrinsically motivated exploration, see Baranes et al. 2014; Mirolli and Baldassarre 2013).

Despite these initial, promising results, a gap exists between the potential of robotic modeling of development as a tool to study cognitive development and its impact on the field. Progress in the field could benefit from more systematic exchanges between the "synthetic" method of robotics and the experimental paradigm (Pezzulo et al. 2011).

Conclusions and Open Challenges

An action-oriented approach has the potential to offer a much-needed unifying framework for the study of cognitive development, as it integrates multiple factors (e.g., sensorimotor learning, sociality, genes) and re-describes them from an action-based perspective (e.g., by asking which genetic biases can be considered to be action-based or useful to acquire SMCs). A research agenda for action-based cognitive development should consider how all these factors (and possibly others) are integrated, how they interact over time (for an initial proposal, see Figure 4.1), and what action-oriented aspects of development explain higher cognitive abilities.

One source of difficulty in research is that the most widely used empirical approaches to the study of cognitive function have intrinsic limitations in the study of action-based cognitive development. Subjects do not simply respond to stimuli predefined by the experimenter but are intrinsically "active" and "exploratory." Moreover, long timescales are involved in cognitive development, as are various interacting factors.

Consider, for example, the learning of a musical instrument. At a certain age, children are able to begin the process of acquiring the skills necessary to play an instrument (e.g., piano or clarinet). Skill learning results from a combination of at least two interacting processes: one domain general, the other context specific (e.g., linked to a specific musical instrument). A domain-general ability for sequence learning permits finger movements to be executed. It is not, however, tied to the mastery of a musical instrument but rather provides the prerequisite "substratum" or "scaffolding" to acquire musical proficiency. Whether a child learns finger movements needed to play the piano or clarinet is determined by the context, as sequence learning is specific to that particular instrument. The study of how a specific ability develops needs to distinguish whether the impact of sensorimotor learning on cognitive development depends on the "scaffold" or on the specific "contents" of the behavioral repertoire that a child learns.

To overcome such difficulties, we need to consider integrating different research traditions and incorporating ideas from dynamical systems and robotic approaches. Care must, of course, be taken to avoid simplifying the phenomenon of interest too much (e.g., taking a too reductionist or behaviorist perspective on what cognitive skills or concepts are and how they are acquired).

Ongoing research is currently investigating the key claims of action-oriented cognition and related embodied and enactivist approaches (Barsalou 2008; Byrge et al. 2014; Engel et al. 2013; Glenberg and Kaschak 2002; O'Regan and Noë 2001). Not all of these studies have investigated the *development* of action-based cognition, and additional research is needed to fill this gap. To begin, several ways that important hypotheses might be tested include the following:

- A fundamental challenge will be to assess the causal relation between action-based processes and development. Action-oriented theories suggest that cognitive abilities and the behavioral repertoire develop in parallel, but these two aspects diverge repeatedly during human development. Initially, infants have little control over their limbs, yet they readily engage in social interactions with other humans. Cognitive abilities seem to be more advanced than sensorimotor skills. There are, however, cases where an expansion of the behavioral repertoire does not readily correspond to an increase of cognitive abilities, such as when children who are supposedly able to *predict* the (sensory) consequences of their actions fail to *understand the implications* of executing an action (e.g., whether the action can cause harm). Do such mismatches in either direction suggest that cognitive development and the expansion of behavioral repertoire occur independently? For example, the ability to judge intentions in young children, contrasts with their use of this ability in laboratory experiments. More empirical research is needed to answer this question.

- After children master abstract concepts, it would be interesting to test whether the sensorimotor conditions under which an abstract concept was acquired continue to drive a child's application of that concept or reasoning with it. Such tests might involve exposure to potentially interfering sensorimotor stimuli (of the sort that facilitated acquisition) to see whether such stimuli continue to affect processing in abstract reasoning tasks and, if they do, to determine the precise nature and extent of their effect. In this way, we might test the extent to which concepts are, even after mastery, partly sensorimotor. This could constitute part of a larger program to test the "detachment" of abstract concepts from all sensorimotor conditions during cognitive development (Pezzulo and Castelfranchi 2007).

- To investigate whether sensorimotor learning merely scaffolds the development of concepts, or scaffolds development and remains an important part of their representation/instantiation, participants (e.g., children) could be given a novel sensorimotor experience after concept acquisition and then tested to determine whether this changes their application of concepts. Another possibility is to study the effects of early appearing motor deficits (e.g., due to prematurity or other pre- and perinatal hazards). Is cognitive development slowed globally by a pre-existing motor deficit, or are there specific effects on some but not all aspects of cognitive development? The ability to control vocalizations, hand movements, or facial muscles (disturbed in the case of Moebius Syndrome) are all known to occur as specific motor impairments from birth and thus might provide useful models for the study of fine-grained development of action-based cognition.

- To test the extent that concepts become "detached" from sensorimotor function, one could look at bimanual amputees, in which a lot of what comprises the sensorimotor contingencies represented at the neocortex is lost. Current evidence suggests that these patients do not have a radically different conceptual world or model of the world (Aziz-Zadeh et al. 2012). To elucidate developmental effects, it would be necessary to observe children born with severe motor limitations, either due to brain abnormality or bodily malformations, such as absence of limbs. If the view of "second-order contingencies" holds, the neural processing of abstract concepts might require the extrapolation and differential activation of nonsensorimotor aspects of experience.

Developmental robotics may offer novel insights into the mechanisms of action-based cognitive development (Cangelosi and Schlesinger 2014). Current studies are providing concrete mechanisms of proximal drivers for exploration, like *intrinsic motivation* or *curiosity*, and showing that they can automatically structure developmental stages in the long term—from the development of sensorimotor affordances to the onset of speech communication (Oudeyer and Smith 2016). Models of intrinsically motivated exploration, which drive the organism to explore through maximizing the reduction of prediction errors, combined with social guidance and motor synergies have demonstrated how such interacting mechanisms can self-organize a "learning curriculum," shaping the steps that progressively lead an organism to learn increasingly more complex contingencies (e.g., in the domain of early vocal development in infants; Moulin-Frier et al. 2014). These models provide precise predictions regarding the nature of these intrinsic rewards, as well as on behavioral consequences of such a self-generated curriculum (Gottlieb et al. 2013). An intriguing aspect of this research is that it might allow us to understand how developmental trajectories are constructed by a cognitive agent over time as a result of autonomous exploration rather than depending on fixed predefined stages. This speaks directly to a view of development as an *active* process.

The pursuit of these and other research directions holds promise to improve our scientific understanding of cognitive development by contributing to a much-needed theoretical synthesis of an action-based approach—one that will impact future education and technology.

Acknowledgments

We wish to thank Larry Barsalou, Karl Friston, Bernhard Hommel, Pierre Jacob, Tony Prescott, Friedemann Pulvermüller, Paul Verschure, Gabriella Vigliocco, and the other participants to the Ernst Strüngmann Forum who contributed to our discussions for their useful insights.

Action-Oriented Models
of Cognitive Processing

5

Can Cognition Be Reduced to Action?

Processes That Mediate Stimuli and Responses Make Human Action Possible

Lawrence W. Barsalou

Abstract

After treating action as peripheral for decades, cognitive scientists increasingly appreciate the fundamental roles it plays throughout cognition. Because action shapes cognitive processes pervasively, some theorists propose that cognition can be reduced to action. This chapter proposes that the central roles of action in human cognition depend on important processes that mediate between stimuli and responses. From this perspective, the unique features of human cognition reflect not only a remarkable potential for action, but also powerful abilities that mediate action in response to the environment. Sophisticated action results from sophisticated mediation; in particular, from mediating processes associated with representation, conceptualization, internal state attribution, affect, and self-regulation. Integrated with action systems, these mediating processes endow humans with unusually flexible and powerful means of shaping their physical and social environments. Without taking these mediating processes into account, it may be difficult, if not impossible, to explain human action. It may also be difficult to explain basic cognitive phenomena associated with memory, concepts, categorization, symbolic operations, language, problem solving, decision making, motivation, emotion, reward, self, mentalizing, and social cognition. Instead of reducing cognition to action, an alternative is to develop a viable theory that does justice to the importance of action in cognition, while integrating mediating processes that complement it.

Introduction

Peripheralizing action is undoubtedly one of the great distortions of traditional cognitive science. Although for decades it has been argued that action is central to cognition, many researchers still view action as little more than making responses, as indicated by the almost complete omission of action from modern cognition texts. Many researchers do not consider action to be a significant factor when theorizing about phenomena of interest or when developing experiments to test their theories. Often the biggest concerns about action are how to counterbalance response handedness and trim reaction time distributions.

A complementary distortion has been to view cognition as an information storage system, which Clark (1998) referred to as the "filing cabinet" metaphor. Accordingly, the primary purpose of the cognitive system is to develop accurate models that represent the world, and then to use these models in reasoning effectively during language and thought. From this perspective, it seems reasonable that the motor system would not play central roles in cognition, but would instead serve primarily to output information as needed, either from the storage system or from operations that act upon it.

Corrective steps have increasingly remedied these distortions. An initial step was the insight that timing mechanisms in the motor system play central roles in cognition (Ivry and Keele 1989). Another was the increasing realization that action is central to cognition (e.g., Glenberg 1997; Hommel et al. 2001; Prinz 1997), as are bodily states (Barsalou et al. 2003) and embeddedness in situations (e.g., Aydede and Robbins 2009; Barsalou 2003; Clark 2008). In addition, researchers have increasingly realized that anticipated action shapes perception and learning ubiquitously (e.g., Clark 2013b; Engel et al. 2013; Friston 2010). Reflecting these trends in the literature, researchers are becoming aware that action is central to cognition. As the title of Engel et al.'s (2013) article states, a "pragmatic turn" is occurring.

Engel et al. (2013), however, argue for a stronger reductive position, suggesting that cognition can be reduced to action (see also O'Regan and Noë 2001). They state, for example, that "cognition is action" and advocate that the field "transform the whole theory of cognition into a theory of action" (Engel et al. 2013:203, 207). They further argue that the construct of *representation* constitutes a further distortion in traditional theories and is unnecessary for explaining cognition. They propose that cognition can be fully explained with *directives*, which are "dispositions for action embodied in dynamic activity patterns" (Engel et al. 2013:206; see also the related construct of sensorimotor contingencies in O'Regan and Noë 2001).

If this account is correct, then the field should focus on action and move forward. Even if it is incorrect, it still might establish a useful dialectic, swinging the field in a much-needed direction before oscillating into a further evolved state. A similar debate occurred in the responses and excellent commentaries to O'Regan and Noë's (2001) target article. Some of the points presented there will be echoed here, but readers interested in these issues are encouraged to examine that discussion more closely.

An even larger debate has addressed the question of whether cognition can be reduced to *sensorimotor* systems in the brain (with reduction to action being a special case). More broadly, can cognition be reduced to *modal* systems, including systems for perception, action, and internal states (e.g., interoception, emotion, reward, motivation, proprioception, taste)? Conversely, this question can be asked in a somewhat different way: Are *amodal systems* necessary for cognition, with modal systems being insufficient?

Here I focus on three topics that may be useful to consider when addressing this family of questions. First, what is the nature of representation, and can reductionist programs afford to dismiss it? Of particular interest are caricatures of representation often used to discredit the construct, together with a modern account better suited for current research. Second, what phenomena must any reductive approach to cognition explain? What do we mean by *cognition*? Of particular interest are the extensive roles that representation has traditionally played in understanding cognitive phenomena. Third, what other basic building blocks besides action might be necessary to create cognition as we currently understand it? Of particular interest are powerful processes in humans that mediate between stimuli and responses.

Misrepresenting and Representing Representation

In many theories, representation is assumed to play fundamental roles throughout cognition. Conversely, representation is often criticized on numerous grounds and is sometimes dismissed as having nothing to do with cognition. Can these perspectives be reconciled?

Caricatures of Representation

One important problem in this ongoing debate is that critics often caricature representation. In perception, representation is sometimes cast as a picture in the head, or as a complete three-dimensional model, viewed by a homunculus. In memory and conceptualization, representation is often cast as a static structure built of amodal symbols residing in a disembodied storage system (e.g., a predicate, frame). Such accounts of representation—associated with traditional and increasingly outdated accounts of cognition—appear unlikely. Indeed, many proponents of representation readily dismiss such accounts and focus on more promising alternatives. Further, proponents note that representation is often dismissed unfairly because of its exclusive association with older accounts.

Broad Views of Representation

At least since Dretske (1995), representation has been defined in ways that make it a broader construct than the narrow subclass that critics often target. Within the broad class of representational systems, traditional representations occupy a relatively narrow corner. Many other more biologically and cognitively plausible forms of representation exist. For example, modern accounts establish representation in neural nets, neural circuits, dynamical systems, situated action, grounded cognition, and so on (e.g., Bechtel 2008; Markman and Dietrich 2000a, b; Prinz and Barsalou 2000; Ramsey 2007; Rupert 2011; Vosgerau 2010). From this perspective, Rupert refers to cognition and the brain as "massively representational."

A simplified form of these technical accounts specifies that A represents B when two conditions are met:

1. A possesses information about B, such that utilizing A makes it possible to interact effectively with B in some way.
2. A was established for the purpose of providing information about B in the service of achieving a goal (e.g., via neural architecture, evolutionary selection, etc.).

Thus, an analog thermometer represents temperature because the mercury level (A) carries information about temperature (B) and was designed to do so. In contrast, a footprint carries information about the size, shape, and identity of a person's foot, but it is not a representation because it was not created for this purpose and serves no goal. Interestingly, sensorimotor contingencies (O'Regan and Noë 2001) and directives (Engel et al. 2013) are representations under this definition (cf. commentary by Scholl and Simons 2001).

Thus, a representation does *not* require the stereotypical properties associated with traditional theories of cognition: It need not contain a complete detailed account of what it represents, nor be a picture in the head, a three-dimensional model, or a predicate calculus structure. It need not contain amodal symbols, reside statically and unchanging in long-term memory, or be viewed by a homunculus.

Consider an example of what might constitute a more plausible form of representation. As people experience hammers, brain areas that process their multimodal aspects become active and associated together (Martin 2007). Specifically, distributed associative patterns become established across fusiform gyrus (shape), premotor cortex (action), inferior parietal cortex (spatial trajectory), and posterior temporal gyrus (visual motion). Following many learning episodes with hammers, an increasingly entrenched associative network reflects the aggregate effects of neural processing in these areas. Based on the definition of representation above, this sloppy, difficult-to-localize network constitutes a representation because it carries information about hammers that can be used to perform situated action with them later (reflecting an evolutionary function for brains to operate in this manner).

A second form of representation results from activating small varying subsets of this distributed network on specific occasions. Upon seeing a hammer (or hearing the word "hammer"), a subset of the hammer network becomes active to represent or "simulate" the processing of a hammer in one of infinitely many ways (Barsalou 1999). Typically, these simulations remain unconscious, at least to a large extent, while causally influencing cognition and action. To the extent that part of a simulation becomes conscious, mental imagery is experienced. Such simulations need not provide complete or accurate representations, but are likely to be incomplete and distorted in many ways, representing abstractions, caricatures, and ideals as well as specific learning episodes.

In a Bayesian manner, the hammer simulated on a given occasion reflects aspects of hammers experienced frequently in the past, together with aspects that are contextually relevant (Barsalou 2011). In other words, the underlying network generates one of infinitely many simulations of a hammer dynamically, each adapted to the current situation. Once this simulation exists, it represents a hammer temporarily in working memory, producing, for example, anticipatory inferences about the object's affordances (e.g., Barsalou 2009; Vosgerau 2010). When the simulation dissipates, this particular act of representation ends.

These two forms of representation diverge considerably from classic accounts in the following ways:

1. There is no permanent static representation of hammers in long-term memory, built from amodal symbols, that is loaded into working memory identically across different occasions.

2. The representation that does reside in long-term memory results from superimposed effects of associative learning distributed across relevant sensorimotor systems, with the resultant network changing constantly after every learning episode (and overlapping considerably with networks for other categories).

3. When this distributed network is accessed, it produces one of infinitely many hammer representations dynamically.

4. These representations serve temporary representational functions by providing useful inferences in specific situations.

What about Cognition Must a Reductive Theory Explain?

Providing a complete account of cognition is *not* the goal of this discussion. Instead, the goal is to describe phenomena that reductive accounts of cognition probably need to explain. As will be seen, representation is traditionally assumed to play a central role in these phenomena. Thus, any theory which aims to explain these phenomena without representation must offer compelling alternative accounts.

Memory

In many memory phenomena, a cognitive/brain state provides information about a situation that is not present. Typically, researchers assume that information about a past event resides in memory, which later becomes active to represent it currently. Often the activation of a memory is assumed to motivate and guide subsequent action, but not necessarily (it could simply evoke emotion).

During explicit memory, recollection of a previous episode becomes conscious, including information about when and where the episode occurred (Squire et al. 2004). Medial temporal structures, such as the hippocampus, integrate memory elements for people, objects, settings, actions, events, self-relevance, and affect (distributed across multimodal brain areas). Not only are these distributed neural patterns often viewed as representations of past situations, the act of conscious recollection is typically viewed as a representational activity. Retrieving a memory provides information about an event that has passed (and thus is not present), representing what the event was like (not necessarily accurately). An explicit memory can similarly provide information about what a similar future event might be like, supporting future action.

During working memory, frontal areas maintain neural activity in posterior sensorimotor areas that became active while processing a stimulus that is no longer present (Levy and Goldman-Rakic 2000). Typically, these distributed representations are viewed as representing the recent stimulus for the purpose of performing a subsequent task effectively (e.g., *n*-back recall). In mental imagery, multimodal states activated in working memory are often viewed as consciously representing something that is not present, such as an object or event, again potentially supporting future action.

For both explicit and working memory, adopting classic forms of representation is not necessary, such as static amodal data structures. Instead, many modern researchers assume that memories result from the dynamic activation of distributed and constantly varying information in neural networks.

Concepts and Categorization

Conceptual knowledge constitutes another fundamental form of memory, often referred to as *semantic memory* (Barsalou 2012). Rather than representing specific events, however, conceptual knowledge represents categories in the world and in experience (e.g., objects, events, settings, mental states, bodily states). Conceptual knowledge is often viewed as representational in several ways. First, the concept that becomes established for a category (e.g., hammers) aggregates information acquired while processing category exemplars, thereby representing information about the category as a whole (e.g., their shape, function, motion). Second, a specific state can be constructed in working memory that represents the category temporarily. Third, these temporary states can be used to perform representational acts, such as imagining what a category exemplar might be like when one is not present or drawing anticipatory inferences about one that is. As described earlier, the representational processes underlying conceptual processing emerge naturally across distributed neural networks that function dynamically.

Categorization constitutes one of the most basic processes that organisms perform. As objects and events are perceived, they are mapped to concepts in the brain, thereby producing rich inferences from associated conceptual knowledge to support situated action (e.g., perceiving an object, categorizing it as a hammer, and inferring that it can be used to connect two boards with a nail). Once a type-token binding becomes established between an exemplar and a concept during categorization, the concept interprets the exemplar in one of infinitely many ways, given that an infinite number of concepts can be used to categorize any given exemplar (e.g., a hammer could be categorized as a tool, paperweight, or political symbol). Notably, such processing goes significantly beyond simple stimulus-response approaches that lack mechanisms for stimulus interpretation. A perceived hammer is not just a sensory state. When reading a manuscript outside on a windy day, a nearby hammer could be interpreted as a paperweight instead of a tool. Typically, stimulus interpretation is viewed as a representational process, with a concept projecting relevant interpretive inferences onto a stimulus once it has been categorized in some way.

Conceptual Operations

For decades researchers have assumed that basic conceptual operations underlie cognition (e.g., Pylyshyn 1973). As just discussed, categorization implements the basic conceptual operation of type-token binding. Although type-token binding could be implemented using amodal symbols in a predicate-inspired formalism, many more cognitively and neurally plausible implementations have been suggested as well (e.g., Barsalou 1999; Pothos and Wills 2011; Smolensky 1990). It is reasonable to assume that conceptual operations exist in any species that performs categorization.

Constructing relational structures is another basic conceptual operation. For example, organisms construct relational structures to represent part-whole relations in objects (e.g., eyes as parts of faces), and to represent relations between agents, goals, actions, objects, and outcomes in events (e.g., people eat soup with spoons). Once a relational structure has been constructed, it is generally assumed to temporarily represent a corresponding configuration of referents in the world, categorizing them via a type-token relation. Across diverse theoretical perspectives, relational structures are assumed to have important computational properties, such as productivity (e.g., Barsalou 1999; Fodor and Pylyshyn 1988). Again, relational structures offer infinite construals about a situation, produce powerful inferences that support situated action, and are likely to arise in any species that recognizes relations between stimuli, goals, actions, and outcomes.

Language

In humans, language offers an unusually powerful means of representing and conveying conceptual knowledge during communication and social coordination. Across diverse sociocultural activities, speakers attempt to convey conceptual representations through language to listeners, who attempt to reconstruct the intended meanings of these utterances. When comprehension occurs successfully, it establishes co-reference to relevant referents in the world and interprets them conceptually via lexical meanings and relational structures.

Similar to memory and conceptual knowledge, language offers another powerful means of representing objects and events not present. Especially important is the ability to represent situations that have never been experienced, or that may even be impossible to experience. Entertaining and working with such possibilities probably requires language and may be argued to be responsible for much of human culture and technology.

Problem Solving and Decision Making

During problem solving, agents attempt to construct a not-yet-realized plan for achieving a goal (e.g., how to paint a room). Often, different plans are formulated before one is executed, so that the possibilities can be evaluated, compared, and tweaked. During the planning process, each plan entertained is typically assumed to represent a possible action sequence in the world, together with outcomes that could result from performing it.

Similarly, in decision making, possible choices are evaluated based on temporarily constructed representations of them (e.g., when purchasing a product or choosing how to act in a social situation). By assessing represented choices before one is selected, their relative merits can be assessed, thereby attempting to optimize desirable outcomes.

Affect

Organisms experience a wide variety of affective states, including core affect, emotion, motivation, and reward. Interestingly, affective states are often associated with salient qualia (at least in humans), implicating their significant status in human consciousness. Indeed, "hot" affective states often trump "cold" cognitive states in many decision-making contexts. How people feel is central to how they act.

Action often originates in affective states (Frijda 1986). Arguably, affective states evolved because of their importance in signaling the necessity of action in certain situations and motivating its successful execution. Can affective states, however, be reduced to action?

An affective state can include activity in the cardiovascular, respiratory, autonomic, endocrine, and immune systems as well as in the musculature.

In general, these states are assumed to result from conceptually appraising an object or event as significant for oneself in some way. For example, appraising an object as "threatening" triggers bodily states that signal danger. Analogously, appraising an object as a "tasty food" triggers bodily states that motivate consumption. As these examples illustrate, an affective state can be viewed as saliently representing the significance of an object or event for oneself.

Once an affective state exists, it often motivates action aimed at resolving the object or event that has destabilized homeostasis. In addition to energizing the motor system, an affective state may initiate a wide variety of actions associated with approach, avoidance, consumption, and so forth, all designed to create change in some (hopefully) useful way. Thus, an affective state, besides representing self-relevance, can be viewed as representing a desire to effect self-relevant change. Because affective states are so salient subjectively, what they represent often comes to dominate the choice of actions, and the intensity with which actions are pursued. Although affect and action are often closely related, affect appears to complement and coordinate action in various ways, rather than being reducible to it.

Self and Mentalizing

What is the nature of the self system that detects self-relevance and initiates relevant actions? Increasingly, self is viewed as taking multiple forms that serve multiple purposes (Damasio 2000; Gallagher 2013). For example, a conceptual form of self is often viewed as representing the identity that one would like (or ought) to be (e.g., traits, roles, values, goals, norms). An experiential form of self is often viewed as representing who someone is at the current moment (e.g., bodily states, affect, thoughts). Clearly, these different forms of self evolve through situated experience of action and play central roles in motivating action. Nevertheless, they also seem to have a representational function of defining how one conceptualizes and experiences oneself.

Just as the importance of the default mode network in the brain has become increasingly apparent, so too has the constant mentalizing and mind wandering that it produces (e.g., Buckner et al. 2008; Gerlach et al. 2014; Smallwood and Schooler 2015). When people are not performing a focused task, they often appear to mentalize about themselves, social connections, daily activities, long-term goals, emotional events, appetitive stimuli, and so forth. Not surprisingly, brain areas that implement mentalizing overlap extensively with areas that process self-relevance (Northoff et al. 2006).

Typically, researchers assume that mentalizing serves a representational function. While mentalizing, for example, one might imagine various forms of one's actual and desired selves. Similarly, one might ruminate about various events and interactions that have implications for oneself. In many cases, representing nonpresent situations in their absence makes it possible to increase understanding of their self-relevance, to regulate affective reactions, and to perform effective future action.

Social Cognition

Humans are adept at representing others as well as themselves. Indeed, many proposals about the evolution of human cognition suggest that its primary adaptations supported revolutionary advances in social cognition, interaction, and communication (Donald 1993; Tomasello 2009). Thus, to understand human cognition, it is essential to understand its social character.

Considerable literature demonstrates that social interaction revolves around attributions about the cognitive and affective states of others. Similar to how individuals represent their own senses of self, they similarly try to represent other people's sense of self conceptually and experientially. In the process, perceivers represent a wide variety of cognitive, affective, and bodily states, as they attempt to explain and predict what people are thinking and feeling, and why they act as they do.

Which Underlying Capabilities Enable Human Cognition?

If we ask ourselves what makes humans so remarkable as a species, would we say that it is simply our ability to act? Certainly, we act in ways that are unique and powerful (at times frighteningly). However, are there other remarkable abilities, that make our actions possible, which cannot be reduced to action in a compelling way?

If one asked cognitive psychologists to provide a value for X in "Cognition is X," what would be their response? I'm betting that many would reply "memory." Arguably, a sensory or motor process becomes cognitive once memory in some form contributes to it through top-down processing. Many cognitive psychologists might also include attention as a key capability. Memory and attention are critical for cognition, not only in humans, but across species, along with sensation and perception (especially in grounded approaches that emphasize the importance of sensorimotor systems in higher cognition). I suspect that most modern researchers would not be comfortable defining cognition solely as action without including these other basic systems. Without them, even simple actions cannot occur, much less sophisticated ones.

In addition to these basic processes shared across species, are there other important capabilities that make the unique and powerful character of human cognition possible? In general, capabilities that mediate sensation and action seem like a good place to look, given that humans are typically viewed as excelling in this regard. Perhaps humans have capabilities, not best understood as action, that explain why human action is so remarkable. Characterizing such capabilities accurately and optimally may require postulating mechanisms beyond those associated with action per se.

Representing Nonpresent States

People represent nonpresent states frequently in episodic memory, working memory, imagery, problem solving, decision making, mind wandering, and social cognition. Donald (1993) argued that humans excel in their ability to represent nonpresent states in the past and future, whereas most other species are largely locked into the present moment, with modest exceptions (see also Prinz and Barsalou 2000). By representing nonpresent situations, humans evolved to achieve diverse goals beyond the reach of other species. From this theoretical perspective, a central computational strength of human intelligence is its ability to represent nonpresent situations.

Again, this ability need not imply static amodal representations in long-term memory. Instead, dynamic multimodal representations temporarily constructed in working memory may typically represent nonpresent situations. Consistent with this perspective, Donald (1993) argued that human fitness increased significantly when written language made it possible to represent conceptual understandings indefinitely, archiving them in a stable and precise manner across generations. On one hand, the value of written records reflects the vagaries of dynamic neural representations that are imprecise and unstable; on the other, it reflects the utility of developing stable representations that are useful.

Representing nonpresent situations appears central to expanding the action repertoire of humans. By examining a represented situation from novel perspectives, diverse actions can be generated creatively that exceed stimulus-response conditioning (see commentaries by Pylyshyn 2001; Van Gulick 2001). By representing geography with maps, navigational capability expands significantly beyond action-based route navigation (Spelke and Lee 2012). The ventral stream represents objects and space in ways that can transcend action (Milner and Goodale 1995). Representing stable objects in perception appears to support effective eye scanning (see commentaries by De Graef et al. 2001; Tatler 2001) and may underlie shape and color constancy. Similarly, representing intended actions perceptually as forward models makes the tracking and successful execution of action possible (see commentary by Gallese and Keysers 2001). It seems difficult to explain how processing false beliefs and counterfactuals, together with the actions they enable (e.g., deception), could occur without representation (Wellman et al. 2001). Finally, establishing principles of the physical world scientifically, not just in human minds but in written records, has led to a spectacular expansion of human action. It's difficult to imagine how modern engineering, technology, and medicine could have developed without explicit attempts to represent scientific principles cognitively during their discovery, documentation, and application.

In general, the utility of representing nonpresent states is consistent with the principle that distinguishing data from process is computationally useful. In a completely compiled procedure, no flexibility exists—data embedded in the procedure always operate under a single task-specific interpretation. In contrast, when data are represented independently, they can be processed more flexibly, allowing novel affordances to be exploited at later times as they become apparent and needed. Because systematic procedures can extract these affordances from data, homunculi are not necessary. Arguably, human cognition capitalized on this computational principle by evolving prolific abilities to represent nonpresent situations. Cognition built solely on action would be much less flexible and adaptive.

Conceptualization under Linguistic Control

Humans excel in their ability to interpret the world conceptually in creative ways that support novel and powerful action. For example, humans develop scientific taxonomies to conceptualize, organize, and categorize domains of experience (Malt 1995). Where would chemistry and biology be without taxonomic classifications of elements and species? On a daily basis, humans perform diverse forms of goal-directed categorization that enable mundane but still impressive forms of action (Barsalou 1991; 2003). By conceptualizing foods as containing calories, fats, and nutrients, for example, it becomes possible to categorize them as healthy and unhealthy, thereby controlling one's weight and longevity. By reappraising a situation as affording a challenge instead of a threat, people can change their emotional reaction from fear to excitement, in turn, changing their action from avoidance to approach.

Representation appears central to the conceptualization of nonpresent objects and events dynamically, creatively, and effectively in the service of goal pursuit. Developing conceptual knowledge about categories enables humans to represent them offline, while trying to understand their nature and select actions for interacting with them effectively in future online situations. In particular, the possibilities entertained, evaluated, and manipulated during planning, decision making, and mind wandering originate in conceptual knowledge that represents the world and human action in it.

Language appears central to people's impressive conceptual abilities. The productive and quasi-compositional structure of linguistic forms is likely to play central roles in manipulating conceptual structures creatively, as people attempt to understand and manipulate the physical world, and to coordinate action within social groups. As a consequence, various forms of abstract thought arise, enabling actions that would not be possible without them (as in science, government, and business).

Internal State Attributions

Attributing internal states to objects and people appears to be an especially important conceptual ability in humans. People naturally attribute essences to objects even when they do not exist (Gelman 2003). Similarly, people attribute internal traits to themselves and others which arguably do not exist (Mischel 1968). Not only do lay people attribute internal states informally, scientists attribute them formally (e.g., the attribution of atomic structure to chemical substances, energy and force to objects, and genetic structures to organisms).

In all cases, attribution of internal states supports action. By attributing traits to people, we can predict their behavior. By attributing atomic structure to chemical substances, we can predict how chemical compounds will react. By attributing self-concepts to individuals, social organizations develop the ability to regulate and coordinate individual action via roles and group identities. The ability to develop, represent, and manipulate conceptual structure—in this case about internal states—enables action which otherwise would not be possible.

Rich Affective Experience

Humans appear to experience richer affective states than do other species. Not only are we motivated to eat, reproduce, and survive, we are motivated to develop artistic skills, become better people, improve our communities, and worship deities. Similarly, people do not only perceive pleasure and safety to be rewarding, they also find it rewarding to succeed professionally, to have others recognize their contributions, and to leave a legacy. Human emotional experience appears unusually rich: not only do people experience fear and disgust, they experience complex social emotions such as gay pride, guarded optimism, and civic responsibility.

From the perspective of appraisal and constructionist theories, these complex forms of affect reflect the impressive conceptual abilities of humans (e.g., Barrett 2006; Scherer 2001; Wilson-Mendenhall and Barsalou 2016). Specifically, the ability to construct complex conceptualizations of affective situations increases the richness of the affect experienced in them. In turn, rich affective states produce unusually complex and nuanced actions aimed at resolving the respective situations.

Self-Regulation

For every cognitive or social ability that has ever been studied scientifically, a dual-process theory has probably been proposed to explain it (e.g., Sherman et al. 2014). Generally speaking, dual-process theories contrast cognitive processes that are relatively fast, implicit, effortless, and habitual with cognitive processes that are relatively slow, explicit, effortful, and regulatory. Although the extent to which implicit habits govern daily life is impressive, so is the human regulatory capacity. Using an expanded executive system together with metacognition, humans regulate perception, conceptualization, affect, and action extensively, at least relative to other species.

Significantly, self-regulation can be viewed as internal action that regulates external action. Nevertheless, the executive system depends critically on representing situations in their absence, establishing conceptual structure that informs sophisticated goal-directed action, and assessing self-relevance and affect continuously. Without these complementary abilities, executive action could not function as effectively as it does in humans. All of these abilities work together, mediating sensation and physical action. In an integrated manner, these mediating abilities endow humans with unusually flexible and adaptive means of designing and selecting effective actions in their physical and social environments.

Implications for Cognition and Action

Based on the mediating processes just reviewed, definitions of cognition and action follow that motivate the central role of action in cognition.

Cognition

Imagine a system that maps sensed stimuli directly onto motor responses in a deterministic manner. Because this system has no processes that mediate between stimuli and responses, it cannot change its responses to a given stimulus adaptively.

Now imagine inserting the kinds of mediating processes just described between sensing a stimulus and responding to it. With mediating processes in place, it becomes possible to change responses to the same stimulus in ways that optimize outcomes for the system in its environment.

This perspective suggests that cognition can be defined as mediating processes between stimuli and responses that yield adaptive action. Under this definition, cognition evolved for the purpose of creating novel actions, selecting among actions, regulating actions, and so forth, thereby increasing the chances that relevant goals in the environment are achieved successfully. Importantly, cognition is not an end in itself, as is often assumed implicitly when its implications for action are minimized or ignored. Instead, cognition exists to support adaptive action.

Action

In a system with no mediating processes, an action simply reflects the stimulus that triggered it. The action is evolution's solution to the stimulus, encoded into a hard-wired system. For all intents and purposes, the stimulus is the proximal cause of the action. In a system with mediating processes, however, the proximal cause of an action is one or more cognitive states, with any eliciting stimulus being a distal cause (except for reflexes). As a consequence, nonreflexive actions reflect contributions from mediating states.

Consider the action of donating to charity over the Internet. The distal stimulus for this action might be learning about a disaster somewhere in the world. Via mediating processes, the disaster activates conceptual knowledge and affect, which in turn motivates making a charitable donation. At the motor level, the act of donating might include eye movements associated with reading a computer screen and finger movements associated with operating a keyboard. At the level of mediating processes, however, the action provides resources to a charity for disaster relief. The important goal of this action is not to produce the supporting motor responses, but to produce sociocultural effects on the environment (i.e., to transfer funds from a person's bank account to a charity to aid victims of the disaster). A tremendous amount of conceptual knowledge about banking, social organizations, disasters, identity, and social responsibility makes this action possible, together with supporting emotion, motivation, and reward. Without these mediating processes, donating money to disaster relief would be incomprehensible, much less feasible. It is difficult to imagine such actions as simply the result of sensorimotor contingencies.

Important human actions generally appear to reflect mediating processes in this manner. Consider the actions that modern humans use to acquire physical resources, including food, shelter, and wealth. We do not simply pick food up off the ground and sleep in caves. Instead, we purchase food from stores, prepare food in kitchens, procure residences, and furnish them. We alter the physical environment extensively (arguably too much), clearing and farming land, damming rivers, and managing wildlife. While acquiring physical resources and altering the environment, we invent and use sophisticated artifacts, including machines, computers, and communication systems. Rather than performing these actions in isolation, we often coordinate them in social groups, relying heavily on language to do so. In performing sociocultural actions, we also rely heavily on institutions and cultural knowledge, including science, technology, ethics, and law.

In general, one can view this remarkable action repertoire as serving niche construction. Over the course of human evolution, we have created physical and social environments that have increased our fitness and, in turn, our ability to harvest environmental resources. Without this action repertoire, such niche construction would not have been possible.

It is difficult to imagine our action repertoire evolving without the kinds of mediating processes described earlier. Sophisticated conceptual understandings of the environment, social groups, tools, and ourselves are essential for formulating actions and representing their anticipated impact on niche construction. Similarly, our affective states, social awareness, communication skills, and regulatory capacities further contribute in myriad ways to our actions and their environmental impact. The mark of a sophisticated human action is the sophisticated cognition behind it.

Clearly, automatized expert performance in many domains benefits from the development of sensorimotor contingencies, making responses to relevant stimuli more efficient through learning. Nevertheless, it is difficult to imagine how the remarkable repertoire of human actions could have developed in the first place without sophisticated mediating processes. Similarly, it is difficult to imagine how we continue to regulate this action repertoire and develop it further without these processes. Even when we perform a relatively automatized action, mediating processes are often available to comprehend, support, and alter it adaptively, should doing so be useful.

Conclusion

Increasingly, researchers appreciate the central roles that action plays throughout cognition. To understand any cognitive process, it is important to understand not only its constant entwinement with action, but also how action contributed to creating it.

Here I have proposed that mediating processes constitute human cognition, including representation, conceptualization, internal state attribution, affect, and self-regulation. Without these mediating processes, human action would not exhibit the remarkable and powerful forms it takes. Although I propose that these mediating processes cannot be reduced to action, I hasten to add that none of these mediating processes could develop without action. Action is necessary for these processes to develop during childhood and is deeply implicated in the forms they take. Furthermore, the expression of these mediating processes occurs through action.

Reducing cognition to action is not only an ambitious project, it is a provocative one. Undoubtedly we can learn much from it. Ultimately, the project could succeed, or at least contribute to dialectic change.

An alternative project might be to sketch the outline of a viable theory that does justice to the importance of action in human cognition, integrating it properly with the powerful processes that mediate sensation and action. No doubt, such integration must also take into account the physical and social situations in which human perception and cognition are embedded as well as the constant and constraining couplings between them all.

Acknowledgments

I am grateful to the organizers of the Ernst Strüngmann Forum for the opportunity to write this chapter, to Esther Papies, Chris Frith, Cecelia Heyes, Marek McGann, Friedemann Pulvermüller, and Gottfried Vosgerau for helpful comments on this chapter, and to all the Forum participants for their inspired and inspiring contributions across our week of meetings.

6

The Mindful Filter

Free Energy and Action

Karl J. Friston

Abstract

This chapter frames key questions about embodied cognition and action in terms of
active inference; namely, the premise that the brain is trying to infer the causes of its
sensory input—and samples that input to minimize uncertainty about its inferences.
This provides a process theory for embodied exchanges with the world that can be cast
as a Bayesian filter—or more simply predictive coding—equipped with classical reflexes. The ensuing (embodied inference) perspective raises interesting questions about
embodiment and enactivism: Can we ever truly observe worldly states? Are there such
things as representations? Can we make inferences about other agents who are making inferences about us? These questions are unpacked in terms of the unobservability
assumption, sensorimotor contingency theory, and the active inference perspective on
mirror neurons and inferring the intention of others.

Active Inference and Predictive Coding

This chapter begins with a brief overview of active inference, with a particular
focus on process theories and implementation in the embodied brain. Predictive
coding will be used as a metaphor for neuronal message passing, as we consider how the basic imperatives of the Bayesian brain (predictive coding) can
be met through active sampling of the sensorium. Having established the basic
idea, I conclude with a statement of the three basic questions articulated above
and the constraints afforded by active inference.

Recent advances in theoretical neuroscience have inspired a (Bayesian)
paradigm shift in cognitive neuroscience. This shift is away from the brain as
a passive filter of sensations—or an elaborate stimulus-response link—toward
a view of the brain as a statistical organ that generates hypotheses or fantasies
(fantastic: from Greek *phantastikos*, the ability to create mental images, from
phantazesthai), which are tested against sensory evidence (Gregory 1968).

This perspective dates back to the notion of unconscious inference (Helmholtz 1866/1962) and has been formalized in recent decades to cover deep or hierarchical Bayesian inference—about the causes of our sensations—and how these inferences induce beliefs, movement, and behavior (Dayan et al. 1995; Lee and Mumford 2003; Friston et al. 2006; Hohwy 2013; Clark 2013b).

Predictive Coding and the Bayesian Brain

Modern formulations of the Bayesian brain—such as predictive coding—are among the most popular explanations for neuronal message passing (Srinivasan et al. 1982; Rao and Ballard 1999; Friston 2008; Clark 2013b). Predictive coding is a biologically plausible process theory for which there is a considerable amount of anatomical and physiological evidence (Mumford 1992; Friston 2008). For a review of canonical microcircuits and hierarchical predictive coding in perception, see Bastos et al. (2012); for a treatment of the motor system, see Adams, Shipp et al. (2013) and Shipp et al. (2013). In these schemes, neuronal representations in higher levels of cortical hierarchies generate predictions of representations in lower levels. These top-down predictions are compared with representations at the lower level to form a prediction error (usually associated with the activity of superficial pyramidal cells). The ensuing mismatch signal is passed back up the hierarchy to update higher representations (associated with the activity of deep pyramidal cells). This recursive exchange of signals suppresses prediction error at each and every level to provide a hierarchical explanation for sensory inputs, which enter at the lowest (sensory) level. In computational terms, neuronal activity encodes beliefs or probability distributions over states in the world that cause sensations (e.g., my visual sensations are caused by a *face*). The simplest encoding corresponds to representing the belief with the expected value or *expectation* of a (hidden) cause. These causes are referred to as *hidden* because they have to be inferred from their sensory consequences. We will see later that these expectations are not limited to beliefs about the causes of sensations but include expectations about how those sensations are sampled through action. In summary, beliefs about hidden causes are engendered by ascending prediction errors, reporting sensory information that has yet to be explained. When these expectations are endowed with dynamics, descending predictions can preempt or anticipate sensory trajectories, thereby minimizing prediction errors as the sensorium unfolds.

Predictive coding represents a biologically plausible scheme for updating beliefs about states of the world using sensory samples (Figure 6.1). In this setting, cortical hierarchies become a neuroanatomical embodiment of how sensory signals are generated; for example, a face generates luminance surfaces which generate textures and edges and so on, down to retinal input. This form of hierarchical inference explains a large number of anatomical and physiological

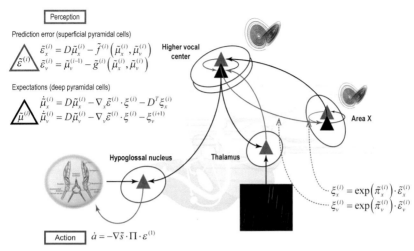

Figure 6.1 Hierarchical message passing in predictive coding using the (simplified) neuroanatomy of a songbird. Neuronal activity encodes expectations about the causes of sensory input, where these expectations minimize prediction error. Prediction error is the difference between (ascending) sensory input and (descending) predictions of that input. This minimization rests upon recurrent neuronal interactions among different levels of the cortical hierarchy. The available evidence suggests that superficial pyramidal cells (red triangles) compare the expectations (at each level) with top-down predictions from deep pyramidal cells (black triangles) of higher levels. Left: the equations represent the neuronal dynamics implicit in predictive coding. Prediction errors at the i-th level of the hierarchy are simply the difference between the expectations encoded at that level and top-down predictions of those expectations. The expectations per se are driven by prediction errors so that they perform a gradient ascent on the sum of squared (precision weighted) prediction error. [See Appendix for a detailed explanation of these (simplified) equations.] Right: scheme of the songbird's auditory system, showing the putative cells of origin of ascending or forward connections that convey (precision weighted) prediction errors (red arrows) and descending or backward connections (black arrows) that construct predictions. In this example, area X sends predictions to the higher vocal center, which projects to the auditory thalamus. However, the higher vocal center also sends proprioceptive predictions to the hypoglossal nucleus, which are passed to the syrinx to generate vocalization through classical reflexes. These predictions can be regarded as motor commands, while the descending predictions of auditory input correspond to corollary discharge. Note that every top-down prediction is reciprocated with a bottom-up prediction error to ensure predictions are constrained by sensory information. The Lorenz attractors associated with higher levels of the hierarchy indicate that generative models in the brain can possess autonomous, and possibly chaotic, dynamics with deep (hierarchical) structure (for details, see Kiebel et al. 2009; Friston and Kiebel 2009).

facts (Friston 2008; Adams, Shipp et al. 2013; Bastos et al. 2012). In brief, it explains the hierarchical nature of cortical connections, the prevalence of backward connections, as well as many of the functional and structural asymmetries in the extrinsic (between region) connections that link hierarchical levels (Zeki and Shipp 1988). These asymmetries include the laminar specificity of forward and backward connections, the prevalence of nonlinear or modulatory backward connections (that embody interactions and nonlinearities inherent in the generation of sensory signals), and their spectral characteristics—with fast (e.g., gamma) activity predominating in forward connections and slower (e.g., beta) frequencies that accumulate evidence (prediction errors) ascending from lower levels.

Precision and the Encoding of (Un)certainty

One can regard ascending prediction errors as broadcasting "newsworthy" information that has yet to be explained by descending predictions. However, the brain has to select the channels it listens to—by adjusting the volume or *gain* of prediction errors that compete to update expectations in higher levels (Clark 2013a). Computationally, this gain corresponds to the precision or confidence associated with ascending prediction errors. However, to select prediction errors, the brain has to estimate and encode their precision (i.e., inverse variance). Having done this, prediction errors can then be weighted by their precision so that only precise information is accumulated and assimilated at high or deep hierarchical levels. As with all expectations, expected precision maximizes Bayesian model evidence (see Appendix). In other words, not only does the brain have to infer the causes of sensations, it also is in the difficult game of inferring the context, in terms of the reliability of prediction errors at each level of the hierarchy and the implicit confidence in associated predictions.

The implicit broadcasting of precision-weighted prediction errors rests on synaptic gain control (Moran et al. 2013). This neuromodulatory gain control corresponds to a (Bayes-optimal) encoding of precision in terms of the excitability of neuronal populations reporting prediction errors (Feldman and Friston 2010; Shipp et al. 2013). This may explain why superficial pyramidal cells have so many synaptic gain control mechanisms, such as NMDA receptors and classical neuromodulatory receptors like D1 dopamine receptors (Goldman-Rakic et al. 1992; Braver et al. 1999; Doya 2008; Lidow et al. 1991). Furthermore, it places excitation-inhibition balance in a prime position to mediate precision-engineered message passing within and among hierarchical levels (Humphries et al. 2009). The dynamic and context-sensitive control of precision has been associated with attentional gain control in sensory processing (Feldman and Friston 2010; Jiang et al. 2013) and has been discussed in terms of affordance in active inference and action selection (Frank et al. 2007; Cisek 2007; Friston, Shiner et al. 2012). Crucially, the delicate balance of precision over different hierarchical levels has a profound effect on inference, and may also offer a formal understanding of false inference in psychopathology (Fletcher and Frith 2009; Adams, Stephan et al. 2013). We will see that it also plays a crucial role in sensory attenuation.

Active Inference and Action

So far, we have only considered the role of predictive coding in perception or perceptual inference, through minimizing prediction errors. However, there is another way to minimize prediction errors; namely, by resampling sensory inputs so that they conform to predictions. This is known as active inference (Friston et al. 2011). In active inference, action is regarded as the fulfilment of descending proprioceptive predictions by classical reflex arcs. In more detail, the brain generates continuous proprioceptive predictions about the expected location of the limbs and eyes, and these predictions are hierarchically consistent with the inferred state of the world. In other words, we believe that we will execute a goal-directed movement, and this belief is unpacked hierarchically to provide proprioceptive and exteroceptive predictions. These predictions are then fulfilled automatically by minimizing proprioceptive prediction errors at the level of the spinal cord and cranial nerve nuclei (see Adams, Shipp et al. 2013 and Figure 6.1). Mechanistically, descending proprioceptive predictions provide a target or *set point* for peripheral reflex arcs, which respond by minimizing (proprioceptive) prediction errors.

The argument here is that the same inferential mechanisms underlie apparently diverse functions (e.g., exteroception in the visual cortex, interoception in the insula, and motor control in the motor cortex). Crucially, because these modality-specific systems are organized hierarchically, they are all contextualized by the same conceptual (amodal) expectations that generate descending predictions in multiple (exteroceptive, interoceptive and proprioceptive) modalities. In short, action and perception are facets of the same underlying imperative; namely, to minimize hierarchical prediction errors through selective sampling of sensory inputs. However, there is a potential problem:

Action and Sensory Attenuation

If proprioceptive prediction errors can be resolved by engaging classical reflexes (action) or changing expectations (perception), how does the brain adjudicate between these two options? The answer lies in the precision afforded to (proprioceptive) prediction errors and the consequences of movement sensed in other modalities. To engage classical reflexes, it is necessary to increase their gain through augmenting the precision of (efferent) proprioceptive prediction errors that drive neuromuscular junctions. However, to preclude (a veridical) inference that the movement has not yet occurred, it is necessary to attenuate the precision of (afferent) prediction errors that would otherwise update kinematic expectations or beliefs (Friston et al. 2011). Put simply, my prior belief that I am moving can be subverted by sensory evidence to the contrary; thereby precluding movement. In short, it is necessary to attenuate *all of the sensory consequences of moving*—leading to an active inference formulation of sensory attenuation. Sensory attenuation is the psychological phenomena where the magnitude of self-made sensations are perceived as less intense (Brown et al. 2013).

In Figure 6.1, the (afferent) proprioceptive prediction error was omitted from the hypoglossal nucleus (for discussion of this omission and the agranular nature of motor cortex, see Shipp et al. 2013). This renders descending proprioceptive predictions *motor commands*, where the accompanying exteroceptive predictions become *corollary discharge*. The ensuing motor control is effectively *open loop*. However, the hierarchical generation of proprioceptive predictions is contextualized by sensory input in other modalities, which register the sensory consequences of movement. These sensory consequences are transiently attenuated during movement. In summary, to act, one needs to temporarily suspend attention to the consequences of action in order to articulate descending predictions (Brown et al. 2013). With this conceptual framework in place, let us now consider some key questions it raises about the nature of representation and mindful action.

Some Key Questions

The first question raised by the above formalism is whether causes of sensations can ever be observed or perceived. Put simply, is perception the same as inference or are they somehow distinct? This argument becomes particularly acute when inferring the mental state of others, where one might call upon generative models of self-made acts to infer the intentions of others (Kilner et al. 2007; Teufel et al. 2010; Brown and Brüne 2012). For example, Bohl and Gangopadhyay (2013) distinguish among several versions of what they term unobservability assumption, as discussed by Michael and De Bruin (2015):

1. It may be understood as a phenomenological thesis: one does not experience oneself as perceiving others' mental states, that is, others' mental states do not (ever?) have the same kind of vividness or experiential presence as colors and shapes. Here, phenomenologists might urge that perceiving someone's sadness, for example, is a qualitatively different experience from merely believing that they are sad, and that this is a distinction to which we should do justice.
2. It may be interpreted as a metaphysical thesis: mental states cannot be perceived because they are immaterial entities.
3. It may be read as an epistemological thesis: perceptual experiences do not ground judgments about others' mental states, at least not without the help of inference and background knowledge.
4. It may be understood as a psychological thesis about the processes by which we come to ascribe mental states to others (i.e., these are not perceptual processes).

Mainstream approaches argue that even if we do perceive (some of) others' mental states, we still need to account for how this is achieved. Mainstream accounts of mind reading attempt to do this. Herschbach (2008), Michael et al. (2014), and Lavelle (2012) argue that mind-reading approaches offer accounts of the subpersonal inferential processes that underlie mental state ascriptions.

A first question, then, is: Just what version of unobservability assumption is affirmed or presupposed by which approach? The basic issue behind this question is whether there is something peculiar about theory of mind that distinguishes it from subpersonal inferences normally associated with perception (e.g., in the context of unconscious inference as proposed by Helmholtz 1866/1962). Clearly, one key difference between inferences about a world that contains other agents is that they have to be hierarchically deeper than the more elemental (possibly unconscious) inferences about, say, visual features. However, one may ask whether this is a sufficient distinction to dissolve questions posed by the unobservability assumption.

A second question is whether the brain entertains representations. In radical versions embodiment, representations per se are precluded. This is nicely articulated by Anil Seth in his treatment of predictive coding or processing (PP) (Seth 2015):

> The notion of counterfactual predictions connects PP with what at first glance seems its natural opponent: "enactive" theories of perception and cognition which explicitly reject internal models or representations (Hutto and Myin 2013; Thompson and Varela 2001). Central to the enactive approach are notions of "sensorimotor contingencies" and their "mastery" (O'Regan and Noë 2001), where a sensorimotor contingency refers to a rule governing how sensory signals change in response to action. Accordingly, the perceptual experience of, for example, redness is given by an implicit knowledge (mastery) of the way red things behave, given certain patterns of behavior. Mastery of sensorimotor contingencies is also said to underpin perceptual presence: the sense of subjective reality of the contents of perception (Noë 2006). From the perspective of PP, mastery of a sensorimotor contingency corresponds to the learning of a counterfactually equipped predictive model, connecting potential actions to expected sensory consequences. The resulting theory of Predictive Perception of Sensorimotor Contingencies provides a much needed reconciliation of enactive and predictive theories of perception and action.

Clearly, there is some consilience between active inference (predictive processing) and sensorimotor contingency theory. The inferences we make about our sensations are (at some level) amodal and entail necessarily predictions of a proprioceptive sort—consistent with our embodied interactions with objects causing sensations. This is true even at the level of visual searches (Wurtz et al. 2011). Do posterior beliefs, then, constitute representations? The active inference story suggests that the answer is yes—although the answer is nuanced to consider probabilistic representations. Does the encoding of expectations (sufficient statistics or posterior beliefs), therefore, violate the tenets of radical embodiment and, if so, what are the implications for this and other theoretical formulations?

A related issue here is whether inferences about hidden states are conscious or unconscious. At what point do unconscious inferences become conscious, and what is the relationship between inference and creature (or indeed state) consciousness? There are clearly some relatively simple arguments about reportability of (state) consciousness content that require inference at a sufficiently deep hierarchical level to generate proprioceptive (or interoceptive) predictions that engage action. It may be that the notion of # may help organize the formal distinctions between conscious inferences and unconscious (subpersonal) inferences.

Finally, we come to questions about theory of mind and the mirror neuron system, which are a necessary aspect of active inference. This is self-evident from the fact that the same dynamics (posterior expectations) of neuronal populations—that occupy a privileged position in the cortical hierarchy—encode predictions about sensations during perception and generate motor commands (proprioceptive predictions) during action. Does this provide a sufficient account of theory of mind, or does it subvert our understanding of agency in some fundamental way? Is it appropriate to associate the same processes underlying perceptual inference with inferences about others? What are the implications of subsuming inference about low-level perceptual features and the intentions of others within the same hierarchical framework?

Sensory attenuation, in particular, requires us to either act or observe—but not do both at the same time. Is sensory attenuation, therefore, an integral part of agency and a sense of self? Can inferences about others be informed by predicting internal (as opposed to external) states of the world?

This brings us to another intriguing area: interoceptive inference and its putative role in inferring the emotional states of self, and possibly others (Seth 2013; Seth et al. 2011). Can one usefully transcribe the principles of active inference in the domain of motor control to emotional pragmatics and shared autonomic states? This is a fascinating area that speaks not just to our sense of self; it may also have had implications for psychopathology, when failing to attend or attenuate interoceptive prediction errors, as has been proposed in autism (Happe and Frith 2006; Lawson et al. 2014; Van de Cruys et al. 2014; Pellicano and Burr 2012).

In conclusion, deep questions arise from the challenges inherent in action and embodiment—questions that are particularly important for the notion of active inference. In subsequent chapters in this volume, various attempts are made to resolve, dissolve, or simply celebrate these questions.

Appendix

This brief description of generalized predictive coding is based on Feldman and Friston (2010). A more technical description can be found in Friston, Stephan et al. (2010). This scheme is based on three assumptions:

1. The brain minimizes a free energy of sensory inputs defined by a generative model.
2. The generative model used by the brain is hierarchical, nonlinear, and dynamic.
3. Neuronal firing rates encode the expected state of the world under this model.

Free energy is a quantity from statistics that measures the quality of a model in terms of the probability that it could have generated observed outcomes. This means that minimizing free energy maximizes the Bayesian evidence for the generative model. The second assumption is motivated by noting that the world is both dynamic and nonlinear, and that hierarchical causal structure emerges inevitably from a separation of spatial and temporal scales. The final assumption is the Laplace assumption, which leads to the simplest and most flexible of all neural codes.

Given these assumptions, one can simulate a whole variety of neuronal processes by specifying the particular equations that constitute the brain's generative model. In brief, these simulations use differential equations that minimize the free energy of sensory input using a generalized gradient descent:

$$\dot{\tilde{\mu}}(t) = D\tilde{\mu}(t) - \partial_{\tilde{\mu}} F(\tilde{s}, \tilde{\mu}). \tag{6.1}$$

These differential equations say that neuronal activity encoding posterior expectations about (generalized) hidden states of the world $\tilde{\mu} = (\mu, \mu', \mu'', \ldots)$ reduce free energy, where free energy, $F(\tilde{s}, \tilde{\mu})$, is a function of sensory inputs, $\tilde{s} = (s, s', s'', \ldots)$, and neuronal activity. This is known as generalized predictive coding or Bayesian filtering. The first term is a prediction based upon a differential matrix operator, D, that returns the generalized motion of expected hidden states, $D\tilde{\mu} = (\mu', \mu'', \mu''', \ldots)$. The second (correction) term is usually expressed as a mixture of prediction errors that ensures the changes in posterior expectations are Bayes-optimal predictions about hidden states of the world. To perform neuronal simulations under this scheme, it is only necessary to integrate or solve Equation 6.1 to simulate the neuronal dynamics that encode posterior expectations. Posterior expectations depend upon the brain's generative model of the world, which we assume has the following hierarchical form:

$$s = g^{(1)}\left(x^{(1)}, v^{(1)}\right) + \exp\left(-\tfrac{1}{2}\pi_v^{(1)}\left(x^{(1)}, v^{(1)}\right)\right) \cdot \omega_v^{(1)}$$

$$\dot{x}^{(1)} = f^{(1)}\left(x^{(1)}, v^{(1)}\right) + \exp\left(-\tfrac{1}{2}\pi_x^{(1)}\left(x^{(1)}, v^{(1)}\right)\right) \cdot \omega_x^{(1)}$$

$$\vdots$$

$$v^{(i-1)} = g^{(i)}\left(x^{(i)}, v^{(i)}\right) + \exp\left(-\tfrac{1}{2}\pi_v^{(i)}\left(x^{(i)}, v^{(i)}\right)\right) \cdot \omega_v^{(i)}$$

$$\dot{x}^{(i)} = f^{(i)}\left(x^{(i)}, v^{(i)}\right) + \exp\left(-\tfrac{1}{2}\pi_x^{(i)}\left(x^{(i)}, v^{(i)}\right)\right) \cdot \omega_x^{(i)}$$

$$\vdots$$

(6.2)

This equation describes a probability density over the sensory and hidden states that generate sensory input. Here, the hidden states have been divided into hidden states and causes $(x^{(i)}, v^{(i)})$ at the i-th level within the hierarchical model. Hidden states and causes are abstract variables that the brain uses to explain or predict sensations—like the motion of an object in the field of view.

In these models, hidden causes link hierarchical levels, whereas hidden states link dynamics over time. Here, $(f^{(i)}, g^{(i)})$ are nonlinear functions of hidden states and causes that generate hidden causes for the level below and—at the lowest level—sensory inputs. Random fluctuations in the motion of hidden states and causes $(\omega_x^{(i)}, \omega_v^{(i)})$ enter each level of the hierarchy. Gaussian assumptions about these random fluctuations make the model probabilistic. They play the role of sensory noise at the first level and induce uncertainty at higher levels. The amplitudes of these random fluctuations are quantified by their precisions that may depend upon the hidden states or causes through their log-precisions $(\pi_x^{(i)}, \pi_v^{(i)})$.

Given the form of the generative model (Equation 6.2) we can now write down the differential equations (Equation 6.1) describing neuronal dynamics in terms of (precision-weighted) prediction errors. These errors represent the difference between posterior expectations and predicted values, under the generative model (using $A \cdot B \triangleq A^T B$ and omitting higher-order terms):

$$\dot{\tilde{\mu}}_x^{(i)} = D\tilde{\mu}_x^{(i)} + \left(\frac{\partial \tilde{g}^{(i)}}{\partial \tilde{\mu}_x^{(i)}} - \tfrac{1}{2}\tilde{\xi}_v^{(i)}\frac{\partial \tilde{\pi}_v^{(i)}}{\partial \tilde{\mu}_x^{(i)}}\right) \cdot \xi_v^{(i)} + \left(\frac{\partial \tilde{f}^{(i)}}{\partial \tilde{\mu}_x^{(i)}} - \tfrac{1}{2}\tilde{\xi}_x^{(i)}\frac{\partial \tilde{\pi}_x^{(i)}}{\partial \tilde{\mu}_x^{(i)}}\right) \cdot \xi_x^{(i)} + \frac{\partial tr(\tilde{\pi}_v^{(i)} + \tilde{\pi}_x^{(i)})}{\partial \tilde{\mu}_x^{(i)}} - D^T \xi_x^{(i)}$$

$$\dot{\tilde{\mu}}_v^{(i)} = D\tilde{\mu}_v^{(i)} + \left(\frac{\partial \tilde{g}^{(i)}}{\partial \tilde{\mu}_v^{(i)}} - \tfrac{1}{2}\tilde{\xi}_v^{(i)}\frac{\partial \tilde{\pi}_v^{(i)}}{\partial \tilde{\mu}_v^{(i)}}\right) \cdot \xi_v^{(i)} + \left(\frac{\partial \tilde{f}^{(i)}}{\partial \tilde{\mu}_v^{(i)}} - \tfrac{1}{2}\tilde{\xi}_x^{(i)}\frac{\partial \tilde{\pi}_x^{(i)}}{\partial \tilde{\mu}_v^{(i)}}\right) \cdot \xi_x^{(i)} + \frac{\partial tr(\tilde{\pi}_v^{(i)} + \tilde{\pi}_x^{(i)})}{\partial \tilde{\mu}_v^{(i)}} - \xi_v^{(i+1)}$$

$$\xi_x^{(i)} = \exp(\tilde{\pi}_x^{(i)}) \cdot \tilde{\varepsilon}_x^{(i)}$$

(6.3)

$$\xi_v^{(i)} = \exp(\tilde{\pi}_v^{(i)}) \cdot \tilde{\varepsilon}_v^{(i)}$$

$$\tilde{\varepsilon}_x^{(i)} = D\tilde{\mu}_x^{(i)} - \tilde{f}^{(i)}(\tilde{\mu}_x^{(i)}, \tilde{\mu}_v^{(i)})$$

$$\tilde{\varepsilon}_v^{(i)} = \tilde{\mu}_v^{(i-1)} - \tilde{g}^{(i)}(\tilde{\mu}_x^{(i)}, \tilde{\mu}_v^{(i)}).$$

This produces a relatively simple update scheme, in which posterior expectations $\tilde{\mu}^{(i)}$ are driven by a mixture of prediction errors $\tilde{\varepsilon}^{(i)}$ that are defined by the equations of the generative model.

In neural network terms, Equation 6.3 says that error-units compute the difference between expectations at one level and predictions from the level above (where $\xi^{(i)}$ are precision weighted prediction errors at the i-th level of the hierarchy). Conversely, posterior expectations are driven by prediction errors from the same level and the level below. These constitute bottom-up and lateral messages that drive posterior expectations toward a better prediction to reduce the prediction error in the level below. In neurobiological implementations of this scheme, the sources of bottom-up prediction errors are generally thought to be superficial pyramidal cells, because they send forward (ascending) connections to higher cortical areas. Conversely, predictions are thought to be conveyed from deep pyramidal cells by backward (descending) connections, to target the superficial pyramidal cells encoding prediction error (Mumford 1992; Bastos et al. 2012).

Note that the precisions depend on the expected hidden causes and states. We have proposed that this dependency mediates attention (Feldman and Friston 2010). Equation 6.3 tells us that the (state-dependent) precisions modulate the responses of prediction error units to their presynaptic inputs. This suggests something intuitive—attention is mediated by activity-dependent modulation of the synaptic gain of principal cells that convey sensory information (prediction error) from one cortical level to the next. This translates into a top-down control of synaptic gain in principal (superficial pyramidal) cells and fits comfortably with the modulatory effects of top-down connections in cortical hierarchies that have been associated with attention and action selection.

Acknowledgments

KJF is funded by the Wellcome Trust.

7

Prediction, Agency, and Body Ownership

Jakob Hohwy

Abstract

The idea that the brain is an organ for prediction error minimization is becoming increasingly influential. Since this idea posits action as playing a central role, it has the potential to reveal new perspectives on troubled notions of action, sense of agency, and body ownership. Elucidating these notions may help ascertain how close this new framework is to contemporary views of embodied, enactive, and extended cognition, which also makes action central to cognition. The prediction error minimization framework suggests novel and somewhat provocative notions of action, sense of agency, and body ownership and, in important respects, it pulls in the opposite direction from the embodied, extended, and enactive approaches.

Introduction

There is increasing focus on the idea that the brain is fundamentally engaged in prediction error minimization (PEM) (Friston 2010; Clark 2013b; Hohwy 2013). If this is true, then it ought to have consequences for embodied, enactive, and extended (EEE) approaches to cognition (Clark 2008; Thompson 2007; Noë 2004; Shapiro 2011).

When PEM is understood specifically in terms of the free energy principle (FEP), there is a central role for action in perception and cognition. Under FEP's notion of active inference, organisms *act* to maintain themselves in their expected states. That is, sensory input is selectively sampled under a favored hypothesis about the state of the organism and the world, increasing its accuracy. In this way, action becomes indispensable for explaining perception. Although there is a central role for action within PEM/FEP, there is ample scope for discussion about the extent to which this affords a good match with traditional accounts of EEE cognition (Bruineberg and Rietveld 2014; Hohwy 2014; Seth 2014).

There has been relatively little focus on what PEM itself might tell us about action, as well as what PEM might tell us about a crucial element of action; namely, the body with which action is performed. If, as PEM would have it, the brain is fundamentally engaged in perceptual inference, active inference, precision optimization, and complexity reduction (Friston and Stephan 2007; Friston 2010), then it may be difficult to find room for traditional notions of what it takes to be an agent. If it is all inference, then where is the agent who acts on desires and beliefs to move the body? How does the agent's subjective sense of body ownership and of agency arise in a PEM brain? What should be said about the basic requirement for action, namely that we are embodied creatures?

In this chapter, I discuss some of the challenges for PEM concerning agency and body ownership, and propose possible responses to these challenges. In the light of this, it becomes tempting to revise somewhat our normal concepts of action, sense of agency, and body ownership. These concepts figure centrally in the EEE approaches to cognition. However, even though PEM accommodates these concepts, PEM's proclivity for internal processing may not be a natural bedfellow for EEE cognition.

Active Inference: Moving to Save the Hypothesis?

According to PEM, there are two main ways of minimizing prediction error of the internal generative model harbored by the brain: Either the model parameters can be updated to such an extent that the prediction error is decreased—this is *perceptual inference*. Or, the sensory input can be changed to fit with the model's predictions—this is *active inference*. In the latter, a model is selected, its predictions generate prediction error, and then the individual changes the states of the sensory organs (e.g., eye movement, palpation, limb movement) until the prediction error is decreased to within expected levels of noise.

In active inference, there is selective sensory sampling to confirm a hypothesis. If the world cooperates and delivers the expected sensory input, then the selected hypothesis is strengthened. Prediction then becomes more *accurate*. Active inference, therefore, increases accuracy. For example, as I actively explore a pipe in my hand by turning it around in my hand and looking at it from different angles, I increase the confidence in the hypothesis that I am looking at a pipe and not merely an image of a pipe. If the world is not cooperating, in the sense that the predicted sensory input does not occur upon executing the selected action, then prediction error will mount. In that case, the system should switch back into perceptual inference and select another hypothesis (e.g., "it is just a picture of a pipe"), under which new and better predictions can be generated.

In this manner, it becomes clear that both perceptual and active inference is needed for a successful PEM mechanism. The system should, on one hand, be open to revising the hypotheses generated under the model and, on the other, be prepared to bet that a particular, reasonably likely hypothesis is actually true and act accordingly. Importantly, the system needs to maintain an optimal balance between these inferential processes.

From this perspective, PEM affords equally substantial roles to "passive" perceptual inference (which minimizes the bound on surprise) and active inference (which maximizes the accuracy of the model's hypotheses). Insofar as EEE approaches exclusively allow action-related elements in cognition, they are thus not compatible with PEM. Furthermore, PEM crucially works with an internal generative model—a representation—of the causes of its sensory input, which some EEE approaches do not countenance at all.

Active inference is important to representation of the world because intervention in the world helps reveal causal structure. For example, passive observation of how frequent certain events occur is often not sufficient to disambiguate between hypotheses that posit causal relations or common cause. Intervening in the world under the assumption that one of these is true can help disambiguation (e.g., varying random variable A and finding no invariant relation to random variable B will suggest that there is a common cause given by some random variable C). This is a lesson that stems from research on causation (Pearl 2000; Woodward 2003), rather than EEE approaches in and of themselves: without active inference, there would only be very shallow, serendipitous, associative representation of the causal structure of the world. In other words, reflection about causal inference on the causes of sensory input compels a move away from pure association learning and toward interventionist, "enactive" approaches to perception.

Under the wider, FEP-inspired perspective, active inference is what maintains the organism in its expected states. Without action, these states would disperse rapidly and the organism would cease to exist. If, however, it manages to sample sensory input that keeps its long-term average prediction error low, then it will survive for a longer period of time. The assumption here is that the organism is a model of its expected states, in particular given in terms of its homeostasis. This model is a probability distribution, which sets out the states in which this particular organism can be expected to be found, which thus can be said to characterize the phenotype of this organism. The organism acts in the world to make sure it does not stray into states where it is not expected to be found; in other words, it acts to avoid surprise. FEP explains how this can occur given that we cannot directly know our expected states: the free energy is a boundary of the surprise. Thus, by minimizing the free energy (or long-term average of prediction error), the surprise is implicitly minimized as well (Friston and Stephan 2007).

The very basic reasoning here is reminiscent of the ideomotor theory of action, according to which action occurs as a result of predicting how being in the desired state would change the received sensory input (Herbart 1816; Lotze 1852; James 1890; Hommel 2013). With PEM or FEP, this idea is writ large in the sense that all movement is relative to an expectation of a set of states. As we shall see, PEM/FEP offers a more detailed account of action and action initiation, which brings the account even closer to some of the tenets of this older theory.

Action and the Predictive Mind: An Attentional Mechanism?

In the overall PEM setup, it is imperative that the system finds a way of avoiding *in*action; otherwise it cannot explore the causal structure of the world, or maintain itself in its expected states. Perplexingly, however, it is not immediately obvious how inaction can be avoided. The problem here stems from the basic requirement (akin to elements of the ideomotor theory) that action occurs when the system prioritizes a hypothesis that is actually false, over another that is actually true. For example, to move my hand to a cup of coffee from its present position away from the cup, I need to select the (false) hypothesis that my hand is at the coffee cup. This hypothesis generates prediction error, since my hand is not actually at the coffee cup. This prediction error is minimized by enslaving the body (via classic reflex arcs) until the predicted state is obtained. This is how movement occurs, according to PEM (Adams, Shipp et al. 2013; Friston, Adams et al. 2012).

The problem is that it seems more reasonable for the system to just revise the selected hypothesis in the light of the *actual* sensory input: that is, revert to the hypothesis that the hand is positioned where it actually is, away from the cup. This would seem a more economical way to minimize prediction error. Unfortunately, it would also lead to inactivity, as nothing would then compel movement.

The solution to this problem is to consider the evolution of the prediction error landscape on the longer term (Brown et al. 2013). Very fundamentally, remaining stationary for too long would cause prediction error to increase: homeostasis will be compromised if I never act to get some sustenance or to explore new hunting grounds. Humans have to keep acting, because we strongly expect that our environment is always in the process of changing (i.e., it is *volatile*). For the particular movement in the coffee cup example, this translates to the idea that the fidelity of the current sensory input should always be expected to deteriorate. In other words, there should be an expectation that precise prediction error will begin to occur under another hypothesis than the currently selected one (for more on this idea, as it relates to temporal phenomenology, see Hohwy et al. 2015). Under this expectation, the gain on current prediction error should be decreased, in line with the idea that the prediction error which is expected to be imprecise should be down-weighted in inference (Feldman and Friston 2010; Hohwy 2012). With this gain reduction (or gating of sensory input), the system is in a position to select a new hypothesis. As the current sensory input is gated, the hypothesis that the hand is in its true position is deprived of evidence. This means that its competitor—that the hand is at the coffee cup—can win, can be selected for active inference, and can drive movement to the point at which the cup is in hand.

Accordingly, action is initiated as a result of preferential weighting of sensory input. Action, and therewith agency, reduces to a process of optimization of the precisions of prediction error. Within PEM, optimization of expected precisions maps onto the functional role for attention (Feldman and Friston 2010; Hohwy 2012); thus action transpires rather surprisingly as an *attentional* effect. It is easy to conceive of attention as a wholly internal mechanism, one that merely processes statistical regularities in the sensory input. As such, the very notion of agency is not a particularly good fit for the more world-involving, nonrepresentational facets of EEE approaches to cognition and perception.

This particular PEM-based account of agency predicts that action initiation is accompanied by decreased gain on current sensory input; that is, when there is movement, current sensory input is first attenuated. This has been suggested as an explanation of our inability to tickle ourselves. Self-tickle requires movement, and movement attenuates input; the prediction error caused by oneself has less precision, and thus will be less able to drive an update of the hypothesis generated by the internal model. The result would be that self-tickle feels less "tickly" than when others engage in this action, since there is no attenuation of sensory input in other-tickle (Brown et al. 2013).

Over the last few decades, Blakemore et al. (1999) have provided the dominant explanation of the self-tickle effect in terms of efference copies. However, this new take on agency furnishes a revised explanation of the effect. It is thus tempting to revisit the tickle effect, since the prediction, contra the traditional forward modeling idea, is that individuals are unable to tickle themselves even when the tickle re-afference is difficult to predict (for initial evidence in favor of this, see Van Doorn et al. 2015).

Apart from these specific predictions concerning this conception of action and agency, there is a broader underlying reason for believing that this is the right way to conceive of these notions, at least if we accept the full FEP framework in the first place. An often-heard objection to PEM/FEP is that if brains are only interested in minimizing surprise, then organisms should be found sitting inactively in dark rooms so that they never encounter anything surprising (Friston, Thornton et al. 2012). This objection overlooks one important point: FEP addresses the long-term average of surprise (given a model) which, for organisms that live in an uncertain world (e.g., humans), is prone to increase when there is too much inactivity. To repeat the point made above: the precision of given prediction error will decrease over time (cf. volatility), meaning that surprise will increase. This compels an organism to act in the world, to seek out predicted high precision prediction error, and thereby to visit its expected states. We only find organisms in dark rooms if dark rooms define their expected states.

This general train of reasoning expresses broadly the argument just rehearsed for action and agency: action occurs as the organism expects current prediction error to lose precision. Contrary to first impressions, PEM/FEP is thus a surprisingly good match for EEE, since it makes agency imperative. However, contrariwise, the way in which it makes agency imperative makes it a less good match for EEE, since agency is just more statistical inference (i.e., optimizing precisions).

The Body in the Predictive Mind: Inferred Ownership?

From the point of view of PEM, the body is a hidden cause of sensory input. Through action, the body changes its position in the world, and this is read-off in terms of changes to the sensory input and the ensuing prediction error. As such, there is no in-principle difference between the body and other hidden causes in the environment that affect the sensory organs. All such causes need to be inferred through the implicit inversion of the generative model as it minimizes its prediction error.

This immediately suggests a fairly deflationary approach to the experience of body ownership. This seems relevant to EEE approaches to cognition, since they typically assign a special role to the body in cognition, although there is no broad consensus on what the notion of embodiment comes to according to EEE (Kiverstein and Clark 2009; Alsmith and de Vignemont 2012). As discussed below, PEM can cast further light on the notion of embodiment. This will bring PEM closer to some tenets of EEE, although again PEM is fundamentally inferential and representational in a way that sits poorly with EEE.

Embodiment

To engage in active inference, it is necessary to have an accurate inferred model of the body. A poor model of the body will yield poor predictions of future sensory input. For example, if I don't know how long my arms are, then the prediction that my hand will be at the coffee cup is not going to be very confident. [In this vein, Gori et al. (2008) suggest that young children rely on haptic rather than visual information for discerning size, when both are available, because their arms have not finished growing yet.] We can assume, then, that organisms which successfully engage in active inference need to model their body too, just as they model the rest of their environment. With active inference we therefore get a notion of "embodied cognition," because minimizing prediction error through action must necessarily rest on an internal model of the body.

This notion of embodied cognition is, however, purely internal and inferential. The model of the body arises as the internal model churns away on the statistical properties of its sensory and active states. In this internalist sense, embodiment arises because organisms have to move around in their environment so as to seek out high precision prediction error. For some strands of EEE, this is not enough to qualify as embodiment since the role of the body is not to usurp and make obsolete internal representational processing (Brooks 1991b; Hutto and Myin 2013).

Sense of Ownership

Body ownership is a key facet of embodied cognition. Not only do I infer that there is a body out there, I infer that this is *my* body. This might mark a difference between the inference of one's own body and inference of other hidden causes (e.g., trees, houses, other people). If we look at this from the internal PEM perspective, a difference between inference of self and another could arise from different statistical patterns. For example, when my body plays a role in modulating my sensory input, there will often be a systematic association between some of my active states and some of my sensory states. If a prediction that my hand will be at the coffee cup is passed to my active states, that may be reliably associated with subsequent input from the coffee cup at my sensory states. This association will not arise when someone else hands me a cup of coffee. So, inferences that involve my own body will, in general, be different from inferences that involve other's bodies. In addition, inferences about my own body might be different from those about the bodies of other people because my priors about my own body, honed over time, are much more precise than my priors about the bodies of others. This may explain in inferential terms the fact that action is associated with a sense of ownership: ownership is inferred when the bodily causes involved in active inference are marked by such active-to-sensory associations, or with high precision priors.

A further aspect of body ownership could be the specific sources of evidence drawn upon in inference to ownership. In general, when I infer that the cause of sensory input is another person, I do not rely on interoceptive and proprioceptive sources of evidence. If I observe someone else reaching for a cup of coffee, I do not get proprioceptive, thermoreceptive, and kinesthetic input. In contrast, all of these types of evidence are available once I reach out for the coffee, lift it up, and feel its warmth on my palm. This suggests that there is plenty of evidence available in the system to ground robust inferences about body ownership.

In this account, body ownership is not, however, a fundamental aspect of cognitive processing. There is no fundamental truth that we have body ownership, which somehow props up cognition, as perhaps some EEE approaches would suggest. Instead, there is more inference, more discerning of relevant associations in the sensory input and other states of the model of the world. Since actions that are felt as "owned" are attributed to the self, the self-other distinction may to some extent be fuelled by the sense of ownership. Together, it follows that organisms do not *need* to have the capacity to distinguish between self and other to function appropriately. That distinction simply falls out of a truthful, inferred representation of the world. Conversely, the self-other distinction may be disturbed simply when the organism has trouble with forming a truthful representation of the world. This speaks to some mental disorders, such as schizophrenia, which often are characterized by self-other disturbances.

Interoceptive Inference

This overall inferential picture of embodiment and body ownership is further enhanced once we consider inference on internal bodily causes. From the inferential point of view of the brain, there is no difference between causes of sensory input that lie beyond the body and causes that lie within the body. That is, just as there is exteroceptive inference of causes in the external world, there is interoceptive inference of causes within the body (Hohwy 2011; see also Aspell et al. 2013; Seth 2013; Suzuki et al. 2013; Fotopoulou 2015). Interoceptive sensory organs are affected by bodily states, and this input is conceived as a prediction error that is minimized in probabilistic inference. Interoceptive perceptual inference thus occurs when internal prediction error (e.g., an increase in arousal) is explained away under some context-dependent hypothesis. This is what leads to bodily sensations and emotions. (In this way, PEM sides with a broadly James-Lange view of emotions; for a key study, see Schacter and Singer 1962.)

The interoceptive aspect implies that even internal states of the body itself are inferred. This puts further pressure on a recurrent theme in EEE approaches to cognition; namely the brain-body barrier is not especially privileged for understanding cognition (Hurley 1998). Instead, with PEM, we get a requirement that there is inference of hidden causes, relative to some sensory veil (or, in terms of causal nets, a "Markov blanket," where a state is knowable once states of the blanket, consisting of its parents, children, and parents of children, are known). This veil can, in principle, be placed in a number of different ways, but there seems good reason to suspect that the sensory epithelia and parts of the spinal cord form our sensory veil (Friston 2013; Hohwy 2014).

At the heart of PEM, and more generally FEP, sits the idea that organisms will act in whatever way to make homeostatic predictions come true. If the organism believes a certain arousal state is unexpected, it will act to bring arousal within expected levels again. This means that acting is, at heart, self-fulfilling prophesying. Under this conception of PEM/FEP, cognition gets a very embodied flavor, since now the ultimate driver for PEM is the imperative to visit only those states that will prevent dispersion of the agent; that is, the states that will allow it to retain homeostasis. In one sense, this means the framework is friendly to some basic tenets of embodiment in EEE (cf. Thompson 2007). However, it is an oddly truncated version of embodiment. The only thing that matters to the organism is its ability to maintain its integrity; it does not matter at all which body or environment allows this to happen. So even though there is embodiment in some sense, there is also an urge to throw away the body and the environment insofar as an understanding of the workings of the biological system is concerned.

If embodiment and body ownership are inferred from the sensory input, rather than being fundamental irreducible elements of cognition, then we should expect that embodiment and body ownership are subject to a range of familiar perceptual distortions. For example, just as our inference of the co-location of a visual and an auditory input may lead us astray in the ventriloquist illusion (Alais and Burr 2004), inference of embodiment and ownership may occasionally lead us astray. This is found in the rubber hand illusion and the full body versions of this illusion (Botvinick and Cohen 1998; Tsakiris and Haggard 2005; Ehrsson 2007; Lenggenhager et al. 2007). By manipulating visuotactile and proprioceptive sensory input in various ways, it is easy to make people infer that nonbody objects belong to their body (e.g., when visual and tactile stimulation occurs in synchrony, it invites the (false) inference that they have a common cause located on the skin). This can be done for many types of conditions (e.g., involving the adoption of strange body sizes as well as supernumerary or invisible limbs). There is also the expected flow-on effect from these illusions to active inference. For example, people will perform scale errors when embodying a very small or very large body (van der Hoort et al. 2011), or they will display subtle differences in subsequent movement (Paton et al. 2011; Palmer et al. 2013).

Sense of Agency: Recessive or Tied to Counterfactual Reasoning?

It is one thing to be an agent and quite another thing to have a sense of agency, to be aware of agency, and represent oneself as an agent. The sense of agency differs from the sense of body ownership, since one can have one without the other, as in when someone pushes what I know to be my arm (for key discussions on the sense of agency, see de Vignemont and Fourneret 2004; David et al. 2008; Tsakiris et al. 2007). Sense of agency seems to be involved in delusions of alien control in schizophrenia: patients attribute agency to other agents even though they have a sense of ownership (see Hohwy and Frith 2004; Gallagher 2000). There is some uncertainty about how "thick" the conscious sense of agency is. It seems that this sense recedes into the background of consciousness unless it is challenged and emerges primarily as a sense of lack of agency.

A classic treatment casts the sense of agency in terms of forward modeling (Wolpert and Ghahramani 2000; Chambon et al. 2014). Sense of agency arises when predictions of sensory consequences from movement match the actual consequences of movement (or when the predicted sensory consequences match the desired state). If there are problems with forward modeling, there might be problems with sense of agency; this has been proposed as an explanation of delusions of alien control (Frith et al. 2000b). Given the manner in which PEM explains action initiation (see above), this account can be revised. In particular, it may be that delusions of alien control arise when occurrent proprioceptive input is not attenuated, and where this causes unexpectedly strong prediction error that is explained away under an alternative (and delusional) hypothesis of external agency being the cause of movement. In other words, in such patients, movement can occur under a false, more complex hypothesis about the world; this posits other agents and thereby leads to delusions of alien control. Conversely, when a patient is able to avoid this false hypothesis, inactivity follows (e.g., in the form of catatonia, or waxy flexibility) since active inference must now be impeded (Brown et al. 2013).

It seems likely that the sense of agency is associated with the attenuation of sensory input as movement unfolds. This would make the concept of sense of agency essentially similar to that of the forward modeling account, even if the underlying mechanisms are different (for PEM, as discussed above, attenuation happens as an element of precision-weighting of proprioceptive prediction error). In both cases, sense of agency occurs when sensory input during self-generated but not other-generated action is attenuated. (Note, however, that the explanations of how delusions arise differ significantly on these two accounts.)

Given this conceptual similarity, we still need to understand how sense of agency can be anything else other than a recessive conscious feeling—one that we have less of, the more control and agency we have. Within PEM, it is possible to identify a further element to the conception of the sense of agency. This sense is associated with the process of model selection that leads up to decision making and movement (Friston et al. 2013). Agents may have learned that it pays to consider a suite of alternative hypotheses rather than jump to action on whatever hypothesis presents itself first. This may be the case, in particular, for actions that will unfold over the medium and long term (e.g., deciding what to have for lunch or what education to pursue), in contrast to actions on short timescales (e.g., reaching for the coffee cup), which may have a restricted repertoire of highly confident hypotheses.

Consideration of a suite of hypotheses calls for counterfactual processing, where the objective is to pick the hypothesis most likely to allow agents to maintain themselves in what they represent as their expected states. Counterfactual processing is the internal generation and comparison of predictions of sensory (exteroceptive and interoceptive) outcomes of hypothetical actions: if I were to do X then I would experience Y, whereas if I were to do X*, then I would experience Y*. The idea is that predicted outcomes are compared relative to the expected state, and the outcome generating the state closest to the expected state is deemed most probable and will thus get to drive action. The inferential process here is minimizing the Kullback–Leibler divergence[1] between predicted outcomes and expected states through variational Bayes.

This process relates to notions of mental time travel (Suddendorf et al. 2009) and decision-making processes in general. It creates a link between deliberation, decision making, and action. It goes beyond basic PEM/FEP since not all organisms capable of minimizing their free energy (e.g., *Escherichia coli*) need be able to engage in counterfactual reasoning. It is a good strategy, however, for minimizing the long-term average of prediction error in organisms (e.g., humans) that model the world in a spatiotemporally deep cortical hierarchy.

With respect to the sense of agency, the idea is that by minimizing the Kullback–Leibler divergence between the predicted and expected states, an agent is able to represent itself because this processing leads to beliefs about the actions about to be performed (for evidence along these lines, see Chambon and Haggard 2012). This does not have the recessive feel of the traditional understanding of sense of agency as attenuation of prediction error. Instead it is associated with explicit, internal representation of oneself as a cause of later sensory input. This process is also associated with active model selection, as different hypotheses are considered and compared against each other; this could speak to the idea that we have sense of agency not only when we are the agents of our movements but also when we are in control of our decision making and execution of action (for a broader view of agency, see Pacherie 2014). Sense of agency is thus essentially related to the individual feeling of being the cause of events and feeling that one could have done something else—a paradigmatic conception of agency, expressed already in Hellenistic philosophy (Bobzien 2006; Frith 2014).

Returning to delusions of alien control, it is tempting to think about these passivity experiences in light of this further, more positive notion of sense of agency. For example, there could be conditions under which counterfactual model selection fails to occur; thus the selected hypothesis is not "vouched for" in the way that gives rise to sense of agency. This may be a worthwhile line of inquiry as such an account might also apply to other passivity experiences, such as thought insertion (the belief that other agents are thinking one's thoughts), that have proven difficult to account for under the traditional account of sense of agency (Martin and Pacherie 2013). For example, if counterfactual model selection failed to occur in patients with thought insertion, we might predict that they should not have anticipated regret. Likewise, patients with schizophrenia show some deficits in future-oriented mental time travel, consistent with this speculation (D'Argembeau et al. 2008).

The overall picture is that PEM has room for some version of the traditional conception of sense of agency. More importantly, PEM may also be able to

---

[1]   The Kullback–Leibler divergence of probability density function Q from the density P, denoted $D_{KL}(P\|Q)$, is a non-negative measure of the information lost when Q is used to approximate P.

accommodate a richer conception of sense of agency. This would play a central role for a PEM system with a deep causal hierarchy, trying to optimize its active inference.

With respect to the aspirations of EEE, it is clear that sense of agency plays a central role for PEM. However, there seems to be a disconnect between the sense of agency and action itself. The sense of agency arises in internal model selection and action itself is a hidden state of the world, which must be inferred. Again we see that PEM addresses the role of action in terms of the workings of internal generative models.

Concluding Remarks

This discussion has focused on action, sense of body ownership, and sense of agency within a framework that primarily views the brain as an organ for prediction error minimization. Some interesting aspects of these notions are that action is an attentional phenomenon, that experience of body ownership is a perceptual inference, and that sense of agency may be related not only to sensory attenuation but also to the ability to reason counterfactually about possible actions. Though PEM is heavily imbued with action and agency, action and our representations of agency are best understood in terms of an inferential, internalist conception of cognitive processing that occurs wholly behind a sensory veil, segregated from the world it is modeling and within which it is acting. This overall focus on agency fits well with contemporary notions of embodied, extended, and enactive cognition. The "enveiled" conception of cognition, however, is anathema to many of these contemporary trends in cognitive science.

Acknowledgments

I thank Chris Frith and anonymous reviewers for many helpful comments on an earlier draft. I am grateful also to the organizers of this Ernst Strüngman Forum.

8

Sensorimotor Contingencies and the Dynamical Creation of Structural Relations Underlying Percepts

Jürgen Jost

Abstract

A cognitive system is coupled to its environment via a sensorimotor loop. It receives signals from, and in turn acts upon, the environment, and the resulting actions influence the structure of future signals. This chapter looks at several possible optimization principles for this loop, all of which have certain shortcomings. It argues for a more refined interaction where the actions create certain structural relations among the sensory data as well as between those data and the system's movements. These relations induce correlations between neuronal activities, and it is argued that these correlations underlie percepts which correspond to a specific spatiotemporal neuronal pattern. Such a pattern is the result of a learning process that transforms correlations into associations.

Introduction

The pragmatic turn in cognitive science (Engel et al. 2013) grounds cognition in interactions with the environment via sensorimotor couplings, instead of internal representations of an external world. This has two important aspects. On one hand, the cognitive system acts on the environment in such a way that it can best extract information from it and can let the environment carry out computations, instead of having to simulate them internally. On the other, actions cause variations in sensory input from which correlations can be extracted via sensorimotor contingencies. Such correlations then underlie percepts. The first aspect has been explored by Clark (2008); the second builds upon the sensorimotor account of vision of O'Regan and Noë (2001). Here, we will focus primarily on the second aspect. Our considerations will be guided and constrained by:

Thesis 1   Models that require very difficult and computationally intensive computations for creating percepts and identifying objects cannot be right. Perception in standard situations is fast, apparently effortless, and simple.

Of course, one may object here that perception only appears simple, and is described as simple in first-person reports, and that the underlying neuronal activity patterns may be substantially more complex. At least the speed of perception is a fact that is independent of subjective experience, and while perception may be more difficult than it subjectively appears, its ease and general reliability do impose constraints on any computational model.

The question then is: Why and how do the sensorimotor loop and sensorimotor contingencies make the implementation of this principle possible? To illustrate, consider the following example: When you want to catch a ball, you could, in principle, attempt to compute its trajectory according to the principles of Galilean and Newtonian dynamics. For that, you would need to have precise information about its initial position and velocity, its mass, its air resistance, and intervening factors such as wind direction and speed. You would then need to solve the corresponding differential equations to find out where the ball will land. To do this to the desired accuracy is very hard, if not outright impossible. In fact, there is an easier well-known solution.[1] Simply run in such a way that the angle under which you see the ball remains constant. Then you will arrive at the right time at the right spot where the ball will land and be able to catch it. This seems much simpler than the Newtonian strategy. The crucial point is that by using the constant angle strategy, you can extract a lot more information from the environment than with the Newtonian strategy, where accurate information is utilized only at the beginning of the process. You would continuously sample information about the position and speed of the ball, and you would adjust your actions so that the information about the viewing angle remains constant. That is, you would let the environment, or in this case more precisely the ball, compute its own trajectory, whereas with the Newtonian strategy, you would have to do that computation yourself. But there is more insight to gain here. Through your actions (i.e., running toward the ball with this strategy) you would generate specific correlations between your own motions and your visual input. The moving ball is correlated in a different manner with your motion than the static environment. This then generates the percept of the flying ball. Of course, this latter effect could also be achieved by actions other

---

[1] Note: this simplified account is not appropriate for all ball-catching activities. For instance, in football, it would not be the appropriate strategy to take, for various reasons. Nevertheless, this strategy is typically used by non-experts and many animals (e.g., dogs) as they attempt to catch a ball.

than running toward the ball, for instance by simply moving your head so that your gaze can follow the ball. Putting this in more abstract terms, your actions generate variations of your sensory input, which you can use to extract specific structures. These specific structures consist in correlations, and once your neuronal system has learned to translate these correlations into associations, it will generate specific spatiotemporal activity patterns in response to such correlations. As shall be argued, these will then underlie the corresponding percepts.

Evolutionary Principles

Before we get to that task, however, let us widen our perspective and view the problem from an evolutionary point of view. A biological organism (or for our purposes, being interested in cognition, more specifically an animal) lives because it is the descendent of ancestral lines whose members all successfully survived and reproduced, in fierce competition with other biological organisms. Its genome was produced by recombining those of its parents, perhaps with some variations. Therefore, we expect that it is also well adapted to environmental conditions and circumstances, not necessarily the current ones, but at least those in which its ancestors succeeded. This is what the term "fitness" attempts to capture, as the actual or expected number of descendents.[2] We also expect that its structures and parts are well suited to cope with its environment. In that sense, loosely speaking, those structures should serve some purpose and contribute as much as possible to the reproductive fitness. Importantly, survival is not an ultimate goal in itself, but is subordinate to that of successful reproduction. Of course, evolution never stops, and biological structures are therefore, in general, not optimal. One could argue that equilibrium theories in biology, and for that matter also in economics, are fundamentally flawed because the competition never sleeps. Nevertheless, in light of the above, it seems insightful to approach them via optimization principles. While the optima may typically not be realized by biological structures, they may often come close to being optimal. More importantly, we may then interpret observed dynamics with the help of the metaphor of transients in a fitness landscape structured by local optima.

Thesis 2   The function of the sensory system of an animal consists in gathering relevant information. Information is relevant when the animal can select actions which, conditional on this information, are useful for the animal in the sense that it increases its fitness.

Action does not need to take place immediately upon the acquisition of relevant information. An animal can gather and store a lot of information as the basis for the selection of future actions. Information can also be indirect. For

---

[2]   This concept is not as trivial as it may appear. For further discussion, see Jost (2003, 2004).

example, you can see that it is raining and seek cover, so as not to become wet, or you can observe a gray cloud and infer that it will rain. Or, you can check the weather forecast.

Another question is whether and how the information is represented. The representation need not be explicit. It can be represented implicitly, and perhaps only partially or incompletely, through memory traces. It can also be stored externally. It only needs to be accessible when needed. In Thesis 4, we shall discuss how learning mechanisms transform the information contained in correlations into internal associations.

The interaction between sensations and actions is more subtle and intricate than expressed by Thesis 2. The activity of the system is also necessary to generate percepts (discussed further below; Thesis 3). In particular, an animal acts not only on the information that it has acquired, it acts to acquire information in the first place.

The Sensorimotor Loop

We begin with a diagram representing the dynamics of the sensorimotor loop.

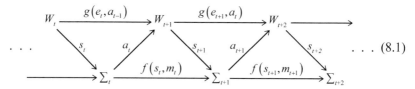

$$\tag{8.1}$$

Here, Σ_t is the system at time t, and W_t is the environment, the external world, at time t. At time t, the environment is in state e_t, and the system in state m_t. The latter receives the sensory signal s_t from W_t and acts via a_t upon the environment. The effect of the action affects the external state e_{t+1}. Thus, we have the state transitions

$$e_{t+1} = g\left(e_t, a_{t-1}\right) \tag{8.2}$$

$$m_{t+1} = f\left(s_t, m_t\right). \tag{8.3}$$

The transitions in equations (8.2) and (8.3) could be deterministic or stochastic, but the fundamental point is that the external state e_t is not directly accessible to the system. Thus, from the perspective of the system, there is a fundamental asymmetry between equations (8.2) and (8.3). The system can only derive some partial, incomplete, and possibly inaccurate information about e_t from its sensory data s_t, and it can partly influence the next state e_{t+1} through its action a_t. Here, we have arranged the relative times so that the environment is always a little ahead of the system, in the sense that the sensory signals are received instantaneously, but the system's actions only become effective at the next time step.

There are some conceptual problems associated with this representation of the sensorimotor loop, which will be addressed below. However, let us first try to extract some insight from this representation.

A similar model could also be formulated in continuous time. In that case, the difference equations (8.2) and (8.3) would be replaced by (ordinary) differential equations. Ordinary differential equations are, in fact, mathematically easier to treat than difference equations, but perhaps it is intuitively easier to grasp the meaning of difference equations.

Optimization Principles

Various optimization principles have been proposed for the sensorimotor interaction of a system with its environment. Optimization, here, can mean either maximization or minimization, and in fact, the same, or a similar, quantity is maximized in some and minimized in other principles. This may sound somewhat puzzling and may make optimization principles questionable, but following previous work (Jost et al. 1997; Jost 2004), it may indeed be plausible to sometimes minimize and sometimes maximize a particular quantity, and in that case, the system will take recourse to schemes that operate on different timescales. The timescale of actual perception, of correlation-based learning, and of biological evolution are clearly separated, but as one knows from many models in physics, a dynamical system can by itself separate into slow and fast dynamics (such an effect leads to a so-called center manifold on which the slow dynamics takes place; see, e.g., Jost 2005). In cognition, sensory inputs are gathered and different inputs need to be compared to detect regularities. The latter should naturally occur on a slower timescale than the former. That is, cognition requires both online interaction with the world and offline processing of data. Therefore, finer distinct timescales may well exist, but here we will not explore these details (for a general discussion of timescales in cognitive and other complex systems, see Jost 2004). Some of the quantities to be optimized involve differences between two terms, and so naturally one of them will be maximized while the other will be minimized. Again, being able to operate on different timescales may help to avoid such conflicts.

There is the basic choice between exploitation and exploration: to utilize what one knows already to its fullest extent, without wasting any energy or incurring any risk by searching for new opportunities, or to search actively for better resources than what is currently available.

Usually, such optimization principles are formulated in information theoretical terms. Informally speaking, the alternative is whether one should go for predictability and prefer a completely black TV screen, which never changes its state or rather go for novelty and appreciate white noise most. Of course, neither option by itself sounds like a very intelligent or useful strategy.

To describe a sample of such optimization principles, we need some basic concepts from information theory. For simplicity, we will only consider situations with finitely many possible states. Thus, let there be states x_i for $i = 1,...,n$ with probabilities $p(x_i) = p_i$ (for short), precisely one of which has to occur at any given time. Thus

$$\sum_i p_i = 1. \tag{8.4}$$

We also say that there is a random variable, called X, whose possible states are $x_1,...,x_n$. Before finding out which one occurs, we are in a state of uncertainty, and when we observe the actual state, we gain information; that is, we reduce our uncertainty. The expected loss of uncertainty is quantified by Shannon's information or entropy:

$$H(X) = H(p_1,...,p_n) = -\sum_i p_i \log_2 p_i. \tag{8.5}$$

(This information is measured in bits, where a bit is the information gained when learning which of two equally probable events occurred.)

When we have another random variable Y, with possible states $y_1,...,y_m$, we can also look at the probability $p(x_i, y_j)$ for simultaneously X being in state x_i and Y in y_j. As in equation (8.4), $\sum_{i,j} p(x_i, y_j) = 1$, and we have the Shannon information of the pair:

$$H(X,Y) = -\sum_{i,j} p_{i,j} \log_2 p_{i,j}. \tag{8.6}$$

Now, X and Y might be independent, or equivalently $p(x_i, y_j) = p_X(x_i)p_Y(y_j)$ for all i, j (where we now use a subscript to indicate the random variable whose probabilities we are taking). In this case, observing Y does not reduce our uncertainty about X. When they are not independent, in contrast, Y contains some information about X. Thus, when the state X cannot be directly observed, but the state of Y is accessible, we can extract some information about the former from the latter. This is quantified by the mutual information between X and Y:

$$MI(X:Y) = H(X) + H(Y) - H(X,Y). \tag{8.7}$$

This quantity is symmetric in X and Y; that is, when Y contains information about X, then X contains the same amount of information about Y. Also, this quantity is always nonnegative; that is, when X and Y are not independent, the entropy $H(X,Y)$ of the pair is smaller than the sum $H(X) + H(Y)$ of the individual entropies. We can then also define the conditional information about X when Y is known:

$$H(X \mid Y) = H(X,Y) - H(Y). \tag{8.8}$$

From equation (8.7) we then have:

$$MI(X:Y) = H(X) - H(X \mid Y). \tag{8.9}$$

In words, the mutual information $MI(X:Y)$ tells us how much the uncertainty about the state of X is reduced when we learn about the state of Y. Again, when X and Y are independent, then $H(X \mid Y) = H(X)$ and consequently $MI(X:Y) = 0$.

The mutual information between two random variables can also be conditioned on a third one:

$$MI(X:Y \mid Z) = H(X \mid Z) - H(X \mid Y,Z). \tag{8.10}$$

Next we introduce the Kullback–Leibler distance or relative entropy between two probability distributions p and q:

$$D(p \| q) := \sum_i p_i \log_2 \frac{p_i}{q_i}. \tag{8.11}$$

(Note that this expression is not symmetric in p and q.) Let now X and Y be random variables on the same space with individual distributions p_X, p_Y and joint distribution p. Then

$$MI(X:Y) = D(p \| p_X p_Y), \tag{8.12}$$

the Kullback–Leibler distance between the joint probability distribution p and the product distribution $p_X \cdot p_Y$ under which X and Y are independent. Once more, this says that the mutual information between X and Y quantifies how far they are from being independent.

Equipped with these tools, we can now formulate some optimization principles for the system S in equation (8.1) that can control its actuators and acquire information through its sensors. Again, we assume that either of them can only be in finitely many states. Let us proceed in steps. For each strategy proposed below, shortcomings will be identified, and this will then motivate the next strategy.

In particular, we shall discuss some caricatures of various strategies proposed in the literature. Often, those strategies involve more than two time steps; that is, they not only optimize some quantity at time t + 1 conditioned on what occurred at time t, but look further into the future and recall longer sequences from the past. As this adds little to the basic principle, we shall suppress this issue systematically. In technical terms, we could say that we assume that all considered processes possess the Markov property; that is, all information from the past relevant for the future is contained in the present state.

Strategy 1   Minimize the surprise; that is, the sensory information $H(S_{t+1})$.

Optimal behavior: Close your eyes.

Strategy 2   Look for novelty; that is, maximize $H(S_{t+1})$.

Optimal behavior: Seek random noise.

Clearly, neither of the preceding strategies is very meaningful. More precisely, the system should try to acquire some information through its sensors, but it should not strive to gather as much information (8.5) as possible because such information may not contain any meaning for the system. Therefore, the following strategies attempt to collect information that is either produced by the system's own actions or is predictable in terms of past sensor information.

Strategy 3   Maximize the empowerment.

$$MI(A_t : S_{t+1}) = H(S_{t+1}) - H(S_{t+1} \mid A_t). \tag{8.13}$$

That is, try to act in such a way that you get as much information as possible about your sensor states at time t + 1.

The empowerment principle states that the system should act such that $H(S_{t+1})$ is large,[3] but $H(S_{t+1}) \mid A_t$ is small. Thus, the sensors should deliver a lot of information in the future, but this should already to a large degree be predictable by the actions carried out at present. As interpreted by Klyubin et al. (2005), empowerment is the amount of information that the systems can inject into the environment via its actuators and recapture through its sensors.

Optimal behavior: Wiggle your feet and look at them.

Strategy 4   Maximize the predictive information.

$$MI(S_t : S_{t+1}) = H(S_{t+1}) - H(S_{t+1} \mid S_t). \tag{8.14}$$

Optimal behavior: Look for complicated situations that you understand well.

Before proceeding, let us analyze the example of saccadic eye movements with those concepts.[4] Here, $H(S_{t+1})$, the information received on the retina after the

---

[3] It is important to note that by the design of the sensorimotor loop (8.1), the actions at time t will influence the state of the external world at time $t + 1$ and thus also the sensory data s_{t+1} that the system will get back at time $t + 1$.

[4] Only an abstract account is presented here, suppressing many important details, to elucidate the underlying principles. For a precise analysis of saccadic eye movements from an information perspective, see Bruce and Tsotsos (2006).

movement, may be quite large. At least when the scene is still, it is, however, essentially determined by the combination of S_t (i.e., what had been received before the movement) and A_t, that movement. Thus, $H(S_{t+1} \mid S_t, A_t)$ is small. In that sense, the quantity $H(S_{t+1}) - H(S_{t+1} \mid S_t, A_t)$ could be large, and we would have an amalgam of Strategies 3 and 4. In any case, this would not amount to true novelty. In fact, for a still scene, the best action would be not to move the eyes at all, because then $S_{t+1} = S_t$ (although for the ball-catching example discussed above, this is no longer so trivial, as there, one piece of sensory information, the viewing angle, should indeed remain constant). Neither does this seem to be the main purpose of saccadic and other eye movements. Instead, one important aspect is that they serve to focus on what is novel and therefore interesting (discussed below, when another important function of such movements, namely the generation of correlations, is identified).

Thus, while the two preceding strategies go in the right direction, they still lead to a somewhat solipsistic attitude, insofar as the system is not really trying to learn something about its environment itself. It would be desirable to include the external states.

Strategy 5   Maximize the nontrivial information closure.

$$MI(S_{t+1} : E_t) - MI(S_{t+1} : E_t \mid S_t) = H(S_{t+1}) - H(S_{t+1} \mid E_t)$$
$$-H(S_{t+1} \mid S_t) + H(S_{t+1} \mid E_t, S_t). \quad (8.15)$$

In contrast to Strategy 4, the system is not just trying to learn something about itself: it also seeks predictable sensory information about the environment. The problem with this strategy, however, is that the system has no means to access E_t except indirectly through its sensory data S_t. Therefore, it is not in a position to implement Strategy 5.

It is important to emphasize that the above only presents caricatures of principles developed in the literature. The actual principles are more refined and often find useful applications. It is not the purpose of this volume to produce comprehensive literature surveys and beyond the scope of this chapter to list all the optimization principles proposed for the information processing by embedded agents. Therefore, a survey of the very extensive literature on those issues is not attempted here; neither shall I mention the many variants of those principles proposed over the years. I would like, however, to list at least those sources that have introduced the quantities discussed above.

The concept of empowerment was introduced in Klyubin et al. (2005); however, in contrast to the above presentation, it was conceived as a multistep principle that is more powerful than the simplified one-step version presented here. In particular, the optimization of the action according to (8.13) was conceived only as an intermediary step for finding the optimal sensory input S_t that enables the highest control over future inputs through own actions. A similar remark applies to the use of predictive information as an optimization principle in Ay et al. (2012). Finally, the nontrivial information closure (8.15) was introduced by Bertschinger et al. (2008) for a different purpose, in the context of a system theoretical analysis, and not as an optimization principle for an agent interacting with its environment.

Looking at the proposed Strategies 1–5, we seem to be at an impasse. In order not to be overwhelmed by meaningless noise, the system can only look at its own feet. The way out may consist in adaptation and learning. Importantly, the system cannot only control its actions a_t, but also modify its internal state m_t. Thus, we must look at the problem from a somewhat different perspective: that of learning theory.

We assume that there is some external probability distribution p that the system tries to infer on the basis of its sensory data. That is, the sensory data are supposed to be random samples of p. (To avoid technical difficulties, we assume here that p is stationary, i.e., does not change in time.) The system then creates a subjective model q which it adapts on the basis of the incoming sensory data stream. Thus, it tries to minimize the Kullback–Leibler divergence (8.11):

$$D(p\|q)=\sum_{s}p(s)\big(\log p(s)-\log q(s)\big).\qquad(8.16)$$

Since the system has only its past sensory data $s_\tau, \tau = 0,...,t$ at its disposal, at each time t, it can adapt q so as to minimize

$$\sum_{\tau=0,...,t} p(s_\tau)\big(\log p(s_\tau)-\log q(s_\tau)\big)\qquad(8.17)$$

within some class of probability distributions that need to satisfy some complexity bound to avoid overfitting (Vapnik 2000). Of course, when t gets large, such a batch learning may become unfeasible. We could introduce some fading memory effect to the extent that signals from the more remote past get lower weights, or even get forgotten. Alternatively, we can perform some stochastic gradient descent, that is, increase the subjective probability of a signal s_τ whenever it occurs as a sample. For such a stochastic gradient descent, the frequency of adaptations depends on the unknown objective probability distribution p, but their magnitude is determined by the current subjective model q.

Of course, since p is not known to the system, but only the samples s_τ drawn from it, the first term in (8.16) or (8.17), the entropy term $-H(p)$, needs to be addressed differently. That is where the actions come into play.

Let us look at a concrete example that will also highlight the conceptual difference to Strategy 3. A visual signal is received when a receptive field on the retina is stimulated. Thus, for this example, the collection of receptive fields is the space of values of the random variable S_t. When the system optimizes (8.16) with regard to q, it simply increases the probability assigned to a receptive field whenever that field gets stimulated. In contrast, when it tries to optimize with regard to p, it would shrink a receptive field that gets stimulated so that its probability of being stimulated is decreased, and enlarge other fields that are less frequently stimulated so that their probability of getting stimulated is increased. This will have the effect of making the probability distribution p of the signals (i.e., the stimulation probabilities of the receptive fields) more uniform. Thus, the entropy of p will increase, and since that entropy occurs with a negative sign in (8.16), the Kullback–Leibler divergence $D(p \| q)$ will therefore decrease. Thus, in contrast to Strategy 3, the system does not try to move into a region of the signal space where the signals are consequences of its own actions, but rather rearranges the signal distribution in such a manner as to increase its entropy.

Of course, the system should try to act in such a manner as to decrease $D(p \| q)$ most efficiently. This also leads to another important aspect. When we discussed the Strategies 1–5, we did not specify how minimization or maximization is actually achieved. In principle, the system could employ a stochastic strategy, along the lines just discussed. More interestingly, it could try to form a prediction (e.g., via a Bayesian estimate) about how the quantities to be optimized change in response to own actions, and then select the action which, according to such a prediction, seems best. As a result of the consequences of such an action, the prediction can then be adapted. This has been explored by Little and Sommer (2013), with a discussion of earlier research in psychology and a comparison with reinforcement learning strategies.

It is important to stress, however, that achieving $D(p \| q) = 0$ (i.e., the equality), $p = q$, is typically not desirable. As the environment is always more complex than the system, the system needs to compress the information obtained through its input, instead of trying to produce a faithful copy. Statistical learning theory (Vapnik 2000) tells us that model constraints are needed to avoid overfitting the input.

In particular, the system needs to classify and categorize, instead of simply reproducing the input. Thus, for instance, it can simply entertain a finite number of hypotheses and then check which of them best fits the data. In addition, the system may typically not be able to check all the details of the data at hand simultaneously, but may need to select certain features that it observes. As has been argued (Jost 2004) and subsequently successfully applied in machine learning (Avdiyenko et al. 2015), the system should then select those features that have highest discriminative power between the competing hypotheses. For instance, when there is a certain number of equally likely hypotheses, one should check a feature that is expected to be present under half of the hypotheses, and absent when one of the other hypotheses is correct.

Only an interpretation that captures the significant aspects and ignores the unimportant parts of the input as noise is capable of establishing meaning for the system. Obviously, going beyond the framework of learning theory, the decision about what is significant and what can be treated as noise is important for the system, but that decision may be carried on a timescale that is different from the online or batch interpretation of the input signals. Previously, I have distinguished between the data or external complexity and the model or internal complexity (Jost 2004). The external complexity is to be maximized when the system wants to extract information from the environment. The internal complexity, in contrast, should be minimized to achieve an efficient representation of the data. This leads to two intertwined principles: (a) the system should gather data so as to make its models more accurate and improve its predictions; (b) the system should construct models that enable it to collect more meaningful and useful data.

In the light of these principles, let us turn now to schemes that do not simply maximize some difference of information theoretical quantities, but create specific structures in the sensory input. Such structures will consist in correlation patterns between motor activities and the induced transformations of sensory input, as well as between different sensory inputs. Thus, the system not only tries to identify and model regularities in the input, it actively creates them in terms of correlations between actions and sensations.

Actions Induce Correlations for Creating Percepts

Percepts are not sensory stimuli, but rather brain states (however defined). Of course, we believe that there has to be some correspondence between percepts and neuronal activities. This does not mean, however, that a percept must necessarily correspond to some activity pattern of a specific group of neurons at a specific time. It could be that some spatiotemporal activity pattern that dynamically extends over time gives rise to some, possibly static, percept in our subjective experience, perhaps in accordance with the proposal put forward by O'Regan and Noë (2001). The question then is: How could such a spatiotemporal pattern possibly be characterized and by what mechanism could it be activated? Proposals about the nature of such patterns include synfire chains (Abeles 1991) and the synchronization patterns of von der Malsburg (1973) and Singer (Singer and Gray 1995). Here, we shall formulate some abstract principles about their nature and induction.

> **Thesis 3**   On the basis of sensorimotor contingencies, actions induce correlations between neuronal activities in the same or different brain regions and thereby induce coherent activity patterns that correspond to percepts.

Let us discuss in more concrete terms how this might work. The claim will be that saccadic eye or head movements induce correlations between the stimuli that correspond to different parts of an object and allow the agent to distinguish or identify that object. It is difficult, and for many animals even impossible, to distinguish a still object from its background. When, in contrast, the object moves, there will be specific sensory correlations between the different parts of that object while those specific correlations will not exist between the stimuli coming from the object and those from the background. In addition, the movements of rigid objects subjected to physical forces and of animals, for instance, are very different from each other; thus, the correlations induced by movements should make a distinction between inanimate objects and animals quick and easy.

Saccadic and other eye, head, or body movements can also produce such specific correlations to distinguish an object from its background, in particular in conjunction with stereoscopic depth perception. In other words, we locate an object by how sensory input on the retina changes in response to our own movements. This is fairly obvious. Also, many of the gestalt laws depend on suitable types of correlations (for a more precise analysis, see Breidbach and Jost 2006). In particular, invariances can be realized by motions. For instance, taking an example from O'Regan and Noë (2001), the sensation caused by a straight line stays invariant under motions in the direction of that line. Only when an endpoint of the line is reached, does the nature of the correlations change. What is crucial is the type of correlation between the movements of the eye or the head (i.e., the self-induced motions and the actions on one side, and the sensory stimulations on the other). In line with O'Regan and Noë's (2001) proposal, that which underlies the percept is nothing but a specific correlation pattern of neuronal activities.

An object, however, is more than a geometric shape. The binding problem, as identified in particular by Singer (2001), concerns the combination of the various features of an object into the perception of an integrated object. Again, as I propose here (inspired by discussions with Wolf Singer), this can be achieved through correlations induced by sensorimotor contingencies. The key observation is that the various features are bound together by their common location; that is, one and the same receptive field or region in the retina receives information about color, texture, brightness, etc. Through a retinotopic map, this then activates the corresponding region in the lateral geniculate nucleus (LGN), which is located in the thalamus. When the position in the retina stimulated by such a spot in the object changes because of movements, then all these feature stimulations simultaneously move across the LGN. The location in the retina and the LGN changes, but various features which belong together always stimulate the same area. The thalamus is connected via reciprocal connections to specific cortical regions, and the different features are processed in different cortical regions. Nevertheless, their joint movement across the retina and the LGN induces correlations between specific areas in the corresponding cortical regions. Via mechanisms of delayed feedback, this may induce specific synchronization patterns between those areas. As proposed by Christoph von der Malsburg and Wolf Singer, these specific patterns may correspond to the percept of the object (for a detailed survey of dynamic coordination, see von der Malsburg et al. 2010). The relevant aspect of such synchronous oscillations is not that different neuronal groups are simultaneously active—after all, there is nothing in the brain to record that, and axonal and synaptic transmission delays prevent any intrinsic notion of simultaneity in the brain[5]—but rather that particular spatiotemporal activity patterns in the brain may have some self-sustaining or self-amplifying capabilities. What is important for the present proposal is that sensorimotor contingencies are needed to induce specific correlations that can become amplified via such a synchronization mechanism, or another scheme that establishes spatiotemporally coherent activity patterns. I propose that without such specific correlations as induced by sensorimotor contingencies, a percept of an individual object could not emerge in the brain.

Such a correlation analysis provides a unifying perspective on the "where" and the "what" aspect of visual cognition. At an abstract level, it suggests a common principle underlying object perception (identification of a coherent object in a visual scene as distinct from other objects or the background), identification (tracking of an individual object that is previously known), and recognition (classification of an object as belonging to some general concept).[6] Of course, the nature of the correlations varies between these different tasks. Also, important differences exist between identification of an object as an individual object with an identity preserved across space and time and its classification as a member of some class regardless of its individual identity. However, this is not addressed here.

If this proposal is feasible, we need to work out the relevant mathematics. This should be a combination of nonlinear dynamics as needed to understand synchronization patterns and of information theory that can quantify the correlations on the basis of the concepts introduced in the previous section.

---

5   This issue needs a more careful discussion than is possible here. Let me only make a few comments. Neurons can fire when they receive simultaneous input from a specific set of presynaptic neurons, and one might therefore argue that at least locally, the simultaneous activity of groups of neurons can be detected. There are at least two problems here. First, the postsynaptic neuron reacts with some delay; it cannot record that a set of presynaptic neurons is active now, but only that it has been active a short while ago. Second, it can at best detect that presynaptic neurons have been active within the same small window of time. To record such simultaneity by suitable intracellular measurements, one has to bin spikes. Let us assume, for concreteness, a bin size of 1 ms. It can then be said that neurons A and B have been simultaneously active if their spikes fall within the same bin. However they could also have been active within less than 1 ms of each other, although their spikes fall into different bins. For instance, one could have spiked at 1.9 ms, the other at 2.1 ms. Of course, such a binning introduces artifacts. Even if we say that two neurons fire simultaneously if they emit a spike within less than 1 ms of each other, then A and B can fire simultaneously in this sense, and B and C can also fire simultaneously, but this does not imply that A and C also fire simultaneously. The notion of simultaneity is not a transitive relation. For the technical aspects, see Grün and Rotter (2010).

6   Here I am employing the terminology used in the machine vision literature. This is not completely compatible with the conventions in the psychology literature, in particular concerning the meaning of "object perception," which seems to include the aspect of identification/classification. The machine vision terminology is concerned with the computational requirements and difficulties involved in those tasks, and this approach also offers useful aspects for the present essay.

In technical terms, there is an important task for dynamical systems theory: How could correlations be detected and exploited in the presence of neuronal transmission delays (i.e., in the absence of any global notion of simultaneity in the brain)?

As argued here, actions generate correlations and there needs to be some action selection principle. Actions should be chosen such that they cause suitable correlations from which percepts can then be formed. This issue is not addressed here, although the framework developed in Jost (2004) should be of some help.

Instead, we turn to the important question of how such correlations can create stable and reproducible percepts. This relationship will be argued to be the result of a learning process.

Thesis 4   The nature of learning is to transform correlations into associations. Thereby it can stabilize the induction of specific neuronal activity patterns in response to specific sensory patterns.

Thus, when experienced often enough, specific correlation patterns (between motor activities and sensory stimulations as well as between different types of sensory stimulations induced by sensorimotor contingencies corresponding to specific classes of stimuli) can cause neuronal dynamics which can be interpreted as associations.

There is, in fact, a specific neuronal learning mechanism, some temporal Hebbian scheme, called the spike timing-dependent synaptic plasticity rule, first introduced by Gerstner et al. (1996). It can be seen as a neuronal version of operant conditioning (Jost 2006). The effect is that when the initiating stimulus for some specific dynamical patterns is presented, that pattern is induced without the subsequent stimuli necessarily coming in as well. In other words, we "see" something when triggered by some specific sensory input, because we have "learned" that this stimulus is typically followed by a specific sequence of further stimuli, and since we "know" this, we no longer need to confirm those subsequent stimuli to "perceive" the corresponding entity. This process can only get disturbed or interrupted by subsequent sensory stimuli that contradict the created percept and which may then induce some other percept in turn. On this basis, interpolations can be made. We only need to sample the sensory data at certain intervals to reconstruct what happened in between. Again, this is not an active process, in the sense that it requires a particular effort. The sensory data coming in at certain intervals are simply sufficient to trigger and maintain suitable neuronal dynamics, as long as there is no mismatch between the ongoing neuronal dynamics and the sensations received.

As an aside, since such specific dynamical patterns only emerge as the result of synaptic learning processes, infants do not have such percepts prior to the establishment of the corresponding dynamical pattern. That is presumably why we do not have early childhood memories of percepts. As I have argued, when a percept corresponds to some dynamical neuronal pattern, and if that pattern only emerges as the result of a learning process, and if memory recall operates by triggering that dynamical pattern and thereby evoking the corresponding percept, then there is nothing to be remembered before the learning mechanism has created the relevant associations. Of course, to elaborate this proposal, one would need to investigate whether, and if so, why, the mechanisms creating the associations in question take hold at just the age at which children begin to form memories. In particular, the transition between the essentially memory-less state and that where a rich set of memories becomes available seems to be relatively sudden, and therefore, the underlying mechanism probably must be of a rather general nature.

The preceding proposal is somewhat similar to that of the predictive brain, but is different at some crucial point. There are no explicit predictions, only at best implicit ones contained in the specific spatiotemporal activity pattern triggered by a specific sensory stimulation. As described, the learning process transforms correlations into associations, so that a sensory stimulus can trigger an autonomous neuronal activity pattern which developed from past experiences in response to an entire sequence of stimuli. After learning, the initial stimulus is sufficient to trigger that pattern, and in that sense, the pattern contains an implicit prediction of the entire stimulus sequence. The rest of the sequence is no longer needed, and we might even experience it when it is not there. Of course, when contradictory sensory signals arrive, the neuronal activity pattern may get disturbed and interrupted. One may then say that the implicit prediction contained in the neuronal activity pattern has not been confirmed. For an analysis of backward visual masking effects, in terms of a conflict between internal predictions generated by the original stimulus and the subsequent contradicting sensory signal of the mask, see for instance Elze et al. (2011). Importantly, according to what is proposed here, there is no need for an explicit prediction. The activity pattern simply unfolds as if that stimulus sequence from the past, which repeatedly followed the initial stimulus, were there, unless too many contradictory sensory signals are received. This brings into question the causality paradigm, which is often applied to decode neuronal responses. This paradigm states that only earlier stimuli can influence a response. In physical terms, this is correct. However, when the response is strongly correlated with sensory input that typically follows, or has repeatedly followed, the first stimulus, a relation exists between a neuronal activity pattern and later stimuli. This relation can then be used to decode the meaning of neuronal activity. Of course, we already know from the experiments of Libet (1985) that the subjectively experienced temporal order may differ from that of the underlying neuronal activities.

As just argued, the sensorimotor contingencies can get internalized as a result of the learning process. This scheme can then be iterated. Correlations between firing patterns in different brain regions could form internal percepts. I would even speculate that much of higher cognition can be captured by such a framework. In particular, this allows for the reflexive nature of consciousness. We perceive that we perceive—Leibniz's notion of apperception. In the framework proposed here, such internal percepts can be internally detected, perhaps by other brain regions, or better, by other processes evaluating those percepts. The insight that sensation is coupled to action may then explain the unity of consciousness. We may have different and perhaps conflicting sensations simultaneously, but, as emphasized in the Supramodular Interaction Theory (Morsella 2005), we need to select a single action at each instance, or at least cannot simultaneously carry out conflicting ones.

Limitations

When one tries to apply a correlation-based analysis to high-dimensional data sets, one quickly realizes that this does not work. An important insight of recent research in machine learning is that such principles need to be supplemented by structural priors. Such a structural prior could be a sparsity assumption. For instance, to analyze an auditory scene, one assumes that there is only a small number of sound sources. This is explored in compressive sensing. Or, one might assume that the data are concentrated on or near a smooth manifold that might stretch in many dimensions, but which is intrinsically low dimensional. This is called manifold learning. One might also make more general continuity assumptions (e.g., to identify movement patterns), or one could assume that the data arise as sums of a few tensor products of vectors in low-dimensional subspaces. Of course, when confronted with a specific data set, the question becomes: What is the most appropriate structural prior? This is somewhat analogous to the problem of finding the best heuristics in an intransparent situation, as discussed by Gigerenzer and Todd (1999). Warglien et al. (submitted) argue that structural assumptions like convexity, monotonicity, or continuity are essential for the semantics of verbs. Likewise, gestalt laws also depend on more specific classes of transformations, rather than on simple correlations (Breidbach and Jost 2006).

One needs structural priors or heuristic techniques, or whatever one wants to call them, to generate some preliminary coarse structure within which a more precise correlation analysis can then be successfully applied. The origin of such structural priors and, in particular, whether they are prewired in our brains or can possibly be learned (and if so, how) constitute areas for future enquiry. In some sense, they might constitute a modern version of Kant's concept of synthetic *a priori* knowledge.

Acknowledgments

This essay is to a large extent based on joint research and discussions with Nihat Ay, Nils Bertschinger, and Eckehard Olbrich. I also wish to thank Wolf Singer for insightful discussions about the role of synchronization. Nihat Ay, Friedemann Pulvermüller, and an anonymous referee supplied useful comments on my manuscript.

9

Language, Action, Interaction

Neuropragmatic Perspectives on Symbols, Meaning, and Context-Dependent Function

Friedemann Pulvermüller

Abstract

Neural mechanisms of cognition are built upon action, action perception, and interaction. This chapter explains how this novel perspective, immanent to the "pragmatic turn" in cognitive neuroscience, is enforced by research on language, semantic concepts, and social communication. Whereas classic approaches attributed these specifically human domains to genetic endowment and encapsulated processes, modern cognitive and brain research has accumulated evidence that mechanisms for speech sounds and symbol forms emerge as a result of sensorimotor functional interaction in the brain, and that conceptual-semantic information is extracted from the interaction of learners with their environment and peers. Correlational Hebbian learning in anatomically prestructured network architectures binds articulatory-motor to auditory-perceptual (phonological) knowledge. This epigenetic neurobiological perspective also explains important aspects of whole form (lexical) storage of symbols and constructions, combinatorial (distributional, syntactic, or grammatical) linkage between stored forms, and context-dependent (semantic, pragmatic) binding between forms, their meaning, and interactive function. Over and above evidence for motor system activation in linguistic and conceptual processing, specific studies demonstrate its causal role for these domains. Thus, action-perception theory offers a novel avenue toward neurobiological explanation of the brain mechanisms for language, concepts, and pragmatic communication.

Introduction

When sensory neurons in an artificial network with random connectivity are repeatedly stimulated, these neurons link up with their connected neighbors to form a strongly connected neuron set (Doursat and Bienenstock 2007). This *neuron set* (also termed cell assembly, synfire chain, neuronal avalanche, or neuronal ensemble) can be seen as a sensory "representation" (Hebb 1949). Strong experimental evidence indicates that our brains carry such sensory representations (see Plenz and Thiagarajan 2007; Singer and Gray 1995). However, representational mechanisms do not need to be restricted to the sensory-perceptual domain. In many cases, a degree of motor activity accompanies novel experience: We startle, smile, explore, approach, or retreat in view of new objects or persons. In such cases, the purely "sensory" learning model is insufficient; motor movements or (when these are led by specific goals) goal-directed actions need to be taken into account (Braitenberg and Schüz 1992; Fuster 1995; Jeannerod 1994; Rizzolatti and Craighero 2004; Pulvermüller 1999). When modeling action-perception contingencies in neurocomputational models structured according to cortical areas and their neuroanatomical connectivity, action-perception contingencies lead, by way of Hebbian learning, to the formation of distributed neuronal circuits. These *action-perception circuits* incorporate neurons in sensory model areas and adjacent "higher" multimodal ones, and reach into prefrontal, premotor, and, ultimately, primary motor areas (Garagnani et al. 2008; Pulvermüller and Garagnani 2014). They provide a mechanism for the cognitive correlates of objects with a regular and specific usage, including tools or food items, as well as for actions the individual can perform, because action performance always implies specific motor movements along with sensory autostimulation. Representations[1] that connect motor and sensory knowledge with each other can be called "pragmatic";[2] the focus on interlinked action-perception representations and circuits characterizes what has been called the "pragmatic turn" (Engel et al. 2013). Such action-perception representations may be more powerful neuronal devices than unimodal sensory representations, because they arise from correlated activity in sensory and motor areas, and therefore, their likely neural mechanisms are large widely distributed circuits spread out over multiple cortical areas.

The postulate that action representations and mechanisms are systematically coupled to perceptual ones has a long history in the cognitive and brain sciences

---

[1] When speaking about action-related engrams, Engel et al. (2013) propose to replace the term "representation," with "directive." However, in theories of language and communication in *linguistic pragmatics* (Austin 1962; Fritz 2013; Searle 1969; Stalnaker 2002; Wittgenstein 1953), directives represent only one subtype of social-communicative action; namely requests, commands, and the like. Thus, the use of the term "directive" might suggest an exclusion of other relevant action and interaction types (e.g., assertions or expressions of feelings, planning, and bargaining) (see, e.g., Searle 1979). Therefore, I will use the term "pragmatic representations" to refer to knowledge about all types of actions (communicative and not) and the perceptions to which they relate.

[2] In this sense, the term "pragmatics" can be used to speak about a range of different scientific schools, ranging from "enactive," "grounded," and "embodied" cognitive theory (Varela et al. 1992; Clark 1999; Barsalou 2008; O'Regan and Noë 2001) to pragmatism or pragmatic philosophy (Mead 1938; Dewey 1896; Peirce 1931, vol. 5) and linguistic pragmatics.

(Jeannerod 2006; Fuster 2003; Braitenberg and Schüz 1998; Pulvermüller 1999; Pulvermüller and Fadiga 2010; Rizzolatti and Sinigaglia 2010; Clark 1999; O'Regan and Noë 2001). This hypothesis is radically different from the classic position that modality-specific sensory modules channel information to central systems for attention, memory, language, concepts, and decisions, which, in turn, drive the motor output (see Figure 9.1; Hubel 1995; Fodor 1983). Instead, cognition is seen as being *built from* action and related perceptions, upon motor and sensory brain mechanisms (Jeannerod 1994, 2006).

Are Action-Perception Mechanisms Sufficient for Cognition?

A wealth of data support this pragmatic position of interwoven action-perception circuits. *Mirror neurons* are active in motor planning and execution as well as in the perception (visual or auditory) of specific actions (Rizzolatti et al. 1996; Kohler et al. 2002; Rizzolatti and Craighero 2004). This dual role may be due to their membership in action-perception circuits that formed as a consequence of cortical anatomy and mapping of neuronal correlations by Hebbian synaptic plasticity. *Memory mechanisms* are known to rely on both posterior (e.g., temporal or parietal) areas but also draw upon prefrontal neurons; the parallel functions of neurons in different lobes may relate to their membership in the same distributed action-perception circuits (Fuster and Alexander 1971; Fuster 1995; Pulvermüller and Garagnani 2014). Mechanisms for mapping motor movements on perceptions are necessary for repetition and imitation of behavior, which seem to play a crucial, though basic, role in normal cognitive and language development (Rizzolatti and Craighero 2004; Pulvermüller and Fadiga 2010). Over and above such (basic) perceptual recognition of movements that individuals have previously performed by themselves, it has been argued that the *understanding* of others' actions may depend, to a degree, on action-perception mapping between different individuals, and that the mental simulation of others' actions may be a main component of the understanding of language, symbols, social-communicative function, intentions, and, more generally, meaning (Kiefer and Pulvermüller 2012; Pulvermüller 2013; Glenberg and Gallese 2012; Barsalou 2008; Meteyard et al. 2012). However, arguably, it is not clear how these aspects of higher cognition emerge from circuits that store action-perception contingencies. Animal and (especially) human cognition as well as social interaction certainly require more than imitation, repetition, and simulation (Borg 2013; Jacob and Jeannerod 2005; Hickok 2009; Lotto et al. 2009; Hickok and Hauser 2010; Csibra 2007). In view of a neuromechanistic theory of human cognition, it is thus crucial to explore which cognitive mechanisms are explained by an action-perception perspective and whether there are natural limits to this line of thought.

Starting from established knowledge about mirror neurons and action-perception coupling in the brains of monkeys and humans, I discuss the idea that

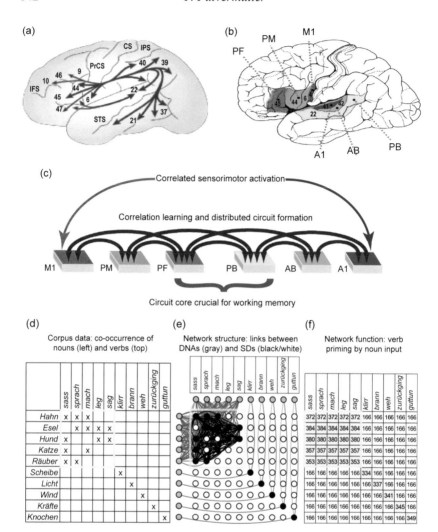

Figure 9.1 Cortico-cortical connectivity, word form circuit formation, and combinatorial semantic learning. (a) Long-range cortico-cortical connections within the perisylvian language cortex and adjacent areas. Abbreviations: IFS, inferior frontal sulcus; PrCS, precentral sulcus; CS, central sulcus; IPS, intraparietal sulcus; STS, superior temporal sulcus; numbers indicate Brodmann areas (Rilling et al. 2008; reprinted with permission from Macmillan Publishers Ltd.). (b) Neuroanatomical subdivision of inferior frontal and superior-temporal cortex into six areas: M1, primary motor; PM, premotor; PF, prefrontal; A1, primary auditory; AB, auditory belt; and PB, auditory parabelt areas (Garagnani et al. 2008; reprinted by permission from John Wiley and Sons). (c) Schematic connection structure of the six areas highlighted. Correlated activation in M1 and A1 during articulations leads to spreading activation in the network and distributed circuit formation for syllables and words. Their richer connectivity determines that PF and PB develop circuit cores (Garagnani and Pulvermüller 2013), where word form circuits link with each other in combinatorial learning.

Figure 9.1 (continued) (d–f) Combinatorial learning of noun-verb co-occurrences in an auto-associative neuronal network model. (d) The matrix shows word pair co-occurrences in a mini-corpus that served as input to the network (verbs in top row, nouns in left column; crosses indicate co-occurrences in text). The matrix section of frequent recombination is highlighted in yellow. (e) Neuronal elements for the same words (gray circles), sequence detectors (SDs) sensitive to specific word pair sequences (white and black circles in square arrangement), and connections between them. Black SDs indicate learning of specific sequences of nouns and verbs previously presented to the network. All word circuits previously involved in combinatorial exchanges are interlinked by way of a conglomerate of heavily interconnected sequence detectors, *the combinatorial neuronal assembly* (black SDs and black between-SD links on top left). Emergence of *generalized links* between those nouns and verbs, which frequently occur in combination with the respective other word group (yellow square), by formation of the combinatorial neuronal assembly is a neuromechanistic result of co-activation of *some* (not all) of the relevant SDs. (f) Result of combinatorial learning for network functionality. After learning, activation of any noun involved in the combinatorial schema (yellow square) primes all of the verbs involved to the same degree, regardless of whether the specific word sequence itself had been subject to learning. The dynamics are discrete in the sense of an all-or-none response. Note the generalization to sequences not previously encountered (Pulvermüller and Knoblauch 2009; reprinted with permission from Elsevier).

pragmatic action-perception representations are the building blocks of higher cognition. The more specific and crucial questions that I will address include:

- Are sensorimotor interactions involved in perception? (See section on "Speech Movement Coupled to Perception.")
- Is there a pathway from sensorimotor coupling to action-perception circuits for understanding meaning? (See section on "From Movement to Meaning.")
- Can (aspects of) abstract meaning be captured by action-perception circuits? (Discussed under "Abstract Meaning.")
- How are communicative actions and intentions realized in an action-perception architecture? (See section "Social-Communicative Interaction.")
- And, generally, given that motor systems are *activated* in perception and cognition, do they also *contribute to* and have a *causal effect on* cognitive processing? (Addressed in all sections.)

These questions have been addressed extensively in the domain of language and communication. This chapter highlights recent work as it pertains to the neuroscience of language.

Speech Movement Coupled to Perception: Mirror Neuron Circuits, Repetition, and Simulation

The correlation of motor activity with sensory input is implied by the very fact that, in the un-deprived individual, movements lead to sensory self-stimulation.

Correlations are mapped in the brain and especially in the cortex, whose neuroanatomical properties seem to be optimal for functioning as an associative
memory (Braitenberg and Schüz 1998). In the human language areas of the left
hemisphere, there are species-specific strong connections through the dorsal
arcuate fascicle between frontocentral areas adjacent to articulatory motor cortex and temporal areas relevant for acoustic processing (see Figure 9.1; Rilling
et al. 2011). These provide the necessary information highway for mapping
of sensorimotor correlations according to Hebbian learning principles (Artola
and Singer 1993; Caporale and Dan 2008). The mirror circuits that map speech
sounds (or phonemes) onto their articulatory motor schemas may therefore be
a result of learning. Note that in view of the variability of phoneme inventories
across languages, an inborn phonological mapping mechanism is insufficient.
The formation of action-perception circuits can be tracked even in adults who
learn novel, meaningless spoken word forms (Pulvermüller, Kiff et al. 2012;
Shtyrov et al. 2010). A range of data support the activation of motor circuits in
speech perception (Fadiga et al. 2002; Pulvermüller et al. 2006) and their relevance for speech sound classification (D'Ausilio et al. 2009; Möttönen et al.
2013). A degree of discussion still surrounds the question whether frontocentral articulatory motor and premotor cortices take a direct causal role in speech
sound processing (Pulvermüller and Fadiga 2010; Möttönen et al. 2013) or
whether their role is restricted to a post-perception task-dependent decision
stage (Venezia et al. 2012). Recent results provided evidence for a direct causal
role of articulatory sensorimotor cortex in semantic understanding of single
spoken words (Schomers et al. 2015).

In neurocomputational studies, action-perception circuits for speech
sounds, spoken word forms, or motor acts can more generally be employed
to activate a specific motor program upon specific sensory (auditory, visual,
etc.) stimulation (Garagnani et al. 2008). Neurodynamically, the strong links
within the action-perception circuits entail a full activation or "ignition" of
the cell assembly after sensory stimulation, a possible brain basis for perception and recognition of objects and word forms. Followed upon ignition of
a circuit, reverberatory activity lasts for some time, thus providing a brain
basis for object- and word-specific working memory. Due to the connectivity
structure of action-perception circuits, memory-related reverberating activity tends to "retreat" to areas where especially strong connectivity to other
areas is present; that is, to higher multimodal cortices (prefrontal and anterior-
temporal). As ignition and reverberation processes provide a neurobiological basis for *recognition* and *working memory* for linked action-perception
information (Pulvermüller and Garagnani 2014), they may be essential for
overt immediate or delayed *imitation* and *repetition* of perceived body actions
or heard words. Since a network can accommodate numerous strongly connected circuits, control mechanisms are necessary to prevent overshooting activity. Such regulation can be model-implemented by local and global inhibition mechanisms, which also provide competition between action-perception

circuits. The degree of regulation and competition between circuits can model aspects of *attention* related to task and context (Wennekers and Palm 2007; Wennekers et al. 2006). Modulation of area-specific inhibition also provides a mechanism for blocking motor output during passive listening and mental simulation as well as for "opening the motor gate" for repetition and speech production. Crucially, regulation and gain control provide mechanisms for *inhibiting "mirror actions" in the social context* (Jacob and Jeannerod 2005; Jeannerod 2006). Because the dynamics of action-perception circuits are under the control of regulation, action-perception mapping cannot be entirely automatic, in the sense that it is not suppressible. Still, mapping from perception to action is automatic in the sense that it rapidly arises in passive perception, even if subjects try to ignore the critical stimuli (Pulvermüller et al. 2003; Pulvermüller 2005; Shtyrov et al. 2014).

From Movements to Meaning: The Case for Action Semantics

The issue of how *meaningful* symbols should be modeled in a pragmatic action-perception network is as yet unresolved. The coupling of knowledge about symbols, including hand gestures, words, and longer constructions, to meaning is achieved through second-stage associative learning. Symbol and action schema become related to each other because different action-perception circuits are being interlinked. In this sense, the comprehension of action words such as "lick," "pick," and "kick" may rely on coupled action-perception circuits for word forms and body part-specific motor schemas (Hauk et al. 2004). One can see this as "mere association" and indeed some researchers chose to move this kind of model in the spiritual neighborhood of behaviorist accounts of meaning and language (Hickok 2010; Mahon and Caramazza 2008). However, the problem that arose from behaviorist approaches to language (e.g., Bloomfield 1933) did not stem from their consideration of behavior as such, but in the static manner in which they construed the relationship between signs, stimuli, and responses (Alston 1964). Semantic links between words and the actions they can be used to speak about are as important for semantic models as are the referential object links immanent to word usage (Chierchia and McConnell-Ginet 2000; Baker and Hacker 2009). These semantic links can, at least in part, be explained neurobiologically by the coupling between action-perception circuits for body actions or objects as well as for symbol form circuits (Pulvermüller 1999). Such learning by correlation mapping is implied by general principles of brain function and, in particular, the cortex's role as an associative memory (Braitenberg and Schüz 1998). In the majority of studies addressing this issue, empirical evidence clearly shows that motor regions (including motor and premotor cortex) are being activated in fMRI experiments on written and spoken action word and action sentence processing (Figure 9.2; for a review, see Carota et al. 2012). Interestingly, words semantically

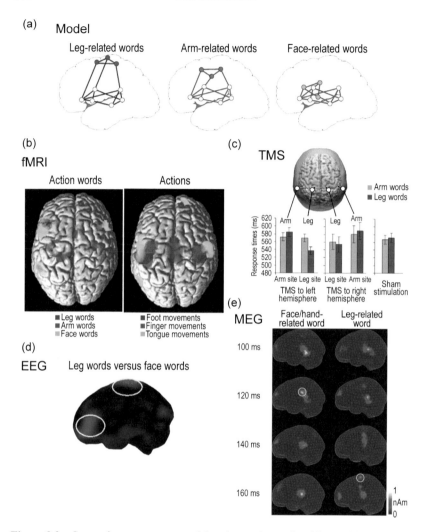

Figure 9.2 Semantic somatotopy model and experimental evidence. (a) Neurobiological model of cortical circuits underlying the processing of words and utterances typically used to speak about actions preferentially involving the face/articulators (e.g., *lick*), arm/hands (*pick*), or leg/feet (*kick*) (Pulvermüller 2001; reprinted with permission from Elsevier). Semantic circuits are postulated in different parts of the motor and premotor cortex. (b) Activation of the motor system, as measured with fMRI, by passively reading *face* (in green), *arm* (red) and *leg* words (blue) and, partly hidden, during motor movements of the *tongue* (green), *index finger* (red) and *foot* (blue) (Hauk et al. 2004; reprinted with permission from Elsevier). (c) Differential facilitation of arm/leg words by magnetic stimulation of the motor cortex controlling the finger/foot using TMS. The brain diagram indicates stimulation loci; bars give average response times of lexical decisions responses expressed by tongue movements (error bars give standard errors) (Pulvermüller, Hauk et al. 2005; reprinted with permission from John Wiley and Sons).

Figure 9.2 (continued) (d) Rapid differential activation of inferior-frontal and superior-central areas by face (in red) and leg words (blue) 200–220 ms after visual word presentation as calculated from EEG recordings (Hauk and Pulvermüller 2004; reprinted with permission from John Wiley and Sons). (e) Rapid activation of frontocentral areas by face/hand (left) and leg verbs (right) 120–200 ms after the recognition point of spoken face/hand-related words (left) and leg words (right), as measured with MEG. Yellow circles indicate early activations of inferior-central areas to face/arm items and, slightly later, dorsocentral activation to leg words; latencies are given relative to word recognition points (modified from Pulvermüller, Shtyrov, et al. 2005; reprinted with permission from The MIT Press).

linked to different body parts preferentially activate the motor representations of these same extremities, so that some (although certainly not all) aspects of action meaning are visibly reflected in the brain response (Hauk et al. 2004; Tettamanti et al. 2005; Kemmerer et al. 2008; Pulvermüller, Cook et al. 2012).[3] Similarly, perceptual circuits for objects linked to word form circuits may underlie referential semantics, although a degree of semantic action-relatedness cannot be denied for many object words (e.g., tool and food words; Carota et al. 2012; Martin et al. 1996).

A range of criticisms have been raised against this pragmatic action-perception perspective on semantic meaning (e.g., Hickok 2010; Bedny and Caramazza 2011). One argument stipulates that word-object and word-action relationships are not 1:1. Some words have different meanings and even within the range of one single meaning, different nuances, or "senses," can be distinguished. This argument, however, does not pose a problem for a neurobiological account, which allows one-to-many relationships to be implemented by interlinking one word form circuit with two or more semantic circuits, and offers regulatory mechanisms to enforce selection between semantic alternatives (Pulvermüller 2002b). Priming in the semantic network—from previously active circuits—contributes to the selection of circuits and circuit parts over competing ones.

A major argument against a pragmatic action-perception mechanism for semantics held that activation of motor systems may be "epiphenomenal" and follow the understanding of action words rather than reflect it. The *epiphenomenality position* implies that some other, truly semantic process precedes (in this case pseudo-semantic) motor system activations, thus predicting that the motor system's response occurs late relative to the truly semantic one. Neurophysiological studies using EEG and MEG could clarify that motor system activation, which reflects the meaning of action words, emerges rapidly. At

---

[3] One study claimed that such "somatotopic semantic" activity does not exist (Postle et al. 2008). However, analyzing the results from their "action observation" localizer with a repeated measures ANOVA with the factors semantic word category (face, arm, leg), motor area (primary, secondary) and somatotopy (face, arm, leg region), yielded a significant interaction of semantic word category with somatotopy, $F_{(4,64)} = 3.8$, GG-eps = 0.64, $p = 0.022$. I thank Greik de Zubicaray for sharing the data.

the earliest latencies, semantic brain responses could be tracked (Pulvermüller et al. 2001; Hauk and Pulvermüller 2004; Pulvermüller, Shtyrov et al. 2005; Shtyrov et al. 2014). A further criticism was that motor system activation might accompany action semantic processing, but may not be *crucial and causal* for it. Meanwhile, a range of data speak against this *afunctionality position*: the causal effect of transcranial magnetic stimulation (TMS) to motor cortex on the processing of specific action-word subcategories (e.g., faster leg word recognition to leg cortex TMS; Pulvermüller, Hauk et al. 2005); work in neurological patients with predominant involvement of the motor system, and consequent action word processing deficit (Bak 2013; Arevalo et al. 2012; Kemmerer et al. 2012); and interference studies in healthy subjects engaging in motor movement and showing body part-specific effects on action word processing (Boulenger et al. 2006; Shebani and Pulvermüller 2013). The epiphenomenality and afunctionality hypotheses could not be confirmed and no principle objections remain against the position that—at least for some symbols, constructions, and meaning aspects—semantics is reflected and carried by the mind and brain's motor system (Kiefer and Pulvermüller 2012; Pulvermüller et al. 2014).

However, meaning is not exhausted by semantic links between language, the world, and the body. *Combinatorial* or *distributional semantic* models construe meaning relationships between symbols in terms of similarities between the contexts in which they frequently occur (Landauer 1999). The limitation of these approaches comes from the fact that they do not cover semantic links between words and the world, so-called *"symbolic grounding"* in objects and actions (Harnad 1990). In contrast, an action-perception account provides not only natural and biologically plausible mechanisms for symbolic grounding, it also accommodates combinatorial semantics. Two complementary mechanisms are offered:

First, a learned symbol with semantic action-perception grounding can frequently co-occur with a novel meaningless symbol. Co-occurrence between simulated semantic activations brought about by the first symbol's circuit with the emerging circuit of the novel word form leads to "parasitic" contextual incorporation of semantic information into the new circuit, which therefore shares semantic neurons with the already established one (see also Cangelosi and Harnad 2001; Pulvermüller 2002a). This mechanism can lead to the coupling of new combinations of semantic features to novel symbols.

Second, symbol forms frequently appearing in sequence can link their circuits sequentially (Pulvermüller 2010; Buzsáki 2010). Neurocomputational simulation studies suggest that this type of combinatorial learning can lead to discrete combinatorial neuronal assemblies (DCNAs), which link together not individual words but, instead, whole classes of semantically and combinatorially similar symbols (Figure 9.1, d–f; Pulvermüller and Knoblauch 2009). Joint and hierarchical activation of sets of DCNAs has been proposed as a brain mechanism for grammar and generation of meaningful sentences and constructions (see Pulvermüller 2002b).

Abstract Meaning: Love, Beauty, Ifs, and Buts

A classic argument against action-perception grounding is based on abstract meaning. Whereas the meanings of the words "eye" and "grasp" can be explained, to a degree, by pointing to similar objects or actions and extracting their common features, those of "beauty" and freedom" cannot. It may be that some common sensorimotor knowledge is inherent in *freeing* actions or instantiations of *beauty* (Lakoff 1987), but it seems likely that additional semantic binding principles underpin such concepts. A remarkable observation has recently been offered: Abstract terms show an over-proportionally strong tendency to be semantically linked to knowledge about emotions (Kousta et al. 2011; Meteyard et al. 2012). This additional embodied-semantic link accounts for advantages in processing speed, which abstract emotional terms show compared with otherwise matched control words (Kousta et al. 2011). In addition, abstract words strongly activate the anterior cingulate cortex, a site known to be relevant for emotion processing (Vigliocco, Kousta et al. 2014). Thus, it appears that at least some abstract words are semantically grounded in emotion knowledge.

If, indeed, abstract emotion words receive their meaning through grounding in emotion (Kousta et al. 2011), it is of crucial relevance to explain how emotion grounding is established. Note that an amodal semantic system account does not address this question. Even if such a system contained an inborn emotion concept of *joy*, it is left unexplained how the learner knows to relate the concept to its corresponding word, and not, for example, to *grief*. The classic answer in semantic theory is that this is possible because abstract emotions and other internal states have characteristic ways in which they manifest in the actions and interactions the learner engages in with speakers of the language (Wittgenstein 1953; Baker and Hacker 2009). Therefore, the link between an abstract emotion word and its abstract concept is by way of the manifestation of the latter in prototypical actions. The child learns an abstract emotion word such as *joy*, because it shows *joy*-expressing action schemas, which language-teaching adults use as criteria for correct application of the abstract emotion word (Wittgenstein 1953; Baker and Hacker 2009). Thus, the manifestation of emotions in actions appears to be the glue between word use and internal state and, hence, between sign and meaning. Only after a stock of abstract emotion words has been grounded in emotion-expressing action can further emotion terms be learned from context.

This action-centered proposal generates further critical predictions that are testable in neuroscience experiments. In particular, over and above activating limbic emotion-related circuits, abstract emotion words should specifically excite the motor system controlling the face and arms, with which emotions are typically expressed. Motor system activation for emotion-expressing body parts was indeed found when adults passively processed abstract emotion words (Moseley et al. 2012); this suggests that for one important class of abstract concepts, semantic grounding in emotion-expressing action is of the essence and can, in part, explain the formation of the link between meaning and symbol. But is this motor activation epiphenomenal? Remarkably, individuals with autism, who are known to be limited in their emotion expression, show reduced motor activation to action and correlated reduced performance in processing action-related words (Moseley et al. 2013, 2015).

However, there are also abstract words that do not heavily draw upon affective-emotional information. Indeed, neuroimaging results suggest very different brain correlates of abstract words and constructions, some of which do not involve emotion-processing centers of the limbic systems (Binder et al. 2005; Shallice and Cooper 2013). A characteristic feature of some abstract utterance is the variability of entities that are typically used. For most concrete object-related words, such as *eye*, the entities used to refer to the object may vary (in size, form, color), but it is normally possible to identify a semantic schema that can be illustrated by a prototype, a typical best representative of the schema (Fillmore 1975); atypical variants may activate the schema representation less than the prototype itself (Rosch and Mervis 1975). For some terms, this model breaks down when their meanings cannot be explained by a prototype, but requires several of them. Consider the case of the word *game* (Wittgenstein 1953; Baker and Hacker 2009; Rosch and Mervis 1975), which can refer to diverse activities ranging from cooperative to competitive, from group to solitary, and from playful to more serious action. No single prototype can represent this space of action schemas and, although prototypical members may be similar (soccer and football), others are very different (soccer and the computer game Tetris). To capture such variable *family resemblance*, semantic representations need to link up with variable action and perceptual schemas. At the neurobiological level, variability means low correlation between word forms and semantic prototypes; this implies that, although word meaning may originally be grounded in specific action and perception schemas, the semantic representation in a sense detaches from specific action-perception knowledge (Figure 9.3). This process of variability-related "disembodiment," implied by correlation learning (the "anti-Hebb" "out of sync-delink" rule), may be effective for many abstract words and concepts. Therefore, abstract words grounded in perceptual schemas (e.g., "beauty") may detach from their perceptual schemas stored in posterior inferior-temporal cortex, thus leaving relatively anterior-temporal representations weakly linked to these concrete instantiations, whereas abstract action terms (e.g., "free" or "game") may show the same process of *variability disembodiment* in prefrontal and parietal areas adjacent to sensorimotor cortex. Weak links between neuronal representations

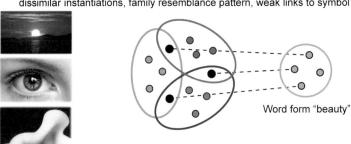

(a) Concrete meaning:
   similar instantiations, semantic feature overlap, strong links with symbol

Word form "eye"

(b) Abstract meaning:
   dissimilar instantiations, family resemblance pattern, weak links to symbol

Word form "beauty"

Figure 9.3 Sketch of putative neurobiological mechanisms for concrete and abstract meaning processing. Both concrete and abstract words as well as constructions can be learned when they are used to refer to real-life events, actions, and objects or their features. A major difference lies in the variability of the sensorimotor patterns that foster semantic grounding, which is typically low for concrete and high for abstract symbols. Assuming Hebbian mapping of correlations, this difference in correlation structure yields different neuronal and cognitive mechanisms for concrete and abstract meaning. (a) Concrete semantics: The concrete word *eye* is used to speak about objects with similar shapes and a range of colors. At the neurocognitive level, this leads to exemplar representations which strongly overlap in their sensorimotor semantic feature neurons, possibly dominated by a frequently processed prototype. Sensorimotor semantic overlap (including, in this case, visual center-surround cells responding to a circle in one color on a background of a different one) to feature neurons more specific to individual exemplars (e.g., to specific color). In concrete semantic learning, neurons of the circuit overlap and frequently occurring prototypical exemplars strongly interlink with the word form circuit due to high correlation of their activations. (b) Abstract semantics: The semantic instantiations of an abstract word such as *beauty* are quite variable, exhibiting a *family resemblance pattern of partial semantic similarity* (Wittgenstein 1953). The diagram illustrates the putative neural correlate of such family resemblance, where sensorimotor semantic feature neurons are only shared between subsets of exemplar representations of variable instantiations of the concept. The low correlation of activations of neuronal circuits for word forms and for each exemplar representation results in weak links between neural representations of sensorimotor knowledge (in modality-preferential areas) and those of verbal symbols (in perisylvian cortex; adapted from Pulvermüller 2013).

of abstract terms and their multiple and variable sensorimotor instantiations may be a hallmark of abstract meaning and key to the focusing of abstract semantic circuits on multimodal prefrontal, parietal, and temporal convergence areas. Extension of activation into one subset of sensorimotor neurons would then depend strongly on priming of some of the many instantiations.

Meaning can be driven by context in the sense that the meaning of a construction is more than that of its symbolic constituents (elementary parts) plus combinatorial regularities. One may speak of a goose as being well-cooked in a *literal* sense, but if somebody states that someone's "goose is cooked," a different *idiomatic* meaning may be relevant. How might this be modeled neurobiologically? In one branch of linguistics, cognitive and construction grammar, whole constructions are assumed to be paired with meanings and stored in a lexicon-like manner (Goldberg 2006; Langacker 2008). Accordingly, the meaning of an idiomatic construction might be distinct and not (or only distantly) related to the meanings of its composite words. However, compositional semantics suggest that sentence meaning is built from word meaning (see, e.g., Davidson 1967), which makes idioms difficult to model. Some proposals, including the neurobiological account, suggest that both views are correct and therefore both single word and whole construction meaning play a role in idiom comprehension. The correlation of the idiomatic, frequently quite abstract meaning with its variable sensorimotor instantiations may draw upon multimodal brain areas removed from sensorimotor systems, whereas the concrete constituent word meanings may engage sensorimotor systems. Indeed, comparison of brain activation maps elicited by idiomatic and literal sentences revealed that multimodal inferior- and dorsolateral-prefrontal, inferior-parietal, and anterior-temporal areas was stronger for idiomatic sentences than for literal ones (Lauro et al. 2008; Boulenger et al. 2009), consistent with a variable abstract semantic pattern. Over and above this idiomaticity effect, some studies of idiom processing reported motor system activation, which reflected aspects of the meaning of action words included in the idioms (e.g., leg motor cortex activity to "Anna ran for president"; Boulenger et al. 2009), suggesting a degree of compositional semantic processing (but see also Desai et al. 2013; Raposo et al. 2009). Because such motor system activation, reflecting the meaning of constituent action words, was already present when idiomatic and literal sentence meaning could first be disambiguated, the data indicate simultaneous construction retrieval and semantic compositional processing of the action meaning of constituent words in idiomatic sentence comprehension. Crucially, precise mapping-in-time using magnetoencephalography (MEG) showed that the brain correlates of abstract idiomaticity and those of action-grounded constituent word meaning occurred at the same time, already 150–200 ms after onset of the critical, sentence-disambiguating words. These results suggest that compositional semantic processing of action-related words (precentral cortex) and non-compositional semantic processing of abstract idiomatic constructions as a whole (prefrontal and anterior-temporal areas) simultaneously and jointly contribute to idiom comprehension (Figure 9.4).

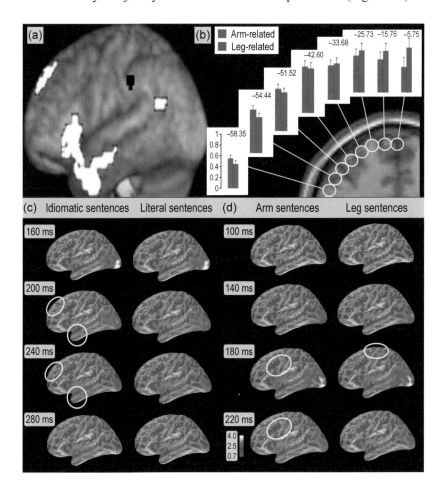

Figure 9.4 Brain activation to idiomatic and literal sentences recorded with fMRI (a, b) and MEG (c, d). (a) Comparison of brain activation elicited during idiomatic and literal sentence processing: white = idiomatic > literal; black = literal > idiomatic (Lauro et al. 2008; reprinted by permission of Oxford University Press). (b) Activation to literal and idiomatic sentences that include arm- (red bars) and leg-related words (blue bars); the red and blue areas indicate where finger and foot movements elicited activity (Boulenger et al. 2009; reprinted with permission from Oxford University Press). (c) Activation time course to idiomatic and literal action sentences (arm and leg sentences collapsed). (d) Activation time course to arm and leg sentences (idiomatic and literal collapsed). Note that constructional idiomaticity and compositional action-relatedness effects were present simultaneously early on (150–200 ms); this suggests that action-embodied compositional and disembodied constructional semantic processes emerge instantaneously at the same time (Boulenger et al. 2012; reprinted with permission from Elsevier).

Social-Communicative Pragmatic Function

Context dependence of linguistic-pragmatic meaning is most obvious in the use of the same utterance (i.e., word or construction) for entirely different action purposes. A word such as *water* can be used in the very same meaning and sense to name an object (i.e., to tell somebody how it is called) or for a request (e.g., to ask somebody for a drink). In pragmatic linguistics, language actions, such as naming and requesting, are called *speech acts* (Searle 1969). If the same utterance is used for different speech acts, the utterance is embedded in different contexts and connected with different intentions and goals. In other words, the utterance is produced with different *predictions* regarding subsequent actions and events, and thus in different *action sequence structures* (Figure 9.5). Neuroscience research has explored the brain basis of different intentions and sequences connected with a body movement (e.g., grasping an object to eat or to place it). Results show that mirror neurons in inferior frontal and parietal cortex indexed goal relatedness, and the proposal is that such neurons index not only basic acts, but their associated *action chains*, including the goal, as well (Fogassi et al. 2005; Iacoboni et al. 2005). Social-communicative interactions are normally more complex than simple action chains (Fogassi et al. 2005) or linearly predictable actions (Pickering and Garrod 2013b) and involve tree-like, sometimes recursive, structures (Fritz 2013; Levinson 2013). Thus, linguistic-pragmatic descriptions take advantage of action tree structures covering the action options opened by a given speech act (Alston 1964; Ehlich 2007; Stalnaker 2002; Fritz 2013; Levinson 2013). Such tree structures also link communicative actions and their goals and intentions manifest in the preferred response actions. A flexible binding mechanism is required to temporarily link the basic action or utterance (e.g., use of the utterance *water*) to the interaction structure of the speech act characterized by the more distant intention (e.g., to be given the object). The relationship between actions is iterative and can have several "layers," because an utterance can be used to request an object, which, in turn, may be performed to please one's companion, and so on (Stalnaker 2002; Fritz 2013).

The brain mechanisms for the postulated action sequence schemas can be assumed to draw upon sensorimotor as well as multimodal cortical areas. When utterances appear in different contexts, their action-perception circuits ignite and bind with the circuits of specific sequence schemas. In the naming context, the circuit for the word form *water* may activate, including its word form part and its object-related referential semantic part. The referential word-object link is known to draw upon inferior-temporal ventral-visual stream circuits (Pulvermüller 1999; Hickok and Poeppel 2007). In contrast, when using the same word to request an object, the sequence schema opens up expectations of a range of partner actions (e.g., handing over the water, or, alternatively, denying the request). Thus, in the request context, motor and action sequence circuits in frontocentral cortex need to be sparked in addition to utterance-related

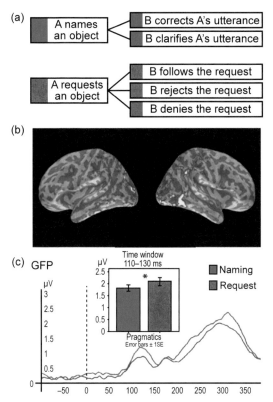

Figure 9.5 Sequence structure and brain correlates of social-interactive communicative speech acts performed with the same words. (a) Action sequence structures for the speech acts of *naming* (in blue) and *requesting* (in red), and results of MEG and EEG experiments on understanding of *naming* and *request* actions. (b) Topographical differences in brain activation 50–90 ms after critical (written) word onset obtained with MEG and distributed source estimation (*request > naming* in red/yellow, *naming > request* in blue). (c) In the EEG response, stronger brain activation was seen in *request* (red line) compared with *naming* (blue line) contexts from 100 ms. Global field power (GFP) is plotted against time (in ms) (after Egorova et al. 2013, 2014).

circuits (Pulvermüller et al. 2014). This neuropragmatic approach predicts that word forms and utterances elicit context-dependent, speech act-specific patterns of activation in motor systems and beyond.

In EEG and MEG experiments, experimental subjects saw actors in communicative contexts, where the same words were used to name and request the same objects. Already ~100 ms after the word critical for understanding the speech acts could be recognized, brain activation distinguished between naming and request actions. There was stronger activation to request and a relevant part of the additional cortical sources were in frontocentral motor systems (Figure 9.5; Egorova et al. 2013, 2014).[4] These results on the brain basis of speech-act processing suggest a local cortical difference which, in part, confirms the above predictions. Such first steps toward understanding the neuronal basis of "how to do things with words" in social communication contexts (Austin 1962) need to be extended in the future, by investigating, for example, populations with deficits in social-communicative interaction with putative relationships to mirror neuron circuits (e.g., autism; Rizzolatti et al. 2009). This topic, under investigated in the neurobiology of language, is at the heart of language as a social phenomenon, as an interactive game activity characterized by action sequences, goals and intentions, commitments about the theory-of-mind assumptions of communication partners, and the use of social-communicative information and knowledge related to linguistic form.

Outlook

A novel pragmatic neuroscience emphasizing the binding between action and perceptual information in the service of mechanisms for higher cognitive processing can draw on a rich reservoir of brain language research supporting this general framework. In particular, it seems feasible to model crucial aspects of semantic knowledge in terms of action-perception circuits specifically linking linguistic symbolic form with meaning grounded in action and perception. Abstract and affective-emotional semantics can be modeled in this framework taking advantage of expression of emotion in action and correlation mapping between symbol forms and sometimes quite variable sensorimotor information, also taking into account combinatorial learning. The emerging picture for semantic circuits is that of richly structured neuronal assemblies, joining together form and meaning information and allowing for fine-grained differential activations reflecting variation in contextual priming and semantic nuances. Temporary binding circuits for meaningful forms into their contextual action-schema networks may account for brain activation, which reflects the communicative context and intentions for which language is used in social interaction. At almost all levels, experimental data indicate that motor systems, along with perceptual ones, are involved in and, critically, exert causal effects on, semantic pragmatic processes.

Acknowledgments

I would like to thank Larry Barsalou, Valerie Keller, Rachel Moseley, Malte Schomers, and Gabriella Vigliocco for comments on earlier versions of this text and for related

---

[4] A range of recent research focuses on neuropragmatic differences in brain activation that distinguishes between speech act types. One research stream focuses on Searle's distinction between direct and indirect speech acts (Searle 1975). There is indication that this distinction may also be manifest in local brain activation (Bašnáková et al. 2014; van Ackeren et al. 2012).

discussions. This work was supported by the Freie Universität Berlin, the Medical Research Council (UK) (MC_US_A060_0034, U1055.04.003.00001.01 to FP), the Engineering and Physical Sciences Research Council and Behavioural and Brain Sciences Research Council, UK (BABEL grant, EP/J004561/1), and the Deutsche Forschungsgemeinschaft, Germany (DFG, Pu 97/15-1 16-1).

First column (top to bottom): James Kilner, Paul Verschure and Jürgen Jost, Larry Barsalou, Moshe Bar, Thomas Metzinger and Jeanette Bohg, Marti Sánchez-Fibla, Jürgen Jost
Second column: Bernhard Hommel, Moshe Bar, Cecilia Heyes and Gabriella Vigliocco, John Tsotsos, James Kilner, Karl Friston, Thomas Metzinger
Third column: Larry Barsalou, Karl Friston, Friedemann Pulvermüller, Alexander Maye, Gabriella Vigliocco, Alexander Maye, Bernhard Hommel

10

Action-Oriented Models of Cognitive Processing

A Little Less Cogitation, A Little More Action Please

James Kilner, Bernhard Hommel, Moshe Bar,
Lawrence W. Barsalou, Karl J. Friston, Jürgen Jost,
Alexander Maye, Thomas Metzinger, Friedemann Pulvermüller,
Marti Sánchez-Fibla, John K. Tsotsos, and Gabriella Vigliocco

Abstract

This chapter considers action-oriented processing from a model-oriented standpoint. Possible relationships between action and cognition are reviewed in abstract or conceptual terms. We then turn to models of their interrelationships and role in mediating cognitively enriched behaviors. Examples of theories or models inspired by the action-oriented paradigm are briefly surveyed, with a particular focus on ideomotor theory and how it has developed over the past century. Formal versions of these theories are introduced, drawing on formulations in systems biology, information theory, and dynamical systems theory. An attempt is made to integrate these perspectives under the enactivist version of the Bayesian brain; namely, active inference. Implications of this formalism and, more generally, of action-oriented views of cognition are discussed, and open issues that may be usefully pursued from a formal perspective are highlighted.

Existing Schema for Action and Cognition

Before considering the form and consequences of models that take an action-oriented view of cognition, it is worth considering how the relationship between these two processes has been described. Here, we consider four schemata which capture different notions of how cognition and action could be coupled (Figure 10.1). It is important to stress that these schemata are not models but

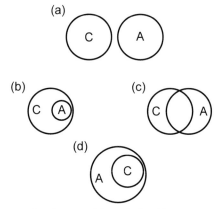

Figure 10.1   Schemata of action-cognition relationships.

rather depictions of different views on cognition and action. Crucially, differences in the schemata do not reflect a fundamental difference in the nature of the coupling but rather in how cognition and action are defined.

At one extreme, in Figure 10.1a, the two processes can be considered to be entirely independent processes: action is simply the behavioral output of the cognitive process (e.g., in the classical sandwich conception of the mind). In this open-loop formulation, action does not, and indeed cannot, influence cognition. Although not explicitly stated, the scheme depicted in Figure 10.1a is implicitly assumed in many models of high-level human cognitive functions used in behavioral and imaging experiments: input is carefully manipulated and output (action) is kept to a minimum (e.g., key pressing) in order to focus on internal processes. Here, action is considered to be a necessary output to disclose internal operations. For example, in a typical experiment that investigates language processing, individuals are presented with written or spoken words/sentences and asked to make a key press decision about them. Such studies, which still constitute the majority of cognitive science and neuroscience studies of high-level cognition, are completely silent as to whether action might play a role.

Alternative approaches (Figure 10.1b–d) consider how action and cognition depend on each other. In addition, action can be explicitly considered to be a subset of cognition or a largely overlapping process (Figure 10.1b and c). This commonly held view defines cognition as information processing à la Neisser. Accordingly, most, but not all, cognition relates to action: cognition influences action, and action influences cognition. This influence is typically, but not always, constructive: cognition and action can exist in harmony without collapsing into one another. They are heavily intertwined, but they are not the same, and none exists solely for the benefit of the other. Just like attention, working memory, and consciousness, cognition and action are intimately related, yet independent. The final scheme (Figure 10.1d) represents a more "enactive" form: cognition is a subset of action and/or cognition subserves action.

In these schemata, action is often only considered as an external goal-directed movement (as opposed to an internal/mental action). Although actions are often defined as goal-directed movements, they could also be considered on a continuum between motor movements and goal-directed actions. For example, the babbling infant may produce the same syllable [ti:] as a one-year-old repeating a word and an older child asking for the beverage. In development, successively more distant goals and predictions (e.g., getting tea, ordering on behalf of someone in a restaurant to please that person) seem to be coupled with the movement representations to yield the cognitive action representation. Because of the relationship between their putative neuronal circuit representations, it may make sense to see movements and goal-directed actions on a continuum.

From a cognitive perspective, an action results from mediating processes that constitute proximal causes of the action relative to the distal stimulus and subsequent sensory processes. An action emanating from a cognitive system has an important conceptual interpretation above and beyond its motor features (often public or social); for example, donating to charity over the Internet. Such an action includes motoric features (reading a screen and typing on a keyboard), while invoking the notion of moving money from one's bank account to an organization to help relieve suffering in a distant group of people. These latter aspects play an important role in conceiving of the action in the first place: donating to charity would not be possible without understanding money, charities, donation, and suffering—and all the causal relations among them—and executing the resultant action effectively and monitoring whether the intended outcome occurred (see Barsalou, this volume). Arguably the most important actions that humans perform are ones that acquire physical resources (e.g., food, shelter, wealth), alter the physical environment (e.g., clearing land, farming), and develop and use technology (tools) to achieve goals as well as establish, maintain, and revise social relations and social status. Crucially, and especially as a result of language and communication, these actions call on cultural institutions, artifacts, and knowledge.

Models of Action-Oriented Processing

Very few models are actually based on a relationship between action and cognition, as illustrated in the third schemata (Figure 10.1c). One important example of an attempt to close the loop connecting brain and environment is the ideomotor theory, as attributed to Lotze (1852), Harless (1861), and James (1890). Its original formulation tried to explain how people acquire voluntary control of their actions, even though they do not seem to have privileged (conscious) access to the motor system. The idea is that agents start interacting with their environment by motor babbling (executing random movements), sensing the reafferent information resulting from these movements (i.e., the self-produced changes in perception), and creating bidirectional associations between the neural pattern producing the movement and the neural pattern representing the reafferent information (Hebbian learning). Given that agents can activate the reafferent codes endogenously by imagining the respective events, these bidirectional associations provide them with retrieval cues to the associated motor patterns. In such a way, the movements can now be executed intentionally. The theory of event coding (TEC) has extended and generalized this approach in various ways (Hommel et al. 2001). First, it claims that perceptual codes and action patterns are represented in a distributed, nonsymbolic fashion. Second, it assumes that perceptual codes and action patterns are integrated into sensorimotor event files; that is, into networks of codes representing the perceptual and movement-related aspect of a given sensorimotor event. Third, it assumes that to represent a given event, the components of a given event file are weighted according to their relevance for the given action goal (intentional weighting; Memelink and Hommel 2013). For instance, when grasping an object, shape, and location, features will be weighted more strongly (and thus contribute more to the representation of the object-grasping event) than color features. Finally, TEC claims that perception and action are the same thing: the process of perceiving an event entails moving in ways that orient one's receptors toward the event of interest, so as to register its perceptual features, and the process of producing an event (i.e., acting) involves moving in ways to generate particular perceptual features which are then sensed and compared to the expected outcomes. That is, both perception and action actively generate reafferent input but the specificity to which this input is predicted is often lower for what we call perception than for what we call action. In essence, TEC assumes that the basis of human cognition is sensorimotor in nature (event files). However, it does not explicitly rule out the possibility of more abstract cognitive codes that are derived from event files.

Other models have been proposed that also make the link between action and cognition explicit. For example, models of sensorimotor representations of grasping movements in frontoparietal cortex can be used to explain the perception of actions as well as the "simulation" or "mentalizing" about actions (Arbib et al. 2000; Jeannerod et al. 1995). A model of frontotemporal cortex shows the emergence of linked action-perception mechanisms from sensorimotor information and functional implications of such learning for working memory (Pulvermüller and Garagnani 2014). Hebb-type learning leads to a strengthening of neuronal connections in a pool of sensorimotor neurons, which implies that activity will be maintained longer in the pool. These models show how higher cognitive functions (mentalizing, memory, and so on) can develop in specific neuroanatomical structures on the basis of associative learning of correlated motor and sensory activity. Thus they provide concrete implementations of the functional parallelism between cognition and action.

Over and above accounting for the emergence of mechanisms for higher cognition, neuroanatomically grounded action-perception models may explain the specific and dissociable brain areas carrying particular cognitive functions.

Here we have provided a few examples of theoretical approaches that link (or unify) action and cognition in conceptual terms. In what follows, we revisit the same ideas from a formal perspective, trying to identify the decomposition of states and their dynamics that constitute the problem at hand. In particular, we consider the optimality principles inherent in ideomotor theory and related developments.

Formal Models of Action and Cognition

To consider the nature and utility of formal models, we start from basic principles and address the usefulness of a formal approach at various levels. In brief, we first appeal to systems biology to identify the sorts of variables (states) that one needs to consider when modeling an agent immersed in its proximate environment. Equipped with a partition of states, we then utilize optimality principles to define classes of state ("as if") theories of action and cognition; where each state theory is defined in terms of the quantity that is optimized. Finally, each state theory entails a series of process theories that hypothesize a particular (computational or physiological) process that realizes the optimization. Having defined a set of process models, it is then possible to test them in relation to the empirical behaviors that each predicts.

An example of an optimality principle would be Bayes optimality (i.e., ideal Bayesian observer assumptions), where the state theory could correspond to the Bayesian brain hypothesis, in which the brain behaves as if it is trying to maximize Bayesian model evidence. The corresponding process models could then include predictive coding or (stochastic) population coding that make very different predictions about the neuronal responses that would be elicited by a stimulus. We use this example (among others) to see how the models could be augmented to accommodate an action-oriented paradigm.

State Spaces and Systems Biology

Figure 10.2 illustrates the partition of states necessarily implied by a system that is acting on its environment. This partition considers the distinction between external states of the world that are hidden from the internal states of an agent—in the sense that external states are hidden behind sensory states. Internal states could correspond to neuronal activity, connection strengths, or any other neuronal states characterizing the brain at one point in time. Crucially, sensory states are caused by external states that subsequently change internal states. Conversely, internal states cause changes in agential states (e.g., actuators or muscles), which then cause changes in external states. Sensory and

$$\partial x / \partial t = f_x(x, a)$$
$$\partial s / \partial t = f_s(x, a)$$

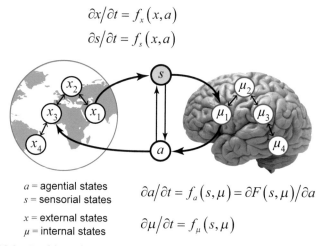

a = agential states
s = sensorial states

$$\partial a / \partial t = f_a(s, \mu) = \partial F(s, \mu) / \partial a$$

x = external states
μ = internal states

$$\partial \mu / \partial t = f_\mu(s, \mu)$$

Figure 10.2 Partition of states necessarily implied by a system that is acting within its environment.

agential states, therefore, couple the world to the brain in a circular fashion, inducing a cycle of action and perception (Fuster 1990). Mathematically, these states insulate the internal states from the external states (and are known technically as a Markov blanket).

The connections or edges in Figure 10.2 denote causal or statistical dependencies and render the graph a Bayesian network or graphical model of the situated agent. These dependencies are most generally described in terms of (stochastic) differential equations. These equations of motion describe the evolution of hidden states, the way that sensory states respond to hidden states and the dynamics of internal states and agential states. The upper two sets of equations constitute a description of the world and how it causes sensory impressions. The lower two sets of equations (for internal and agential states) can now be regarded as a formal model of perception and action.

Optimality Principles

Optimality principles are so ubiquitous in the physical sciences that it is difficult to think of an example in physics that does not rely on an optimality principle. Important examples include Hamilton's principle of least action, which underlies all classical motion and thermodynamic laws, which underlie the behavior of systems at thermodynamic equilibrium in statistical physics. In our case, we can define a state theory of action and perception by casting the equation of motion for internal and agential states as a gradient descent (or ascent) on some function of internal and sensory states. This presupposes that the dynamics of the system or situated agent—say, the sensorimotor loop and/or the neuronal system—can be derived from a variational principle; that is, it assumes we can identify some quantity that the system is trying to optimize. The existence of this quantity (known as a Lyapunov function) is generally guaranteed for the sorts of systems in which we are interested. Identifying the quantity (or quantities) that the system is trying to optimize is crucial because the implicit dynamics (action and perception) will lead to fundamentally different sorts of behavior.

Which Optimality Principle?

There are a range of optimality principles or objective (Lyapunov) functions one might consider. These range from information theoretic quantities based upon the principle of maximum information transfer, or minimum redundancy, through to functions that explicitly accommodate goals, such as utility functions. Some of the more prevalent information theoretic functions are reviewed by Jost (this volume). Most suppose that internal states should possess the greatest mutual information with the hidden or sensory states. In other words, one should be able to predict the external states, given the internal states. An important aspect of these optimality functions is the constraints or priors under which mutual information is maximized. In Pezzulo et al. (this volume), we see a ubiquitous constraint; namely, sensations are caused by a small number of external states at any one time. This *sparsity assumption* can be used to optimize the form of interactions among internal states using a variety of schemes that lead to receptive field properties and architectures that are remarkably reminiscent of functional anatomy.

In optimal control theory and reinforcement learning, the objective function is generally cast as a utility or reward function, also referred to as negative cost. This optimality function speaks more to action than perception, but perception is usually deployed in the context of some state estimation that implicitly appeals to information theory or the Bayesian brain.

The Bayesian brain hypothesis assumes that the optimality function is either Bayesian model evidence or approximations such as variational free energy. Variational free energy is an approximation to the evidence for a model implicit in the internal states that is evident in sensory states at any given time. This means that if internal states minimize variational free energy, they are implicitly maximizing an approximation to Bayesian model evidence and will look as if they are performing (approximate) Bayesian inference, so that internal states come to represent external states.

Many different optimality principles have been proposed to understand human and animal cognition, and the question of which optimality principle to adopt may appear unresolved. Here, we address some of the key issues in this area. For instance, in some schemes, surprise (i.e., prediction error) is minimized, as in the Bayesian brain hypothesis, whereas other proposals emphasize the resolution of uncertainty by maximizing Bayesian surprise or information gain. In robotics or machine learning settings, for example, exploratory actions can be cast as satisfying curiosity (Schmidhuber 1991b). More generally, optimization principles could emphasize the gain of relevant information or the accuracy and scope of predictions. Either could be subordinate to the other. An agent could try to gather information to improve its predictions, or it could build predictions to acquire new information. Jost (2004) argues that a system could satisfy these two goals by addressing them on different timescales. Evolutionary thinking may help us understand the hierarchical relationship between different goals. Short-term predictions provide a reference or set point for homeostasis (or allostasis), whereas relevant information could provide opportunities for reproductive success in the long term, which is not a homeostatic affair. According to evolutionary biology, fitness (as expressed by the actual or expected number of descendants) is the most basic principle. In this setting, survival of an individual (a homeostatic affair) is necessary for, but subordinate to, reproduction. Therefore, from this perspective, the gain of relevant information should be the overarching principle, and homeostasis should be subordinate.

The optimality principles considered above all have particular biases toward perception or action. For example, the Infomax principle and the Bayesian brain hypothesis do not accommodate action, whereas optimal control theory and reinforcement learning do not easily accommodate perceptual inference. Is there any way to integrate these optimality principles into a common framework or state theory? One approach is to recast the optimization of action in relation to reward or utility functions as an inference problem. This is known as *planning as inference*. The advantage of this is that one can gracefully subsume action or policy selection (i.e., planning) and state estimation (i.e., perception) within the same optimality principle; namely, the Bayesian brain. Furthermore, it is relatively easy to show that optimizing internal states with respect to the evidence for a *generative model* implicitly maximizes the mutual information between internal and external states, subject to (prior) constraints inherent in the generative model. This brings us to *active inference* described by Friston (this volume), which is closely related to the notion of *empowerment* (an information theoretic principle that explicitly conditions mutual information on action).

Active Inference

Active inference can be regarded as the action-oriented (enactivist) version of the Bayesian brain hypothesis; it requires *both action and perception* to optimize Bayesian model evidence. Bayesian model evidence (the log probability of some sensory states under a generative model) is also known in information theory as (negative) surprise. This means that approximate Bayesian inference is another way of saying agents act to avoid surprising sensations (i.e., homoeostasis). Clearly this provides an impoverished account of goal-directed and mindful behavior. However, surprise rests on the violation of predictions, where predictions imply prior beliefs about what will happen. Formally, the generative model is specified in terms of likelihood and prior beliefs. Put simply, this means that there are as many classes of optimality as there are prior beliefs an agent might entertain about hidden states and their sensory consequences (e.g., garnering information that may enhance reproductive success). In other words, we can now express any optimality function from the previous section as a prior belief. The only thing that has changed is that the optimality principle is described in terms of a generative model, allowing the same objective function (Bayesian model evidence or its free energy approximation) to be optimized in all cases. Casting optimal control or Infomax principles in terms of active inference does not fundamentally change the principle; it just provides a common framework in which to model the optimization dynamics that underlie action and perception. In short, instead of asserting that agents maximize utility, we can say that agents believe they will maximize utility and then realize those beliefs through action. This emphasizes the fact that utility functions and prior constraints—necessary to define information theoretic imperatives—can be formulated as part of the generative model embodied in the agent's functional anatomy.

One of the most promising and generic priors arises in optimal control theory and decision theory. Known as KL or risk-sensitive control, it simply states that "I act (believe I will act) to minimize the probabilistic difference between preferred outcomes and those I predict given current evidence about the state of the world." The nice thing about this prior is that it gracefully accommodates reward (or utility) and epistemic value (or information gain). In other words, optimizing internal and agential states under this prior leads naturally to a Bayes optimal mixture of explorative and exploitative behavior.

Implications of the Modeling Approaches for a Paradigm Shift

Building on these examples of both formal and informal modeling approaches, we address the implications of these accounts for the role of action in cognition, both conceptually and practically. Areas are highlighted where the modeling approaches provide an account or insight that may differ from other accounts of an action-oriented cognition.

Implication 1: From Open-Loop to Closed-Loop Theories

The first implication that is clear from the system depicted in Figure 10.2 is that the formal account describes a closed-loop scheme. In other words, the external states change the sensorial states, which change the internal states, which change the agential states, which change the external states, etc. In the closed-loop account, perceiving the world produces states that produce action which in turn change the world, causing the loop to iterate. This closed-loop account is distinct from an open-loop account, where the external states change the sensorial states which change the internal states. Note that an open-loop account can be accommodated by the scheme in Figure 10.2 simply by removing the agential state. This has important implications, which we return to later. In the open-loop account, the primary interest is in how the individual responds to the world. The effects on action are largely irrelevant as are effects of action on the world. Action can occur but only as a response that serves to provide data on underlying cognitive mechanisms. The consideration of whether the system is closed or open loop has implications for how cognitive scientists design and interpret data from experiments. In practice, most research in cognitive science and cognitive neuroscience adopts the open-loop approach, in the sense that the paradigms are not constructed to assess closed-loop performance and effects. If cognitive mechanisms have evolved and developed to support closed-loop performance, as suggested, then theories assuming only open-loop processes may be misguided. One potential paradigm shift would be to study cognitive mechanisms in closed-loop paradigms. Although this will present major methodological challenges (e.g., considerable increases in complexity for design, control, data, analysis, etc.), we might find that our understandings of cognitive mechanisms change considerably. One cognitive domain in which such a change from an open-loop to a closed-loop perspective may be especially important is language, which we use to illustrate some key points.

Language as Action

In linguistics and psycholinguistics, language is often considered to be a system of lexical entries and rules that determine the hierarchical structuring of morphological, lexical, and phrasal units (as is evident in speech or text), distinct from its actual implementation in action and perception. Moreover, language comprehension has traditionally been studied separately from production (and the majority of studies concern comprehension with only a smaller number of studies concerning production). However, it is crucial to see that similarly complex combinatorial schemes may be at work in the linguistic and general action domains (Jackendoff 2011; Pulvermüller and Fadiga 2010) and that, when language is used and learned, it is produced and understood in face-to-face communicative contexts, where production and comprehension are intertwined (e.g., gestures, facial, and bodily movements) between conversational partners to fulfill specific communicative goals. These actions complement speech in real-world contexts and are part and parcel with the linguistic signal. The difficulty in separating the linguistic from communicative information becomes especially clear when we consider languages that can only be transmitted in a face-to-face situation, such as sign languages, but it is just as relevant for spoken languages (Vigliocco, Perniss et al. 2014).

   Finally, language can be viewed as a tool for communicative action and interaction (e.g., Austin 1975; Searle 1969), and it is indeed the communicative and social context that determines whether an utterance such as "water" functions as a naming action or as a polite request bringing about a desired response. Recent work demonstrates that these different social-communicative functions of language have different brain-mechanistic bases (see Pulvermüller, this volume).

From the perspective of an action-oriented view of cognition, we see immediately that a traditional linguistic definition of language is too narrow. Viewing language as action has a number of important consequences and benefits. First, taking a closed-loop approach to language processing, we can consider language production and comprehension as forms of action and action perception. For example, Pickering and Garrod (2013a) specifically argue that in language processing, speakers construct forward models of their actions before they execute those actions, and perceivers of others' actions (listeners) covertly imitate those actions and construct forward models of those actions. Further, Pickering and Garrod have shown how such a closed-loop approach to language production and comprehension can account for a number of psycholinguistic phenomena, in terms of how speakers/listeners interweave production and comprehension processes as well as how production-based predictions are used to monitor the upcoming utterances in dialogue.

Second, and more broadly, an action-oriented closed-loop view of communication opens up new directions in research that use real-world stimuli where the linguistic content expressed in speech but also co-occurring hands, facial, and body actions are part of the communicative actions. Regarding the role of these additional actions such as co-speech gestures, there is evidence that they are integrated during online spoken comprehension (Kelly et al. 2010; see review by Ozyurek 2014). It has further been shown that they provide a critical cue in vocabulary acquisition such that pointing to objects by infants is a precursor to learning objects' names (Özçalıskan and Goldin-Meadow 2009).

These different approaches all appeal to the notion that the object of investigation (language) cannot be defined in the traditional reductionist way (rule governed concatenation of symbols). Instead, they should be seen in the context of (and serving the function of) communication. Thus these approaches call for new methods to study language in real-world contexts.

Implication 2: Mental Representations as
Inferences about External States

From the biological system account described above and shown in Figure 10.2, one can see immediately that internal states must represent external states. This is because prior preferences about outcomes can only be caused by agential states. However, agential states are only functional in sensory and internal states. This means that internal states must stand in for or represent (in some sufficiency sense) external states. This simple observation dismisses radical accounts of enactivism that preclude (implicit) representations and leads us to a formal account of enactivism: namely, agential stages enact the predictions represented by internal states, where internal states correspond to the production of (conscious or unconscious) inferences about external states based on sensory evidence. A convincing demonstration of the role of generative modeling in perception can be found in Nair and Geoffrey (2006). In brief, they show that when building a model capable of classifying handwritten digits, the inclusion of a generative model of how handwritten digits are created greatly improves classification performance (i.e., perceptual inference). This example speaks directly to the embodied nature of generative models the brain might employ to make sense of the sensations generated by (oneself and) others.

Implication 3: The Key Role of Agential States

From the system biology approach, one can see easily the distinction between enactivist and cognitivist formulations by considering the graphical formulation with and without agential states. If one simply removes the agential states (or action) from the system, one can see that the system is still capable of producing lots of interesting inference and (deep) learning. This would be consistent with the vast literature on perception and cognition that does not rely on active sampling. However, as soon as we place agential states into the mix, we now have the interesting issue of how a Bayesian brain would cope when it can choose the sensory evidence to sample. A key aspect of this is that cognitive attributes, such as the value of information, curiosity, and intrinsic reward, only have meaning in the enactivism paradigm. For example, to address the exploitation-exploration dilemma, one has to account for action. In a similar vein, visual search paradigms would not have any meaning from a purely perceptual or cognitivist perspective. Although it is clear that some inference and cognition can be performed by the system without action, speaking against a pure enactivist account, it is also evident that action has the potential to alter perception and cognition radically. This is perhaps most evident in the ability of the agent to conduct an active search to explore the environment during learning, maximizing the information gain from the senses through acting and moving in the environment. In other words, with action, the agent is able to explore the environment, altering the information about the environment that the internal states can access through the sensorial states. Indeed, there is a large literature from robotics that demonstrates this to be the case (Tsotsos 1992; Shubina and Tsotsos 2010).

There seems to be little disagreement that action is important for endowing artificial agents with cognitive capabilities. Several ideas about how to gain additional information from movement have been explored during the last decades. One example is *active vision*, in which the combination of sensor readings from different viewing angles allows higher recognition accuracy than using each of the single readings (Tsotsos 1992; Shubina and Tsotsos 2010). In robotics, the improvement in performance with active visual search is an existence proof that the action-oriented approach is feasible. More importantly, the behavior of these robots shows what can be accomplished within this paradigm. An earlier and equally impressive example is that of the object recognition strategy implemented by Wilkes and Tsotsos (1992). Here, origami objects piled in a jumble can be individually recognized by a camera mounted on a robot arm that can purposefully move about the pile, selecting viewpoints and object characteristics that are used to isolate and identify them. Related approaches in robotics and computer science could generate new predictions with respect to the paradigm shift toward a more action-oriented view of cognition. For example, it has been shown that the ability to associate behavior with a stimulus is intractable in the general sense without attention (Tsotsos 1995, 2011). This suggests that future theories of attention must be broad enough to handle the requirements of an action-oriented paradigm shift.

Demonstrations of the importance of action for learning and cognition are not limited to robotics and computer vision; they have been also demonstrated in humans and animal models. In humans, recent work on vocabulary acquisition has shown that the learning of labels improves more when the infant actively explores the object being named by a caregiver, than when the child simply looks at the object without actively manually exploring it (Yu and Smith 2013, see also Dominey et al., this volume). In animals, the well-known experiment on vision performed by Held and Hein (1963) dramatically demonstrates the importance of active vision. In this study, a pair of kittens was harnessed to a carousel: one was harnessed but stood on the ground and was able to move around by itself, whereas the other was placed in the gondola and was only able to move passively. The point of this experiment was that both kittens learned to see the world, receiving the same visual stimulation. The difference was that the one could move actively, while the other was moved passively. According to Held and Hein, only the self-moving kitten developed normal visual perception. The other, which was deprived of self-actuated movement, could not develop depth perception. In short, self-movement was necessary to the development of normal visual perception with depth. Our movement in the world, the movement from here to there or there to here, gives the dimension of depth to mere visual sensations. The conclusion is that movement is the key to understanding vision.

Future Opportunities

Real-World Experimentation in Humans

New theoretical frameworks require new experimental paradigms and novel analytical methods. Our predominant methods for studying the brain (e.g., the subtractive approach in fMRI) associate particular areas of the brain with particular functions, but are less informative with respect to how regions form networks and how various networks interact. Methodological approaches need to be formulated so that we can study the activation of simultaneously active neural circuits in the brain in response to naturalistic stimuli. Other necessary advances include:

- Software for making and annotating naturalistic stimuli
- Virtual reality to allow more naturalistic interaction while maintaining experimental control
- Use of mobile measures (eye-tracking, NIRS, EEG)
- Analytic tools for studying interacting brains with fMRI and MEG, and data-constrained modeling based on this data

Experimentation in Robotics

One subtle implication of the formulation offered above is that maximizing expected utility (through pragmatic actions) or epistemic values (through epistemic actions) can be cast as a pure inference problem (using standard Bayesian techniques). This naturally prescribes a space of process theories, each based on different forms of (approximate) Bayesian inference. Practically, this also allows robotic research to avail itself of mature algorithms and schemes that have been considered in great depth over the past decades in statistics and machine learning.

Action-Oriented Understanding
of Consciousness and the
Structure of Experience

11

Extending Sensorimotor Contingencies to Cognition

Alexander Maye and Andreas K. Engel

Abstract

An emerging view in cognitive science considers cognition as "enactive" (i.e., skillful activity involving ongoing interactions with the external world). A key premise of this view is that cognition is grounded in the mastery of sensorimotor contingencies (i.e., the ability to predict sensory changes which ensue from one's own action). It is proposed that the learning of sensorimotor contingencies serves basic sensorimotor processing and that it can also be used to establish more complex cognitive capacities, such as object recognition, action planning, or tool use. Recent evidence from robotics and neuroscience supports this claim and suggests that "extended" sensorimotor contingencies might be a viable concept for pragmatic cognitive science.

Introduction

An "action-oriented" approach to cognition holds that cognitive processes are closely intertwined with action and that cognition is best understood as "enactive" (i.e., a form of practice itself). Accordingly, cognition is grounded in a pre-rational understanding of the world—one based on sensorimotor acquisition of skills for real-life situations.

Long before the emergence of modern cognitive science, philosophers emphasized the active nature of perception and the intimate relation between cognition and action. In 1896, the American pragmatist John Dewey formulated an influential sensorimotor approach to perception:

> Upon analysis, we find that we begin not with a sensory stimulus, but with a sensorimotor coordination and that in a certain sense it is the movement which is primary, and the sensation which is secondary, the movement of the body, head and eye muscles determining the quality of what is experienced. In other words, the real beginning is with the act of seeing; it is looking and not a sensation of light.

With striking convergence, the same concept was expressed, more than sixty years later by the French phenomenologist Merleau-Ponty (1962):

> The organism cannot properly be compared to a keyboard on which the external stimuli would play. Since all the movements of the organism are always conditioned by external influences, one can, if one wishes, readily treat behavior as an effect of the milieu. But in the same way, since all the stimulations which the organism receives have in turn been possible only by its preceding movements which have culminated in exposing the receptor organ to external influences, one could also say that behavior is the first cause of all stimulations. Thus the form of the excitant is created by the organism itself.

Merleau-Ponty strongly advocated an anti-representationalist view by emphasizing that the structures of the perceptual world are inseparable from the cognitive agent.

Most motifs of the "pragmatic turn" addressed throughout this volume can be traced back to these two philosophers. Drawing on the pragmatist and the phenomenological traditions, numerous authors have explored the implications of defining cognition as embodied action (Varela et al. 1992; Clark 1998; Noë 2004; Pfeifer and Bongard 2006; Thompson 2007; Engel 2010; Engel et al. 2013).

The notion that cognition can only be understood by considering its inherent action-relatedness is a key ingredient in the sensorimotor contingency (SMC) theory put forth by O'Regan and Noë (2001). Accordingly, an agent's SMCs are constitutive for cognitive processes. In this framework, SMCs are defined as law-like relations between movements and associated changes in sensory inputs that are produced by the agent's actions. "Seeing" (according to the SMC theory) cannot be understood as the processing of an internal visual "representation"; instead, "seeing" corresponds to being engaged in a visual exploratory activity, mediated by knowledge of SMCs. The active nature of sensing has been advocated by other approaches as well. For example, active vision in robotics is often considered to be a sensorimotor approach in which action plays a constitutive role for perception. In the majority of cases, however, the different views captured by the robot's camera at the scene are analyzed without considering the actions involved in bringing about these perspectives. Thus, active vision approaches attempt to compute a veridical representation of a scene by effectively stitching together perceptions from different perspectives. Whereas it has been shown that this improves the reliability of the scene segmentation, the approach still hinges largely on action-ignorant methods for analyzing the individual camera images. The key concept of SMC theory is more radical in the sense that it considers action a necessary component of perception: action does not merely support or interact in some way with perception. Thus, instead of analyzing images individually for their perceptual content, an SMC theory-based approach searches for regularities in the changes between camera images brought about by the specific actions with respect to the objects in the scene.

Although increasing evidence from work in robotics, psychology, and neuroscience support the SMC theory perspective, few attempts have been made to extend these ideas into a more comprehensive framework for cognitive science, and to derive their implications for understanding more complex cognitive capacities. Here, we propose extending the concept of SMC theory and suggest that SMCs be used to define object concepts and action plans, and that the mastery of SMCs could lead to goal-oriented behaviors. Our proposal implies that the notion of SMCs could be expanded into a more generalized concept of action-outcome contingencies, and we use the term "extended" sensorimotor contingencies (eSMCs) to denote this generalized concept.

Limitations of Sensorimotor Contingency Theory

SMC theory provides a fresh approach to explain perceptual awareness as well as a potential alternative to cognitivist theories of consciousness that consider cognition as computation over internal, observer-independent representations. O'Regan and Noë (2001) developed the basic idea and related it primarily to visual awareness, but its pertinence to other sensory modalities is straightforward. In a number of respects, SMC theory reveals limitations that call for further development of the concept as well as its application in empirical research.

Clarification Is Needed

A central claim of SMC theory is that conscious perceptual experience requires attentional exercise to master SMCs (O'Regan and Noë 2001). Building on the intuitive understanding of the term, "mastery" may work to explain the very concept. However, to assess mastery of SMCs in nonhuman agents (in particular, artificial agents), details are needed as to what constitutes mastery and the mechanisms by which an agent can actually achieve it. Similarly, in SMC theory, the process of "attunement" to an environmental feature has not been clearly defined. Attunement can describe the exercise or deployment of already mastered sensorimotor knowledge in a particular context (O'Regan and Noë 2001). However, we do not know whether attunement is a deliberative process which consciously weighs different possible contingencies to be deployed or an automatic process that selects the right set of contingencies in response to a particular situation.

This is complemented by an unclear attitude toward representationalism. O'Regan and Noë (2001) explicitly refute the view of the brain as a world-mirroring device. Perhaps the most interesting proposition of SMC theory is that perceptual experience cannot be equated with the activity of neurons or neuronal populations in specific sensory modalities. Instead, the regularities in the sensorimotor interactions (i.e., the SMCs) give rise to the different qualities of perceptual experience in different sensory modalities. These contingencies, however, need to be acquired, memorized, and maintained. This may create the impression that SMC theory has simply replaced classical concepts of representation with the notion of SMCs, which largely subserve the same role in the overall framework. Even if knowledge of SMCs is declared implicit, by emphasizing the continuous nature of the processes by which an agent attunes itself in its entirety to the environmental structure to fulfill its demands, functions for the discovery of regularities and policies that regulate the plasticity of this knowledge need to be specified.

Notions of Action and Normativity Are Unclear

SMC theory explains perceptual qualities in terms of the regularities in the changes of sensory signals caused by actions. Whereas examples of squashing a sponge or stroking a surface (O'Regan and Noë 2001) may explain the emergence of the corresponding perceptual experience in an intuitive way, closer inspection calls for further elaboration of SMC theory's concept of action. Why can we squash the sponge and stroke the surface and not vice versa? Given that our action space is infinite, why should we select exactly these two movements? How do actions which convey perception get along with actions designed to achieve goals?

   These questions suggest that some type of normativity is involved, reflecting the aptitude of each action for achieving perceptual discrimination and other goals. Likewise, actions can succeed or fail in yielding the intended outcome. Consequently, the question becomes: How do such norms arise? Other enactivist approaches (Thompson 2007; Di Paolo et al. 2010) approach this question by pointing out that the body is not only an apparatus that mediates perceptual experience, but that its self-maintenance (homeostasis) constitutes a source of basic norms by distinguishing actions that promote or break self-maintenance.

   Another direction of development must be to clarify what constitutes "action." In its original form, SMC theory seems to comprise more than just movements or motor activity, but it does not offer a clear distinction between action and movement. The extended version of SMC theory, which we propose here, interprets "action" as being neither coextensive with that of behavior nor with that of movement (Mead 1938); "action" also includes acts that do not involve any overt movements (e.g., thinking, calculating, imagining, deciding). The description of acts or actions typically makes references to goals, whereas behavior can be described without making any reference to mental states.

Generalizing with Sensorimotor Contingencies

To recognize environments and objects contained therein, current SMCs must be compared to previous experiences drawn from a repository of learned contingencies. The process of recognition does not necessarily involve movement at all, but can rely solely on sensory information to match prior experiences. This allows, for example, quick recognition of one's own cup from the pattern it leaves on the retina, instead of the time-consuming process of reenacting some of the contingencies that were acquired when it was seen for the first time. In addition, after some experience with cups, any new one can be recognized as such without fully exploring the contingencies of this particular instantiation. Both aspects require some form of generalization. Whereas learned contingencies may be the basis for recognizing a particular instance of an object ("my cup"), generalized contingencies allow the recognition of a broader object class. Such a generalization process needs to consider the relevant aspects in the typically high-dimensional sensorimotor interaction patterns and reduce them to a basic set of properties that pertain to the whole object class. Relevance, however, is largely contingent on context. Thus, any generalization schema is likely to operate dynamically. How relevant subsets of eSMCs are determined is currently an open issue.

Extending Timescales

Explanations for the basic concept of SMC theory generally employ a single context: gazing at an object, its haptic exploration, the attunement of an automatically guided missile, or driving a Porsche (O'Regan and Noë 2001). It seems straightforward, however, to assume that there are regularities in sensorimotor interactions that extend across different contexts and over longer timescales. Deployment of contingencies over extended timescales appears necessary for certain actions, such as driving to work or preparing a cake. Going to the movies and growing flowers involve regularities that extend over the course of a day or weeks, respectively. Pursuing a doctoral degree to achieve a fascinating position in science may represent an example that plays out at a lifetime timescale. As we suggest below, the concept of SMC theory might be extended to what we call "intention-related" eSMCs, which capture long-term regularities in action sequences and constitute our conscious experience beyond the timescale of object perception. It may even be interesting to ponder how eSMCs connect individuals on a social level, perhaps on timescales greater than an individual's lifetime.

Disrupted Contingencies and Altered Perception

If SMCs shape perceptual experience, their disruption should yield altered perception. This obvious conclusion needs elaboration and experimental validation.

Changes in body morphology, the skeleton-muscular apparatus, or the transmission characteristic of the sensory organs will activate a learning process for the respective altered contingencies. Conscious experience of altered perception is possible as long as the agent has access to both original and altered sets of contingencies. Without the possibility to reenact the undistorted contingencies, they are likely to be extinguished over time. To address these issues, it would be highly interesting to study perceptual alterations in patients with long-term impairment of the ability to move without concurrent cognitive deficits. Targeted changes of SMCs in virtual environments may offer a promising strategy to augment the experiential quality in such patients.

Extending Sensorimotor Contingency Theory

Our approach explicitly departs from the notion that perception is generated by an internal representation of the outer world. Instead, it accounts for the crucial role of action in the process of developing cognitive capabilities, an idea that is immediately appreciated if one thinks about the development of human infants. Our approach turns the classical view, in which sensory information initiates actions, upside down. The agent initiates actions to receive changes in its sensory stimulation, and it learns associations between them. The initiation of actions seems to emerge from spontaneous behaviors of biological organisms, as in the motor-babbling behavior of infants (Westermann and Miranda 2004; Natale et al. 2007).

   Below, we distinguish three types of eSMCs that comprise contingencies at different levels of complexity, from sensorimotor coordination to action-reward contingencies. This suggests a multilevel architecture, where different types of eSMCs are acquired and deployed to implement cognitive processes of increasing complexity. A key hypothesis in our proposal is that a consideration of eSMCs at different scales should help unravel the emergence of cognitive capacities at different levels of complexity and contribute to an understanding of how these may be grounded in basic sensorimotor processes.

Modality-Related Extended Sensorimotor Contingencies

This first type captures the specific changes of the sensory signal in a single modality, depending on the agent's action. Examples include the different perspective distortions that result from eye movements and locomotion, sound pressure profile changes when the head rotates, and the dependence of the force feedback from the force exerted by finger movements. This most basic type of eSMC distinguishes the qualities of sensory experiences in the different sensory channels (e.g., "seeing," "hearing," "touching") and was addressed by SMC theory in its original formulation (O'Regan and Noë 2001). We suggest, however, that the original idea be broadened to account for the full vector of multisensory inputs, together with intrinsically generated normative feedback. This notion is reflected in the computational model of modality-related eSMCs that we describe below.

Object-Related Extended Sensorimotor Contingencies

The next eSMC type relates to the effects on the sensory system that are specific for each object under consideration, and these effects are inherently supramodal. They describe the multisensory impression that an object leaves on a set of actions of the agent. An example is given by the different visual and force feedback signals received when touching a sponge, a piece of cardboard, or a piece of wood. Object-related eSMCs define the object under consideration, and exercising actions from a set of object-specific eSMCs corresponds to the perception of this object. One of the fundamental claims of our approach is that the observed relations between actions and sensory changes are sufficient for recognizing a particular object. Object-specific eSMCs are more numerous and more complex than modality-specific eSMCs.

Intention-Related Extended Sensorimotor Contingencies

The third type denotes a further level of generalization of the concept of SMCs and considers the long-term correlation structure between complex action sequences and the resulting outcomes or rewards, which the agent learns to predict. We propose that intention-related eSMCs capture the consequences of an agent's actions on a more general level as well as on extended timescales. These complex eSMCs include contingencies that are cognitively simulated by the agent and do not relate to factual movement. After learning, intention-related eSMCs could be used to predict whether an action will be rewarding or not and to rank alternatives. At the same time, intention-related eSMCs may provide the basis for action plans that involve several steps to reach an overall goal. In this way, anticipation and anticipatory behavior as well as the sense of agency might be grounded in eSMCs.

Extended Sensorimotor Contingencies in Natural Cognitive Systems

A substantial body of neurobiological findings is compatible with the idea that SMCs are used by the brain, and that different types of contingencies play a key role in natural cognition. Here we review neurobiological evidence that supports the eSMCs concept before considering recent attempts to model and test eSMCs in robot experiments.

Basic Importance

The action-oriented view of cognition advocated here is supported by findings on the role of exploratory activity and sensorimotor interactions for neural development and plasticity. For a long time, developmental processes in the nervous system have been known to be activity dependent. For instance, development of neural circuits in the visual system and acquisition of visuomotor skills critically depend on sensorimotor interactions and active exploration of the environment (Held and Hein 1963). Even in the adult brain, there is considerable plasticity in cortical maps (e.g., in the somatosensory and motor system) that has been shown to depend on action context (Blake et al. 2002). These studies suggest that appropriate action, allowing exercise of relevant eSMCs, is necessary throughout life to stabilize the functional architecture in the respective circuits.

If guidance of action is a dominant function of the brain, one would predict that neuronal response profiles in sensory or association regions should strongly depend on action context. Indeed, clear evidence exists for such an action-relatedness. Activation of visual cortical neurons changes profoundly if self-induced movements are permitted, as compared to passive viewing of stimuli (Niell and Stryker 2010). Gain modulation of neural responses is abundant in the nervous system (Salinas and Sejnowski 2001), demonstrating that sensory activity patterns are always, to a considerable degree, related to or modulated by action. In premotor cortex, the spatial profile of multimodal receptive fields depends on body and limb position (Graziano and Gross 1994). Tactile and visual receptive fields of premotor neurons are in dynamic register and seem anchored to body parts, even if these are moving; this suggests that such polymodal neurons support predictions about expected changes in sensory input (Figure 11.1a). Action-related changes of sensory response properties of polymodal neurons have also been observed in studies involved in tool-use learning (Figure 11.1b) (Maravita and Iriki 2004). It is tempting to speculate that assemblies of such polymodal neurons encode eSMCs.

Important evidence regarding the neural mechanisms of eSMCs comes from research on "corollary discharge" or "reafference signals," which are necessary for an organism to distinguish self-generated sensory changes from those not related to its own action (Crapse and Sommer 2008). Supporting the SMC theory, this research shows that predictions about the sensory outcome of movement are critical for the basic interpretation of sensory inputs. The importance of reafference has been shown in the context of eye movements as well as grasping or reaching movements. Interestingly, similar principles of predicting sensory inputs seem to play a key role in more complex cognitive processes, such as language comprehension or predictions about sequences of abstract stimuli (Schubotz 2007). In all of these cases, activity of motor planning regions seems to be involved in generating the prediction about sensory events, possibly by modulating neural signals in sensory regions. Malfunction

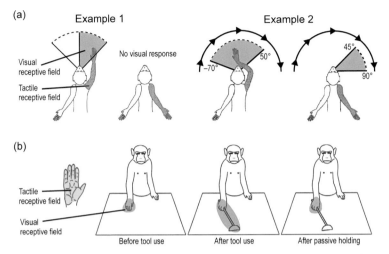

Figure 11.1   Dependence of multisensory receptive fields on motor state and action context. (a) Two examples of neurons recorded from ventral premotor cortex in the monkey. Neurons showed bimodal responses with both tactile (blue) and visual (red) receptive fields. In both examples, the visual response depended on the arm position of the animal. The panel illustrates results from Graziano and Gross (1994). (b) Recording of bimodal intraparietal neurons with tactile (blue) and visual (red) receptive fields. The visual receptive field showed adaptive changes as the animal used a tool to retrieve food, which expanded to include the entire length of the tool. Under control conditions with passive holding of the same tool, this expansion did not take place. The panel illustrates the work of Maravita and Iriki (2004). Reproduced with permission from Engel et al. (2013).

of such modulatory signals and associated disturbance of forward models have been implicated in the pathogenesis of psychiatric disorders such as schizophrenia (Frith et al. 2000b).

Object- and Intention-Related eSMCs in the Human Brain

Our view implies that procedural knowledge is fundamental to acquire object concepts. Thus, storing information about events and objects should generally involve action planning regions. In line with this prediction, neuroimaging studies show that object concepts in semantic memory do not only rely on sensory features alone but, critically, also on the motor properties associated with the object's use (Martin 2007). If subjects are trained to perform functional tasks on certain objects, premotor regions become active during visual perception of these objects. Another intriguing finding is that motor and premotor systems are also active during mental simulation of events (e.g., during mental rotation of objects). Research on the mirror neuron system provides strong support for this view (Rizzolatti and Craighero 2004).

A fruitful approach for investigating the relevance of object-related eSMCs in human cognitive processing is to study the dependence of object recognition on exploratory eye movements during free viewing of images. Using ambiguous images, a recent study showed that eye movements performed prior to conscious object recognition predict the later recognized object identity (Kietzmann et al. 2011). Other studies on natural vision have also shown that eye movements are highly predictive in nature (Hayhoe et al. 2011). These findings suggest that eye movements (like other movements) express the mastery of object- and intention-related eSMCs.

Patients with dysfunctional motor circuits provide the possibility to investigate the functional role and putative mechanisms of eSMCs. Our view predicts that motor dysfunction in such patients should result in a disruption of learned eSMCs and a lack of adaptivity and acquisition of novel eSMCs. This, in turn, should become manifest in altered perceptual processing and altered cognitive capacities. These hypotheses can be tested, for example, in patients with Parkinson disease. In these patients, modified eSMCs could be studied both as a function of dopaminergic medication (with and without levodopa) or as a function of deep brain stimulation in patients who have undergone surgical treatment. Indeed, substantial evidence suggests deficient acquisitions and calibration of eSMCs in patients with Parkinson disease. Thus, altered modality-related eSMCs have been observed in tasks involving rhythmic movement (Gulberti et al. 2015). Alteration of object- and intention-related eSMCs is suggested by difficulties in action naming observed in patients with Parkinson disease (Herrera and Cuetos 2012).

Intention-related eSMCs are also reflected in the "sense of agency," which refers to the experience of oneself as the agent of one's own actions and is, as such, a central constituent for human self-consciousness. Our sense of agency relies heavily on the experience of SMCs or action-effect couplings, but adds to these an additional, more complex level involving intention, experience, and identification. A large body of evidence suggests that the sense of agency depends on the degree of visuomotor congruence or congruence between predicted and actual sensory consequences of an action (David et al. 2008). Only in the case of congruence will an agent register a sensory event as caused by itself; incongruence would lead to the registration of the event as externally caused (Frith et al. 2000b). The more systematic the congruence between action and action effect, the better the agent's capacity to differentiate between self-produced and non-self-produced actions. Thus, we hypothesize that the emergence of the experience of agency is directly related to the mastery of action-effect couplings at the level of intention-related eSMCs.

Extended Sensorimotor Contingencies in Artificial Cognitive Systems

Implementing the concept of eSMCs in artificial agents provides the opportunity to verify the theory and challenge its limits as well as to explore extensions.

In return, robotics may benefit from the virtues of the eSMCs account of cognition. Learning and using the structure of sensorimotor dependencies could endow a robot with an understanding of its different sensory modalities and an action-based perception of objects in its environment. Among other advantages, the object recognition and manipulation capabilities of eSMCs-controlled robots would be based on the experience that a robot would acquire with an object rather than the experiences of a human programmer, thus allowing the robot to discover relevant features itself and, hence, adapt better to unexpected changes in the environment or sensor failures. This synthetic approach requires casting the theory into computational models that are suitable for controlling robots. In the following, we introduce some of these approaches.

Modality-Related Extended Sensorimotor Contingencies

A general formulation of modality-related eSMCs can be given in computational terms by considering the probability of observing a particular sensory input after executing an action given the previous sensory state. This notion is described mathematically by the conditional probability distribution $P\big(o(t+1)\big|a(t),o(t)\big)$ over future sensory observations $o(t+1)$, when the current sensory state is $o(t)$ and action $a(t)$ is executed. Recently, we used an extension of this idea to study the acquisition of modality- and object-related eSMCs and to show how they can be used to control an agent's behavior (see Figure 11.2; Maye and Engel 2012). The robot observed the dependencies between changes in the accelerations and power consumptions of the motors and actions that led to changes in the movement direction. Using reinforcement learning, the robot preferentially reenacted those patterns which maximized a given utility function and thus learned to move smoothly and in an energy-efficient manner.

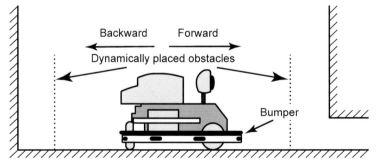

Figure 11.2 Experimental setup to study learning of eSMCs and deploying them for controlling behavior in an artificial agent. In the exploration phase, the robot developed knowledge of the size of its confinement, appropriate movements to escape collisions and energy-efficient movement sequences between the walls. This knowledge allowed the robot to react properly to obstacles that were placed dynamically in the arena. Reproduced with permission from Maye and Engel (2012).

It would be interesting to explore to what extent minimalistic systems with a Braitenberg vehicle-like control architecture could enact modality-related eSMCs. This architecture is characterized by a fairly direct coupling between the sensors and actuators of the agent, which is adjusted with respect to a goal function. As an example, O'Regan and Noë (2001) likened this to a tactical missile that is "tuned" to the eSMCs of airplane tracking, a task easily solved by a Braitenberg vehicle. The vehicle has extensive knowledge of possible input-output relations in its domain and uses this information to adjust its behavior to maximize the goal function. Even if the robot malfunctions, it still experiences sensorimotor regularities, but it cannot effectively exploit them to optimize its behavior.

Object-Related Extended Sensorimotor Contingencies

Object recognition poses a key challenge for artificial agents, for which there is currently no general solution. Classical attempts to solve this problem have aimed at recognizing objects in static images (e.g., snapshots from a camera). The pragmatic viewpoint invites us instead to consider vision as a mode of exploration, as an active process of making sense of the visual input. A paradigmatic example for applying this concept to robot vision is given by Bergström et al. (2011), in which a robot arm interacts with objects in a scene to disambiguate camera images of the scene acquired by the robot. A model of the visual appearance of the scene was generated using an expectation-maximization approach and belief propagation. This model produced a weak hypothesis about the centers of potential objects in the scene. The robot arm then pushed one of these centers, and the ensuing motion flow was analyzed for compatibility with the presence of one or more rigid objects. The result was fed back to the appearance model to improve the segmentation.

A similar concept was used by Björkman et al. (2013) to determine the shape categories of household objects. Object shapes were modeled by implicit surfaces that were adjusted by a regression of Gaussian processes. Visual information from a stereo camera was used to initialize the surface models. Regions of high uncertainty were then touched by a robotic hand, and the models were updated using tactile information. If the robot hand touched not only the object, but pushed it across the ground, information in the resulting motion flow could be used to classify the functional properties of the inspected objects and to group them according to their affordances (Högman et al. 2013).

In one of our own studies (Maye and Engel 2011), we addressed two of the aforementioned limitations of SMC theory: action selection and normativity of eSMCs. We developed a scenario in which object-related eSMCs were not only used to recognize objects but also to let the robot show corresponding behavior. We used a Markov model as described for modality-related eSMCs but extended it by attaching to each individual eSMC a utility value which captured the feedback that the robot received in the respective context. By giving positive feedback for correct actions and punishing wrong actions (and neutral feedback if the action was indifferent), the robot was trained to associate the presence of different object types with specific actions (e.g., sorting cans and boxes by pushing them in opposite directions). The robot modeled the utility of these actions, $a(t)$, by observing reward or penalty probabilities conditional on the sensorimotor context, $c(t)$. This was done by counting the number of rewards received in a situation given by sensory observation $s(t + 1)$:

$$p_{reward}\left(s(t+1)\big|a(t),c(t)\right) = \frac{N_{reward}\left(s(t+1),a(t),c(t)\right)}{N\left(a(t),c(t)\right)}. \qquad (11.1)$$

Another histogram was used for computing the probability of punishment, $p_{penalty}$.

Assigning reward probabilities to eSMCs can be regarded as a method for structuring the sensorimotor knowledge that the agent acquires. It establishes relations between eSMCs that have a similar ecological value. Sensorimotor knowledge could be further structured by considering temporal information, for example, by grouping eSMCs that have been observed in the same time interval and which are therefore probably related.

In the terminology of SMC theory, viable actions result when the agent exercises its mastery of eSMCs. However, SMC theory does not suggest a strategy for choosing which set of learned eSMCs to activate. This is why our model uses the value system described above, which structures and weighs eSMCs according to their expected reward. The decision schema (Figure 11.3) constitutes a simple switch from exploration to exploitation: If there are actions that have never been tried before in the current context, one of these actions will be

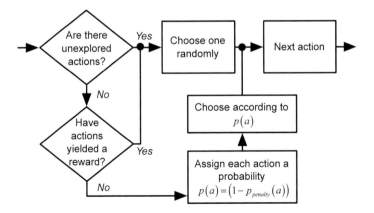

Figure 11.3 Action selection schema that optimizes the robot's behavior based on the observed utility of different eSMCs. Reproduced with permission from Maye and Engel (2011).

chosen. Otherwise, an action which most likely will yield a reward is chosen. If only actions which yielded no or negative rewards (i.e., penalties) have been observed in the past, these actions are assigned probabilities according to their least negative effect, and one of them is chosen according to this probability distribution.

Intention-Related Extended Sensorimotor Contingencies

Temporally extended sensorimotor interaction patterns are not only suitable characterizations of the agent's perceptual experience, they also enable the prediction of future experiences and the planning of proper action sequences. The basic idea is to search for partial matches between learned eSMCs and the current sensorimotor context, and then build a probability distribution over future experiences from these eSMCs beyond the point to which they match. We explored this idea in a study where we intentionally introduced a delay between issuing a motor command in a robot and showing the corresponding change in the movement direction (Maye and Engel 2013). This required the agent to predict potential collisions and to adjust its motor commands accordingly.

Taken together, the approach to model eSMCs using Markov processes of different history lengths led us to the hypothesis that the different levels of eSMCs may involve different timescales over which regularities in the sensorimotor patterns are captured (Maye and Engel 2012). Accordingly, from a modeling perspective, there would be no principle differences between the three eSMCs levels introduced here. Rather, these levels would describe different ranges on a continuum of sensorimotor context scales. This would allow eSMCs to be nested, with eSMCs that span shorter timescales forming the building blocks for eSMCs with longer timescales (Figure 11.4).

Intention-related eSMCs can also be considered from the perspective of the related concept of affordances (Gibson 1979). Knowing an object's affordances allows continuous sensorimotor interaction to change the state of the environment in a predictable and goal-directed manner. A primary application scenario for this approach is tool use, which was explored by Sánchez-Fibla et al. (2011) using a robot that learned to push objects of different shapes along a predefined trajectory or to a given target position and orientation. Sánchez-Fibla et al. introduced the concept of affordance gradients, which are object-centered representations of the consequences that an agent's actions may have.

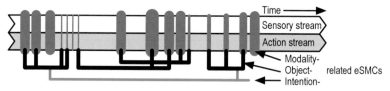

Figure 11.4   Schematic of the timescales that are captured by different types of eSMCs.

These representations are mathematically formulated by vector fields which allow the robot, by means of interpolation, to make predictions about the consequences for actions that have not been tried before. In the case of push actions, affordance gradients can be formally defined by a triplet consisting of a gradient representing the object's shape and two vector fields describing the displacement and rotation of the object when pushed from a given position and in a given direction. This triplet is learned in an exploration phase in which the robot can try out the consequences of pushing the object from various positions and at different angles. The forward model that the robot acquires in this phase can then be used iteratively to develop an action plan to achieve a goal and to update this plan during execution (Sánchez-Fibla et al. 2011).

Major challenges for eSMCs-based approaches to robot control result from the size of sensorimotor spaces and the complexity of the relations between actions and sensations. Modern robots feature many degrees of freedom and large numbers of sensory channels, composing potentially infinite sensorimotor spaces. This is an issue for any robot control architecture. Squashing action space and quantizing sensory signals are frequent solutions for curbing the sensorimotor space. The second issue, the complexity of sensorimotor relations, is more specific to eSMCs-based approaches. These relations can be simple co-variational or causal; they can span different timescales, be described at different levels of abstraction, and be hierarchically nested. Another challenge may be the development of neurobiologically plausible eSMCs models which would link insights from empirical research with results from synthetic approaches. Corresponding models are only beginning to emerge.

Outlook

The approach that we have pursued here departs from the classical notion that presumed "higher" levels of cognition might be fundamentally different from presumed "basic" levels of sensorimotor coordination. It also rejects the notion that architectures embodying complex cognitive functions (e.g., "higher" processing centers in the human brain) differ in principle from modules for more basic functions. Conceptually, the eSMCs approach diverges from much of what has been characteristic of the classical cognitivist framework, moving toward an enactive, embodied view of cognition and a much more dynamic and holistic view on the underlying architectures.

The central role that this approach assigns to action places it in the family of embodied approaches to cognition. It shares ideas with embodied approaches in cognitive linguistics, autonomous robotics, and ecological psychology as well as with enactivist models. Like SMC theory, enactive approaches to cognition (Varela et al. 1992; Thompson 2007; Di Paolo et al. 2010) conceive of cognitive states as interaction processes of embodied agents and the environment rather than brain states. Enactivism, however, entertains the notion of the body as a self-constructing process that gives rise to norms and the concept of autonomy. The act of striving to maintain autonomous identity can explain why agents have interests in the world and provide a touchstone for developing a concept of sense-making. In this respect, enactivism may be able to provide an answer to the question that SMC theory seems to leave open; namely, what constitutes mastery of eSMCs and how can an agent achieve it?

We suggest that the eSMCs approach may provide new perspectives for robotics. In contrast to top-down architectures used in industrial robot applications, where the relevant knowledge is programmed into the robot, our approach advocates letting the agent acquire knowledge by embedding it in an environment which it can explore and manipulate. A consequent interpretation of the eSMCs approach suggests that the discovery and appropriate utilization of regularities in the robot's sensorimotor interactions with the environment would vest it with a primitive form of cognitive states. Notably, the relevant sensorimotor dependencies comprise properties of the embodiment as well as the environment. As such they can be seen to be a control mechanism for morphological computations (i.e., cognitive functions which are carried out through particular properties or suitable arrangements of body parts) as well as for an extended mind that employs the body and the environment to realize cognitive functions. This places our approach in the perspective of active externalism, as proposed by Clark and Chalmers (1998).

One of the challenges for SMC theory is to understand how abstract concepts might be grounded in sensorimotor processes. Embodied theories of cognitive linguistics consider the human body as a source of meaning, concept articulation, and even reasoning (Gallese and Lakoff 2005). Rather than abstract rules of logic, thought processes employ comparisons to sensorimotor experiences, thereby grounding our understanding of the world and the contents of our verbal communication in the physical reality in which we are embedded. Solving an exercise, for example, is thought to involve patterns used to overcome obstacles, such as considering the problem from different perspectives, selecting an appropriate solution method, progressing along the action plan, or determining the suitability of the result. This idea is compatible with an extended version of SMC theory in which eSMCs define the contents of perceptual experience and constitute the substrate for higher-level cognitive processes. This requires learning mechanisms that enable the cognitive agent to generalize and abstract from sensorimotor interaction patterns.

A highly interesting issue is whether the approach proposed here can be extended even further to ground basic aspects of social cognition. It seems plausible to assume that agents deploy learned action-effect contingencies in social contexts to predict outcomes of their own and other's actions. This would allow an agent to coordinate with actions of other agents and enable effective coupling of agents in social contexts. Accordingly, social interaction may strongly depend on the dynamic coupling of agents, and this interaction dynamics may provide important clues to social cognition. This view shares aspects with the interactionist concept of social cognition proposed by Di Paolo and colleagues (e.g., Di Paolo and De Jaegher 2012). They argue for an extension of an enactivist position and emphasize that sense-making in a social environment occurs in a participatory manner and that central aspects of cognitive performance are inherently relational. Furthermore, our concept agrees well with the joint action model by Knoblich and colleagues (Knoblich and Sebanz 2008), who predict that shared intentionality can arise from joint action.

Finally, it might be interesting to consider implications of our approach for understanding consciousness (cf. Seth et al., this volume). In the original version of SMC theory, a key hypothesis is that SMCs can account for qualia as the basic features of phenomenal experience, and that differences in the qualitative character of perceptual experiences result from differences in the relevant SMCs (O'Regan and Noë 2001). What SMC theory leaves open is how more complex contents of conscious experience can be aggregated, selected, and structured. We believe that it would be worth exploring whether an extended sensorimotor account, as discussed here, might help to account for the structure of conscious experience beyond the basic level of sensory qualities.

Acknowledgments

This work has been supported by the EU (FP7-ICT-270212, ERC-2010-AdG-269716).

12

What's the Use of Consciousness?

How the Stab of Conscience Made Us Really Conscious

Chris D. Frith and Thomas Metzinger

Before the birth of consciousness,
When all went well,
None suffered sickness, love, or loss
None knew regret, starved hope, or heart burnings.
— Thomas Hardy, *Before Life and After* (1909)

Regret is the most bitter pain, because it is characterized
by the complete transparency of all one's guilt
— Søren Kierkegaard, *Either/Or* (1843/1992)

Abstract

The starting assumption is that consciousness (subjective experience), rather than being an epiphenomenon, has a causal role in the optimization of certain human behaviors. After briefly outlining some of the critical properties of consciousness, this chapter reviews empirical studies that demonstrate how much can be achieved in the way of action and decision making in the absence of relevant conscious experience. Thereafter, it considers, in detail, the experience of action and suggests that this has two key components: the experience of being an agent, which causes events in the world, and the belief that we could have done otherwise. Such experiences enable us to justify our behavior to ourselves and to others and, in the longer term, to create a cultural narrative about responsibility. Finally, the experience of regret is explored (i.e., the recognition that one could and should have acted otherwise). Regret is a powerful, negative emotion that is suggested to integrate group norms and preferences with those of the individual. The transparent and embodied nature of the experience of regret ensures that cultural norms become an inescapable part of the self-narrative. The conclusion is that conscious experience is necessary for optimizing flexible intrapersonal interactions and for the emergence of cumulative culture.

Introduction

What's the use of consciousness? By asking this question, we indicate that we approach the problem of consciousness mainly via biology and evolutionary theory. Our first assumption is that the appearance and maintenance of the phenomenon of consciousness in humans and other animals implies that there is some continuing evolutionary advantage to consciousness. If we assume that consciousness evolved, then it is also reasonable to assume that some creatures, such as humans, are more conscious than others. It also follows that, at least in our world and under the laws of nature holding in it, we do not believe in the possibility of zombies, those philosophical constructs, functional isomorphs that behave exactly like humans, but in the absence of consciousness. There are some things that such zombies would not be able to do. Our task is to identify these things.

Much previous work on consciousness has concentrated on perception, especially vision (Crick and Koch 1995). The natural antidote to this biased perspective requires that we cease to focus on perception, or action, or cognition in isolation. Thus, to answer the question "What's the use of consciousness?" we need to relate consciousness research to the underlying principle that connects all three elements: the action-perception loop. If there was a formal framework capable of unifying all three aspects under a common principle, and if that framework turned out to be empirically plausible, then it would be natural to describe conscious experience by using the conceptual tools offered by it. For example, conscious experience could then be a single, generative model of reality including a model of the self as currently acting, perceiving, and thinking (Friston 2010).

If consciousness gives an advantage to humans, then it must causally enable humans to achieve more optimal behavior. What class of optimization problems does consciousness enable us to solve? What new types of action does it enable? These potential uses of consciousness are particularly relevant to the current pragmatic turn in cognitive science.

Consciousness and Its Properties

Levels and Contents of Consciousness

At this point we need to make a gesture in the direction of defining consciousness. One distinction is between *levels* of consciousness and *contents* of consciousness. Levels of consciousness relate to the distinction between being awake or asleep as well as between being a man or a mouse. Consciousness comes in degrees (e.g., of wakefulness and alertness). We can ascribe the property of "consciousness" to whole persons or biological systems, and we might distinguish such systems according to their overall level of wakefulness, the presence of an orientation reaction, etc. This is sometimes called *creature consciousness* (Metzinger 1995a). Consciousness, however, can also be viewed as a property of individual states (e.g., representational states in the central nervous system of an organism). This is sometimes called *state consciousness* (Rosenthal 1986). The content of consciousness refers to what our conscious experience is *about*, what we are currently conscious *of*. This relates, for example, to the distinction between being conscious (aware) of the face in our field of view and not being conscious of that face (important in the search for the neural correlates of consciousness; see, e.g., Beck et al. 2001), while the overall level of consciousness does not differ between these two states of the subject in the experiment. Accordingly, we can look for the global neural correlate of consciousness (i.e., the set of physical or functional properties corresponding to the totality of an individual's experience) or the correlate of specific kinds of content; for useful conceptual distinctions, see (Chalmers 2000).

Properties of Conscious Experience

In terms of the contents of conscious experience, three properties are of particular relevance (for further details, see Metzinger 1995b):

1. There is a pure subjective experience, the *phenomenal content* of our mental states. These subjective states have a certain *feel*: there is something *it is like* to be in these states.
2. These phenomenal states are frequently *transparent*. We do not experience these states as representations of reality; we experience them directly *as* reality.
3. Conscious experience is always perceived as part of the current moment, whereas phenomenal experience is characterized by the subjective character of *presence*. What is present is always a whole situation or single, unified world model.

In addition, under standard conditions these states and their contents are experienced from a *first-person perspective*: they are the inner experiences of an individual and seem to be private. This last property raises well-known epistemological problems for the scientific study of consciousness (Jackson 1982; Levine 1983): How can scientific objectivity be applied to something that is subjective and only available as an individual first-person perspective (cf. Nagel 1974, 1986)? This problem reveals an interesting paradox: my conscious experience appears to be private and inaccessible to anyone else, yet it is the only aspect of my mental life to which I have seemingly direct access, about which I possess maximal certainty, and which I can potentially report to others. Perhaps the core problem in consciousness research consists in finding out what exactly a "first-person perspective" is and if it can, at least in principle, be naturalized.

The Problem of Report

In practice the study of subjective experiences depends on the report of the person having the experience. Strictly speaking, there is no such thing as "first-person data." Scientifically we can only access first-person reports, but never the experience itself, in its subjectivity (for further discussion, see Metzinger 2010). These reports need not be verbal; some experiences may be too novel or complex for suitable words to be available. Further, *if* verbal reports are available, they are in constant danger of being "theory-contaminated"—a process by which subjective reports are influenced by the scientific or philosophical theories the subject believes to be true, by specific psychological needs, by social context, or by cultural background assumptions. For the subjective experience of agency, which is the primary focus here, we believe this point to be of particular relevance.

A Novel Proposal

We believe that conscious content may have played a decisive role in the emergence and stabilization of complex societies. This is one prime example for a function of consciousness. To ground our proposal below, we will look at a range of biological and cognitive functions for which conscious processing is not a necessary prerequisite. Then we will consider the experience of action and introduce the notion of "regret," first describing its phenomenological profile, then offering a brief representationalist analysis, and proceeding to isolate its hypothetical function. Finally, we will present a brief sketch of an argument as to why this specific kind of phenomenal experience would have been advantageous under an evolutionary perspective.

Functions for Which Consciousness May *Not* Be Necessary

There is so much empirical evidence in favor of consciousness, viewed by itself, as having little role in our behavior, that we might conclude that it is no more than an epiphenomenon. Huxley (1874) proposed that consciousness, although real and created by the brain, was an epiphenomenon with no influence on behavior:

> Consciousness…would appear to be related to the mechanism of the body simply as a collateral product of its working, and to be as completely without any power of modifying that working as the steam-whistle which accompanies the work of a locomotive engine is without influence upon its machinery.

Humans, he suggested, are "conscious automata." Even if we accept that consciousness does have a role in decision making, there are many cases where better decisions are made when people forgo conscious control (for a review, see Engel and Singer 2008). What, then, is the use of consciousness? Before presenting our own speculations, we will consider and dismiss several candidate processes for which it has been proposed that consciousness is necessary.

Sensory Integration and Global Informational Access

Here is a good candidate for a function of consciousness: Our conscious experience of the world typically involves objects and actions, rather than isolated sensations and movements, and it is plausible that consciousness is necessary for integrating information (Tononi 2008) and for broadcasting information between different processing modules (Baars 1988; Dehaene and Naccache 2001). However, we are doubtful. There is increasing evidence that sensory integration happens even at the earliest stages of sensory processing (e.g. Watkins et al. 2007; Lemus et al. 2010); even high-level, cross-modal integration of symbols can occur without awareness (Faivre et al. 2014). For many activities there is clear need for "global availability" of information. But why should this global access be associated with subjective experience? Furthermore, decisions which require the integration of many sources of information seem to be better made in the absence of conscious reflection (Dijksterhuis and Nordgren 2006), perhaps because appropriate weighting of multiple sources of information is disrupted by conscious deliberation (Levine et al. 1996; Engel and Singer 2008). A similar phenomenon can be observed in the performance of highly skilled acts (Beilock et al. 2002).

Sophisticated and Flexible Top-Down Control

Many accounts suggest that consciousness is necessary for a high-level supervisory system that modulates lower-level automatic processes, especially when unexpected problems arise and when novel skills must be developed (e.g., Norman and Shallice 1986). Again, we find this to be a very plausible suggestion. However, given that there is a hierarchy of sensorimotor control (e.g., Friston 2005), this formulation requires that we specify at which level in this hierarchy consciousness emerges. While there is, as will be described below, good evidence for a role for consciousness in top-down control, we suggest that the level in the hierarchy, at which this operates, is higher than previously supposed. Control of considerable sophistication and flexibility can occur at lower, "automatic" levels. For example, it is well established that much low-level control of action can occur without awareness. This is true for hand movements (Fourneret and Jeannerod 1998) as well as for locomotion involving the whole body (Kannape et al. 2010).

Consider two examples in which monitoring and control occur at an even higher level in the absence of awareness. In a study of walking (Varraine et al. 2002), people were given arbitrary and unpracticed instructions as to how they should change their walking pace when they detected a change in the responsiveness of the treadmill upon which they were walking. Remarkably, they changed their pace correctly for about six seconds before they reported detecting the change. In another study, skilled typists slowed down after they had made errors, but not after they experienced errors inserted by the experimenters, even though their verbal report showed that they were not conscious of the distinction between these types of errors (Logan and Crump 2010). Here, monitoring and control (metacognition) of the low-level process of typing occurred outside consciousness.

Emotion and Motivation

Does the motivation created by affective states, such as pleasure and pain, depend on awareness of these states? Perhaps emotions do not have to be conscious to make people act in particular ways. The following study gives an example of an unconscious emotion: Smiling faces, presented subliminally, caused thirsty people to pour and consume more drink, even though they were unaware of any change in their emotional state (Winkielman et al. 2005). It is obvious that consciousness is required for us to *talk about* an emotion, and further research is needed to identify those aspects of emotion that enable functional availability for verbal report (Metzinger 2003). However, certain kinds of conscious emotion, such as regret, do have effects on behavior (Filiz-Ozbay and Ozbay 2007). Unlike more basic emotions, such as happiness and anger, regret involves counterfactual thinking: "Things would be different, if only I had behaved differently."

Representing the Mental States of Self and Others

Do we need consciousness to account for the mental states (e.g., beliefs, perceptions and intentions) of others and of ourselves (Humphrey 1999; Graziano and Kastner 2011)? We certainly need consciousness to *talk* about our mental states, but can we take account of the mental states of others without awareness? For example, our behavior is affected, automatically, by the action goals of others (Sebanz et al. 2003) as well as by the perceptual knowledge of others (Samson et al. 2010). In Samson et al.'s study, they showed that it took people longer to report the number of targets when another person with a different viewpoint saw a different number of targets, even when this was entirely irrelevant to the task being performed. This effect was not altered when a cognitive load was applied (Qureshi et al. 2010), suggesting that the process of taking another person's knowledge into account was automatic and unconscious.

Do We Need Consciousness to Make Free and Flexible Decisions?

The ability to make free and flexible decisions is a role for consciousness that is most relevant to action. Yet ever since Benjamin Libet's classic experiment, doubt has been cast on this role (Libet et al. 1983). Results from his studies, replicated more recently using fMRI (e.g., Soon et al. 2008), suggest that the awareness of initiating an action comes too late to have any causal role in the decision. From fMRI data, patterns of brain activity, occurring well before a participant reports making the decision, can be used to determine, somewhat better than chance, which action will be chosen.

One problem with these studies is that the decision (e.g., whether to move the left or the right finger) is neither taxing nor of much relevance to real life. However, the discovery of choice blindness, by Johansson et al. (2005), confirms the fragile relationship between decisions and awareness in situations of much greater ecological validity (e.g., Hall et al. 2012). In these studies, participants could be persuaded to justify a decision that they had not actually made. They seemed unaware of the decision that they had actually made.

Being in Control: The Experience of Agency

A striking paradox is revealed by the above-mentioned studies: awareness of decision making seems to have little or no role in causing decisions, yet the vivid feeling of being the author of one's own actions—the sense of *agency*—is a large component of conscious experience. Indeed, it is only because we have this clear experience of being in control that experiments like those of Libet are possible. People have no problem when asked to report the precise moment at which they made a decision. There is strong awareness of mental agency, yet, at the same time, very little awareness of bodily agency. Why should awareness of mental agency be given such salience, unless it has some function?

Since at least the time of Epicurus (Bobzien 2006), the experience of being in control of our actions, our sense of agency, is considered to have two key components: (a) the sense that it is I that am doing a particular action (i.e., I am in control) and (b) the sense that I *could* have done something else (i.e., the counterfactual element). The latter component is critical for our experience of regret: I would have done better, if I had chosen the other option.

Below we outline several aspects of the experience of agency before addressing the question regarding the salience of mental agency awareness.

Intentional Binding

Research on the sense of agency received a dramatic boost from the discovery of intentional binding by Haggard et al. (2002). Libet's technique was used to indicate the subjective timing of an action: the initiation of the action (a button press) and the outcome (a sound) of the action. For an action in which the person was the author (i.e., a deliberate button press), the subjective time between these two events was shorter than the physical time. For an action in which the person was not the author (e.g., finger movement is caused by transcranial magnetic stimulation) the subjective time was longer than the physical time. This suggests that (a) our experience/perception of actions is composed of two components (the initiation and the outcome) and (b) the subjective time between these two components is a marker of intentionality.

The Experience of Being in Control

As is generally the case with perception (Kersten et al. 2004), our experience of action depends on our expectations as well as on evidence from our senses. Thus, in the case of action, there are prospective and retrospective influences on the degree of intentional binding (Moore and Haggard 2008). Prospective influences (i.e., expectations) can arise from learning about the probability that an outcome will follow the movement. Retrospective influences arise from the nature of the outcome. As a result, the time at which a person experiences initiating an action can be influenced, retrospectively, by whether or not the outcome occurs (see also Lau et al. 2007).

These results indicate a considerable malleability for our experience of being in control. As with other kinds of perception, illusions of control can arise, typically through manipulation of expectations. Such illusions have been documented in detail by Wegner (2003) and include people believing that they were controlling an action when they were not, and vice versa. Beliefs about control, caused by instruction, can also alter intentional binding (Dogge et al. 2012).

Responsibility and the Sense of Agency

Whether or not the conscious experience of agency (being in control) has any well-circumscribed causal role in the action currently being performed, the experience has an important role in culture. For example, verbal reports about this specific type of phenomenal experience can now become "theory-contaminated" and begin to drive cultural evolution. As we pointed out above, "theory-contamination" is a process by which subjective reports are influenced by the scientific or philosophical theories the subject believes to be true, by social context, or by cultural background assumptions. Here, our point is that this obvious fact is not only a deep epistemological problem for the philosophy of mind, but that it can also figure in the scientific explanations of the formation of "sociocultural priors" (i.e., the emergence of *new* cultural background assumptions). Cultural beliefs about the nature of agency, such as "free will is an illusion" or "self-control is like a muscle," affect not only our experience of agency, they also impact our behavior (Job et al. 2010; Rigoni et al. 2013).

In addition to specifying the key components of agency, Epicurus believed that agency was the basis for moral responsibility (Bobzien 2006). Critical to this aspect of agency is the extension of the self across time: responsibility cannot be denied simply because an action was carried out in the past. Today, our beliefs about free will are intimately connected with the idea of responsibility (Nahmias et al. 2005). When behavior is caused by conscious states, people tend to judge that the agent acted freely. In contrast, when behavior is caused by unconscious states, people judge that the agent did not act freely (Shepherd 2012). We can only be held responsible for our actions if these have been chosen freely.

The concept of responsibility has a major role in Western legal systems. If we are capable of controlling our actions, then we are responsible for these actions. If, by reason of mental illness, for example, we are not capable of controlling our actions, then our responsibility is diminished. Young children and animals are also generally considered unable to exert control and are therefore not considered responsible for their actions. However, it is very difficult to judge when and to which extent they can control their actions and must take responsibility. Public views vary and have changed over time. On occasions, animals have been tried in court (Humphrey 2002), and the age at which children become legally responsible for their actions varies widely, even within present-day Europe (Hazel 2008).

What Use Is the Ability to Detect Agency? How Does It Influence Our Social Lives?

The importance of beliefs about agency for social cohesion has been explored in the laboratory. Experimental studies of economic exchanges show how easily cooperation within groups can be subverted by the appearance of free riders, people who benefit from the willingness of others to share resources, while not sharing themselves. Cooperation can be maintained by the introduction of sanctions through which free riding is punished (Fehr and Gächter 2002). Furthermore, people prefer to join institutions in which such sanctions are applied (Gürerk et al. 2006). Importantly, however, punishment is only applied when it is believed that free riders are acting deliberately of their own free will. Punishment (or reward for good behavior) was not applied to people believed to be behaving in accord with instructions given by the experimenter, even though the consequences of their behavior was no different (Singer et al. 2006). Here then is an experimental demonstration of a link between perceived responsibility, derived from the perception or belief of deliberate agency, and contingent social regulation. Furthermore, this responsibility is associated with identifiable individuals rather than acts. The experience of agency and responsibility can optimize personal-level interactions between individuals within groups.

Regret

Individual perception is critical for the human experience of regret: I would have done better, if I had chosen the other option. The experience of regret has several important implications for our understanding of consciousness, especially self-consciousness. First, the experience of regret implies an extension of the self across time: backward in time, because the action I am regretting happened in the past, as well as forward in time, because my anticipation of regret will affect my actions in the present (Filiz-Ozbay and Ozbay 2007). Second, the experience of regret emphasizes the importance of cultural factors for consciousness. It may have exactly been the emergence of the specific form of self-conscious suffering, which today we call "regret," that opened the door from biological to cultural evolution. Interestingly, feelings of regret are especially intense when the chosen action has flouted some cultural norm. This normative aspect of regret reminds us that, pre-Descartes, the concept of consciousness was to a large degree synonymous with the concept of conscience.

The Phenomenology of Regret

Regret is a form of suffering (Metzinger 2016). The first defining feature is that regretting something is a distinctly negative form of phenomenal experience, one that we will try to avoid and which we will try not to repeat or intensify (Reb and Connolly 2009). Second, regret is an *embodied* form of conscious experience: phenomenologically, it is predominantly an emotional experience (Gilovich and Medvec 1995) possessing aspects like despair (what has been done can never be changed), shame (one would like to conceal what one has done from the public or one's conspecifics), and guilt (one is acutely aware that one's past actions are immoral in the sense of having caused concrete suffering in others or having violated group interests). Very little is known about the physiological correlates, but the phenomenology of regret itself is frequently described as having interoceptive components (e.g., it can be *heart wrenching*). Third, it typically involves a cognitive aspect as well: a consciously experienced element of understanding or a sudden insight into the inadequacy of one's own past behavior. Fourth, the phenomenology of regret is always one of acutely enhanced self-awareness. In regret we experience ourselves as attaining a form of self-knowledge, which we previously did not have: we have done something morally wrong (or stupid) in the past, and we had the choice of doing otherwise. Interestingly, while the sense of agency is represented as something we possessed in the past, the state of regret itself does not itself involve a sense of agency. While the phenomenology of ownership is crisp and distinct (I *identify* with my regret, it is an integral part of *myself*), regret itself is not an action. It is a kind of inner pain that simply appears in us. This, therefore, is the fifth defining characteristic.

The phenomenology of regret can be described as a *loss of control* over our personal narrative, and in this sense it is also a threat to our integrity. It is a threat to the integrity of our autobiographical self-model, because, on the personal level of description, we become aware of an irrevocable damage to our life narrative. Because it is an emotional and frequently also an embodied experience, perhaps with heart-wrenching qualities, we cannot *distance* ourselves from it—another important way in which regret involves a loss of control. This is not only about our autobiography, but also about our current and future inner life. In this sense, the cognitive aspect mentioned above is "counterfactually rich": if I necessarily will regret what I have done for the rest of my life and if, therefore, I will try very hard never to have this experience again, then a very large number of possible and future states of myself are automatically affected.[1] Regret is something that can overshadow or "color" all other phenomenal experiences that a human being can have.

The Representational Content of Regret

In regret, we have a transparent self-model, whereby the system necessarily identifies with its content. First, this self-model portrays the organism as an *agent* in a strong libertarian sense. It can initiate and control actions, and it can deliberately choose an action on the basis of its desires and values. Second, if such desires and values are represented in the transparent part of the conscious self-model, then the organism necessarily *identifies* with these values and desires, leading to the distinct phenomenology of ownership sketched above. Third, many conscious "acts of deliberation" just appear in the conscious self-model, without any introspectively available precursors. That is, there are specific action-related representational states (e.g., dynamic, conscious goal-representations) which are portrayed as spontaneously occurring and subjectively experienced as *uncaused mental events*. Fourth, there is, therefore, a phenomenology of ultimate origin (free will) grounded in the self-model, depicting the organism as having a certain, crucial *ability*: the ability to initiate spontaneously new causal chains and thus to do otherwise. The individual self is represented in the brain as possessing a plurality of futures open to it, which are fully consistent with the past being just as it was. Fifth, there is a strong representational fiction of *sameness across time*. The agent, as consciously portrayed, possesses a sharp transtemporal identity, it is always the *same* entity that acted in the past, which acts now, and which will act in the future.

---

[1]   It is interesting to note how, phenomenologically, feelings of regret are highly "present": they are hard to suppress and continuously *re*-present themselves to the subject of experience. For the case of conscious perception, Seth (2014) proposed that "counterfactually rich" generative models encode sensorimotor contingencies related to repertoires of sensorimotor dependencies, with counterfactual richness determining the degree of perceptual presence associated with a stimulus. It is intriguing to extend his idea to the emotional layer of the self-model: the greater the counterfactual richness of an emotion, the greater its experiential degree of "presence."

Therefore, action consequences will always be attributed to one and the same entity: the fictional self is *responsible* for its actions.[2]

In addition, there is a novel, and much stronger representation of the social dimension: Other agents exist who have preferences too, which can be frustrated, for example, by actions for which one is responsible. These agents are also sentient, and they have the ability to suffer in many ways. In particular, a *group* exists, one's own group, and this group possesses interests and preferences as well. The group is not a sentient being, but it is a superordinate entity to which *preferences* can be attributed. There is a representation of group interests, which can be violated by individual agents, and of group preferences, which may stand in conflict with individual preferences and can accordingly be frustrated by individual actions.

In departing from theological and ancient philosophical models of regret, we propose that the representational content of regret may result from a failed integration of group preferences and individual preferences. Obviously, we also have the capacity to regret having been the cause of individual suffering, and it is often the case that the individual in question is identical to ourselves. Nevertheless, regret always has to do with conflicting sets of preferences and its representational content is inherently social. In essence, regret results from applying mechanisms of social control to oneself, namely, retribution (self-punishment) and reputation loss (self-blame). Societies are complex, self-modeling systems too, which self-regulate their activity via distributed control mechanisms that include many individual agents. Every good regulator of a social system must be a model of that system (Conant and Ashby 1970; Friston 2010; Seth 2015).

Importantly, for any organism that has acquired the capacity to feel regret and whose behavior is determined by this very special form of conscious content, the self-model and the group-model have become functionally integrated in a much stronger way. As soon as desires and values of the group are represented in the transparent part of the conscious self-model, the organism necessarily *identifies* with these values and desires (Metzinger 2003). This enables an organism to suffer *emotionally* from a self-caused frustration of group preferences. This further creates a permanent and never-ending source of conflict in its inner life. However, this source of conflict simultaneously acts as a strong source of motivation to strive continuously for social cohesion in one's own group. We believe that the conscious experience of regret marks out a critical transition in the internal dynamics of our model of reality: A functional platform for automatic self-punishment has been created. The

---

[2] The term "virtual identity formation" was introduced to refer to this process (Metzinger 2013:5) and it is speculated that one function of mind wandering is the constant creation and functional maintenance of the representation of a transtemporal, fictional "self." Only if an organism simulates itself as being *one and the same* across time will it be able to represent reward events or the achievement of goals as happening to the same entity, as a fulfillment of its *own* goals.

group-model has invaded the organism's self-model to such a degree that the conflict between group and individual interests is now *internally* modeled in a way that includes (a) *sanctions* by the group (regret is internal self-sanctioning) and (b) dynamic *competition* between group and individual interests, which takes place not only on the level of overt, bodily actions but also on the self-model of the individual. In this way, social interactions and group decisions are optimized.

The Causal Impact of Conscious Processing

Viewed in isolation, the conscious experience of agency seems to occur too late to have any causal role in the action with which it is associated. Nevertheless, experience in relation to action can now affect future choices of action, as with anticipated regret. Personal-level experience, therefore, does appear to have a role beyond an individual action. It affects cultural practices, such as moral codes and laws, and shapes the sense of self, by generating beliefs about self-control, thus giving rise to concepts such as responsibility, intentionality, accountability, culpability, and mitigating circumstances. These cultural beliefs are fed back to influence the behavior of the person. This suggestion raises the interesting possibility that the sense of agency and the idea of voluntary action are acquired through cultural learning. The causal link between the group level and the individual level is constituted by the conscious self-model, in which group preferences are increasingly reflected as social complexity increases.

Wolf Singer (pers. comm.) has made the interesting observation that were this cultural learning process to take place before the formation of autobiographical memory, it would appear as "*a priori*": agency and responsibility would appear as a simple, given property in the child's autobiographical self-model as it matures. Taking this point further, we could describe the experience of agency and responsibility as an "abstract prior," a stable hyperprior guiding the process of conscious self-modeling. We treat children as responsible by rewarding and by punishing them. They grow up embedded into a cultural practice of being *held* responsible. Accordingly, their self-model always predicts that they, themselves, will be held responsible, because their autobiographical narrative unfolds in a cognitive niche which assumes that they are in control of their actions and have the ability to do otherwise.

We have already mentioned the wide cultural variations in beliefs about the age at which responsibility should be assigned. There is also some evidence for variation in beliefs about the relevance of self-control and their effect in cultural practice. Among the Mopan Mayas of Central America, perpetrators of crimes are punished according to the degree of damage that they inflicted rather than the degree to which the act was committed intentionally (Danziger 2006). As a result, the defense "I didn't mean it!" is considered irrelevant, and therefore seldom attempted. In terms of legal preconditions of criminal guilt and liability to punishment, this culture has adopted a "consequentialist" (as opposed to a "deontological") approach to justice, in contrast to the test for mental competence and a "guilty mind" (*mens rea*) that is typically applied in "developed" societies. Most types of deontology hold that choices cannot be justified by their effects at all. No matter how morally good their consequences, some choices are morally forbidden, and what makes a choice the right choice is its conformity with a moral norm. Moral norms, very simply, are to be obeyed. This example again illustrates one of our main points: the phenomenal experience of agency becomes theory-contaminated by the way it is verbally described; different meta-ethical theories lead to different "sociocultural priors" that determine which action counts as a *good* action and which agent counts as a *moral* agent. Suprapersonal models of moral agency (those shared by a society) then exert a top-down, causal influence on personal-level behavior by shaping the self-model of individual group members.

Effects of Cultural Beliefs on the Experience and Control of Action

It is unknown whether the unusual beliefs about responsibility of the Mopan Mayas have had an impact on their personal experience of action or on empirical measures such as intentional binding. However, many experiments show how manipulation of beliefs about agency can alter behavior in the laboratory.

In these studies, some participants are presented with statements such as "most rational people now recognize that free will is an illusion" (Crick 1994), while others see statements that do not involve free will. Participants who are led to doubt the existence of free will show increased aggression and reduced helping behavior (Baumeister et al. 2009). They are also more likely to cheat in exams (Vohs and Schooler 2008). Effects can be observed even on more basic aspects of action. It is well established in reaction time tasks, where participants have to be as accurate and as fast as possible, that response times increase immediately after an error (for a review, see Dutilh et al. 2012). This post-error slowing is reduced in participants who have been led to doubt the existence of free will (Rigoni et al. 2013). Furthermore, the amplitude of the brain's readiness potential, measured with EEG, which precedes voluntary responses, is reduced (Rigoni et al. 2011).

Empirical studies of the effects of regret are still in their infancy. Regret can lead to ruminative thoughts and is associated with anxiety and depression (Roese et al. 2009). Furthermore, the experimental activation of regret can lead to delayed sleep onset and insomnia (Schmidt and Van der Linden 2013). It is not surprising, therefore, that we will take action to avoid regret (Reb and Connolly 2009). When we consider the options before us, we will factor in how much regret we anticipate feeling if any particular choice turns out to be suboptimal. This anticipated regret affects our choices (Filiz-Ozbay and Ozbay 2007).

The Connection between Consciousness
and the Evolution of Regret

We have discussed how various types of report about subjective experience can serve as data for developing an empirically constrained theory of consciousness and how such reports can be strongly "theory-contaminated." For many centuries, Western theories about regret had to do with purifying the inner life of the soul, with philosophical self-knowledge, and with man's relation to God. In the Greek philosophical and biblical tradition, important technical concepts were "compunction, "contrition," and "repentance." For example, the experience of regret could be something that leads a human being to a specific type of social action, called "confessing her sins." Here, by considering regret, we want to show how a fresh perspective of these concepts can be gained by connecting a data-driven (socio)biological approach with the more general question of what the central evolutionary functions of conscious experience might have been.

In the history of ideas, we find two main themes dominating theories of consciousness: *integration* (e.g., consciousness as a mental function that creates a union of the senses) and *higher-order moral knowledge* (inner knowledge about one's own bad actions and desires). Interestingly, the first semantic element has been strongly preserved in current research on consciousness (Metzinger 1995b; Tononi and Edelman 1998) whereas the second meaning of "conscious awareness" is almost completely absent.

In more than twenty centuries of Western theorizing on consciousness, an extremely interesting connection is found between phenomenal experience and moral cognition. The English word "conscience" is derived from the Latin *conscientia*, originally defined as jointly knowing, knowing together with or co-awareness, as well as consciousness and conscience. Here, the first point of interest is that throughout most of the history of philosophy, consciousness had a lot to do with conscience. Descartes was the first to separate conscience and consciousness and to constitute the modern concept of consciousness in the seventeenth century. Before modern times, being unconscious meant lacking a conscience. Even today, most people believe that moral considerations should only be applied to acts that are consciously intended (Shepherd 2012). The Latin term *conscientia,* in turn, stems from the Greek term *syneidesis*, which refers to moral conscience, co-awareness of one's own bad actions, inner consciousness, accompanying consciousness, joint knowledge, or disconcerting inner consciousness. Early thinkers were always also concerned with the *purity* of consciousness, with taking a normative stance, and especially with the existence of an inner witness. Democritus and Epicurus philosophized about inner torture associated with the bad conscience (Bobzien 2006) and Cicero formed the matchless term, *morderi conscientiae* (Hödl 1992): in English, the pangs of conscience (agenbite of inwit; Joyce 1922) or, in German, *Gewissensbisse.*

Even before Christian philosophy, the idea existed that conscience is a form of inner violence, a way to persistently hurt oneself.

In many early writings, consciousness as *conscientia* is part of the conscious person as an inner space, into which sensory perception cannot penetrate. It is an inner sanctum which contains hidden knowledge about one's own actions and private knowledge about the contents of one's own mind. Importantly, it is also a point of contact between the ideal and the actual person. In Christian philosophy, this contact is established by testifying or *bearing witness* to one's own sins. All of these concepts from early philosophy suddenly sound completely different when they are not read from the perspective of the later addition of the Christian metaphysics of guilt, but rather when they are read in a fresh and unbiased manner from the perspective of an evolutionary approach to consciousness.

A second interesting idea, found in many early philosophers, is that agents share their knowledge with an ideal observer, typically God. Never, however, was there a convincing argument for saying that this ideal observation is necessarily conducted by a person or one kind or another of individual self. Here, we propose that the "ideal observer"—which lies at the origin of moral cognition and moral behavior—is a *mental representation of group interests*. This is the emergence of a "first-person plural perspective" (Gallotti and Frith 2013). Self-consciousness served as a functional platform for the representation of group preferences in the brain of individual organisms. Upon this platform, individual and group interests could compete. The mechanisms which constitute self-consciousness are often subpersonal; the representational content is suprapersonal.

Consciousness as the Interface between the Person and Culture

Our actions and the brain systems through which they are implemented depend on a hierarchy of top-down control (Felleman and Van Essen 1991; Friston 2005; Koechlin and Summerfield 2007). This hierarchy of control, however, does not stop inside the person. In the examples given above, and, indeed, in most experiments, the highest level of top-down control comes from the instructions given to the participant by the experimenter (Roepstorff and Frith 2004) and, ultimately, from culture.

In many experiments, including those discussed above, instructions are designed to manipulate the beliefs of participants. For example, in economic games participants learn, by trial-and-error, that some of their partners can be trusted to make fair returns of the money invested in them, whereas other partners cannot be trusted. Participants also learn about the trustworthiness of information given by the experimenter (Delgado et al. 2005) or through gossip from other participants (Sommerfeld et al. 2007). Such information changes the participants' behavior, even though there is no actual difference in the behavior of their partners.

In these examples, acquiring a new model of reality (or in traditional Bayesian terms, a *belief* about the world)[3] causes changes in behavior, even when it is false. Interestingly, this can still count as an example of mental causation, because the representational content of the self-model accounts for the shift in behavioral profile, and also because conscious experience itself has the critical role of causally enabling the transfer of a model from one mind (the experimenter's) to another (the participant's). In other words, change in behavior is a causal consequence of shifts in the functional profile of the participant's phenomenal self-model brought about, in this case, by the instructions of the experimenter.

Mechanisms of Suprapersonal Top-Down Control

The learning process that occurs in trust games is nicely captured through a Bayesian mechanism. When we invest money in a partner, we can predict how much of our money will be returned on the basis of our degree of trust (a prior belief). If we get more than expected (positive feedback), our degree of trust increases. If it is less (negative feedback), our degree of trust decreases. However, if we are given prior information about trustworthiness, much greater weight is given to the prior information than to direct experience. This effect has been observed in terms of brain activity (Fouragnan et al. 2013) as well as behavior (Sommerfeld et al. 2007).

We suggest that beliefs arising from instructions, or from culture more generally, exert their effects by modifying prior expectations at the highest level of the personal hierarchy of control. Effects of these modifications demonstrably penetrate deeply into the hierarchy of control, affecting the monitoring of low-level cognitive processes (Rigoni et al. 2013) and associated brain activity (Rigoni et al. 2011).

A similar process might explain the effects of manipulating (or first installing) beliefs about free will. Our basic urge, we believe, is to be selfish, to gain advantages at the expense of others. This is one of those "abstract priors" that emerges through very early cultural learning. To overcome this urge we have to exert self-control (Metzinger 2015). Free will is necessary to exert such control (Nahmias et al. 2005). It is this intentional, top-down control that enables us to behave in a moral fashion. Without such top-down control, we might as well give in to our basic urges and gain all the (short-term) advantages that this

---

[3] It is important to note how the largest part of our model of reality cannot be adequately reconstructed as a set of beliefs (where, according to the standard definition, a belief would be the relation between a person and a proposition). Neural representations in human brains do not come in a propositional format, as they do not have the necessary properties of systematicity and productivity—the information expressed by a Bayesian model in the biological brain is a *subsymbolic* representation of probability distributions.

might bring. This leads to an increase in cheating and general antisocial behavior. Ironically, telling people that there is no free will alters their very behavior, thus providing another example of mental causation and the effective role of conscious self-representation.

Sharing Experiences

We have discussed how instructions and culture influence the person, but there is, of course, traffic in both directions (Sperber 1996). The explicit metacognitive mechanisms that enable us to be influenced by the ideas of others also allow us to influence them. This permits control, not just at the personal level, but also at the suprapersonal level (Shea et al. 2014).

In the choice blindness paradigm discussed above (Johansson et al. 2005), participants are easily persuaded to accept that they have made a different decision from the one they actually made. This phenomenon is part of a larger set of examples showing that we have remarkably poor access to the mental processes underlying our behavior (Nisbett and Wilson 1977). In spite of such meager knowledge, people are more than happy to talk about and justify the decisions they have just made.

Although the conscious experience of agency may have little causal role in the action with which it is associated, the experience will be very relevant to any attempt to justify the action after it has been made. We would be able to claim, for example, that our action was accidental rather than deliberate. By justifying our actions and discussing with others why we do things, a consensus is built about the mental basis of action. Whether or not this is a *true* account of the mental processes, such consensus is likely to be an important basis for cultural norms about responsibility. Thus, consciousness of action enables us to develop a folk psychology critical for the regulation of social behavior (McGeer 2007).

We not only tell each other about our experiences of action, we also share our perceptual experiences. In a series of experiments, Bahrami et al. (2010) have shown how such discussion can create group advantages. In these studies, two people jointly perform psychophysical signal detection tasks. After giving individual reports about the presence of a signal, disagreements are resolved by discussion, leading to a joint decision. If the abilities of the partners are roughly equal, then the joint decision is consistently better than that of the more skillful person working alone. Discussion is crucial for optimizing this group advantage and requires that the partners talk about their confidence in their experience of the signal (Bang et al. 2014). Through such discussion they develop a verbal scale for rating their levels of confidence. Group advantage depends on the development of such a scale (Fusaroli et al. 2012).

We suggest that these group advantages, which depend on the experience of and ability to report confidence in a perception, constitutes another case where consciousness has an important and possibly necessary function. Transparency is the phenomenological equivalent of maximal confidence. In this case, the explicit report of confidence enables optimization of joint decisions. It remains to be seen if elegant new paradigms can be created to show that even these aspects of our mental lives may occur without conscious awareness. For example, do some phenomena associated with hypnosis (e.g. Smith et al. 2013) indicate that instructions can have their effects without awareness?

We conclude that, in relation to both action and perception, a particular kind of self-consciousness arises at the point in the hierarchy of control where the person interacts with other minds. This is the level at which instructions work. At this interface, between the person and culture, there is two-bidirectional traffic (Sperber 1996), such that the person can be influenced by other minds and the person can in turn influence others. So, what use does consciousness fulfill? We propose that at least one kind of consciousness functions to enable explicit communication about subjective experience. This, in turn, causally influences behavior and enables the growth of cumulative culture. This growth is dependent on the development of norms about appropriate behavior. This kind of consciousness creates the social cohesion and cumulative culture that has proved such an immense advantage to humans.

Regret and Regret Prediction: The Argument from Transparency and Modal Competence

Regret is a very specific kind of representational content: it carries information that a biological organism can utilize to optimize future decisions and enables group preferences and norms to have a direct influence on the behavior of individuals. Returning to the question posed in the introduction, what is it that, in our world, a zombie could never do? In our world, a maximally similar but unconscious creature could never be a true functional isomorph. Why? Because it would lack the representation of "realness," and thus it could not compare real and counterfactual states of self and world, and because it would not possess the enormous motivational force that comes from identifying with the contents of one's self-model.

Regret carries self-related information, which often refers to specific *social facts*. The evolutionary advantage of representing this information under the very specific, neurally realized data format of a transparent, egocentric model of reality, as described above and elsewhere (e.g., Metzinger 2003, 2009) is that it forces a biological organism to:

1. Experience the relevant kind of fact about the world as irrevocably *real* (e.g., damage to the interests of its own group, or itself, has been done, the organism itself was the cause of this damage, and it could have done otherwise). Let us call this the "principle of phenomenally transparent representation."

2. Identify with this damage by integrating it with its internal self-representation. We could call this the "principle of transparent self-modeling."

Regret is a particularly powerful form of conscious experience, because it represents the group's interests *in* the individual's transparent self-model, thereby
creating a new form of suffering from which the organism cannot distance
itself—for the simple reason that the relevant form of representational content
has now been functionally integrated with an internal representation of *itself*.
The sense of agency is the decisive causal prerequisite, because it introduces
the phenomenal experience of "I *could* have done otherwise!" (whether true or
not) into the self-model.

Let us define "modal competence" as the ability to represent mentally the
operators of modal logic and their function: □ (It is *necessary* that...) and ◊
(It is *possible* that...), but also F (in deontic logic, It is *forbidden* that...) or P
(in temporal logic, It *was the case* that...). Modal competence is a naturally
evolved form of intelligence, which comes in many degrees. In our context,
the mental ability to represent successfully some things as *possible* and other
things as *real* (i.e., as actual facts) is of highest importance. If a biological organism is to develop higher forms of intelligence like episodic memory, future
planning, or counterfactual reasoning, it needs a simple form of modal competence. To develop these forms of intelligence, it needs a functional mechanism
that reliably distinguishes between what is real and what is only possible or
what happened in the past; for example, the animal must avoid episodic memories from turning into hallucinations and manifest daydreams, or, as in future
planning, it must find an optimal trajectory from a model of the world reliably
marked out as "given" into a second model of the world portrayed as "possible
and desirable." Only conscious representation has this remarkable functional
property and, on the level of self-representation, it is exactly this property that
causally enables the phenomenal experience of "I *could* have done otherwise!"

Under the Bayesian predictive coding framework, we assume unconscious
inferential processes which lead to a continuous, dynamic representation of
probability distributions (Friston 2010). Only conscious experience, however,
can represent something as *real* and as taking place *now* (Metzinger 2003;
Lamme 2015a, b; Melloni 2015), and only self-consciousness provides a singular unit of identification. There could be unconscious models of the organism as a whole, of individual and group preferences, and so on, and they could
certainly be characterized by a high degree of Bayes optimality. But only misrepresenting the probability of a hypothesis as 1.0 and simultaneously flagging
it as a fact holding *now* via a window of presence turns a possibility (or a likelihood) into a reality. This is what makes the zombie conscious. The argument
from transparency is that conscious experience must be exactly the functional
mechanism that "glosses over" subpersonal Bayesian processes by assigning
"realness" to them—that is, by misrepresenting them as exemplifying an *absolutely* maximal likelihood or maximum posteriori probability. It is this step

that turns a process into a thing, a dynamical model into an internal reality, and a self-model into a self.

Therefore, it is only conscious experience that enables suffering and the enormous motivational force that comes with representing something as an irrevocable and untranscendable fact and at the same time as a threat to one's *own* integrity. We believe that it is the conscious self-model that causally integrates the continuous, low-level biological process of sustaining the organism's existence with a specific dynamic representation of the system, namely, a generative self-model that continuously strives to find evidence for the system's very existence (Hohwy 2014; Friston 2013). If this internal self-model has the capacity to integrate social facts (e.g., the frustration of group preferences) then it creates a new biological phenomenon: the causal integration of the individual's striving for self-sustainment with the group's need for cohesion and stability. This is a culturally shaped form of self-consciousness, linked with the idea of identity (see Kyselo 2014). It enables new types of actions aimed at the satisfaction of group preferences, because it makes a new set of facts globally available for introspective attention, verbal communication, and behavioral self-control.

At the outset, we also asked: Which class of optimization problems does consciousness enable us to solve? A well-known neuroscientific concept is "reward prediction" (Hollerman and Schultz 1998; Schultz and Dickinson 2000; Tobler et al. 2006). We want to point out that in complex biological nervous systems the opposite capacity might also exist, and we dub it "regret prediction." If a system has the capacity to distinguish between its own actual and possible future states, then it could also begin predicting future regret (Filiz-Ozbay and Ozbay 2007; Coricelli et al. 2005). It could simulate future states of the self-model that resemble the current one. If it has a self-model that misrepresents it as possessing a precise transtemporal identity, then it will also represent such future regret events as potentially happening to the *same* biological system, to itself. The prediction of future suffering of the kind we have sketched in this chapter allows for the comparison of future states with present states, and opens the possibility of seeking trajectories into more desirable situations. We believe that this new biological capacity—regret minimization—will dramatically have increased the motivational force behind prosocial behavior. The search for one's own coherence turns into the search for group coherence.

The experience of regret is intimately associated with the experience of agency: the experience that I did it and that I could have done otherwise. In closing, we wish to draw the reader's attention to a specific logical possibility. Ultimately, regret, like the experience of being an agent, may be a form of self-deception, a naturally evolved, but functionally adequate form of misrepresenting reality. Exactly this form of "theory-contaminated self-deception" may have provided a mechanism for cultural evolution and the sustaining of social cohesion, therefore providing advantages for the group as a whole (von

Hippel and Trivers 2011; Trivers 2011). Kierkegaard (1843/1992) made a similar point in *Either/Or*: "The deceived is wiser than one not deceived."

Ackowledgments

CDF acknowledges support from the Wellcome Trust and Aarhus University. We are grateful for comments from Cecilia Heyes, Holk Cruse, and Marek McGann.

13

Pragmatism and the Pragmatic Turn in Cognitive Science

Richard Menary

Abstract

This chapter examines the pragmatist approach to cognition and experience and provides some of the conceptual background to the "pragmatic turn" currently underway in cognitive science. Classical pragmatists wrote extensively on cognition from a naturalistic perspective, and many of their views are compatible with contemporary pragmatist approaches such as enactivist, extended, and embodied Bayesian approaches to cognition. Three principles of a pragmatic approach to cognition frame the discussion: First, thinking is structured by the interaction of an organism with its environment. Second, cognition develops via exploratory inference, which remains a core cognitive ability throughout the life cycle. Finally, inquiry/problem solving begins with genuinely irritating doubts that arise in a situation and is carried out by exploratory inference.

Introduction

This chapter examines the conceptual background to "the pragmatic turn," particularly by articulating some of the central principles of a pragmatist approach to cognition and experience and showing how they are relevant to contemporary cognitive science (and its pragmatic turn). The main theme of a pragmatic cognitive science is that "cognition is for action." The "pragmatic turn" provides a framework for understanding cognition that is distinctively different from a traditional framework that takes cognition to be structured by computations on rich representational contents (Engel et al. 2013). I propose that the new pragmatic turn in cognitive science is conceptually grounded in the work of the classical pragmatists, particularly Charles Sanders Peirce and John Dewey.

Engel et al. (2013:202) have formulated the pragmatic turn as follows:

> In cognitive science, we are currently witnessing a "pragmatic turn," away from the traditional representation-centered framework toward a paradigm that

focuses on understanding cognition as "enactive," as skilful activity that involves ongoing interaction with the external world.

The pragmatic turn which they envisaged is largely a matter of sensorimotor interactions. While there is a good case for arguing that cognition is grounded in such interactions, it is not obvious that all of cognition can be exhaustively described in such terms. One of my aims in this chapter is to show that pragmatists do not have to rely solely on a behavioral account of sensorimotor interactions with the environment. Peirce, for example, gave a detailed account of representation, or sign action as he preferred to call it, and did not think of these as exclusively action-oriented in Clark's sense (Clark 1998).

To explain how we think by interacting with the environment, classical pragmatists develop this idea by framing the nature of the interactions in terms of exploratory inferences. The child develops cognitively by actively exploring the local environment (usually under supervision), often learning about phenomena by physically manipulating them. Thus, the primary argument in this chapter is to refocus the pragmatic turn into the kinds of exploratory and inferential actions that are core to the pragmatist project and to deny that this constitutes a slide into neobehaviorism.

I structure the discussion around three principles of a pragmatist approach to cognition, derived from the work of the classical pragmatists and contemporary pragmatist approaches to cognition:

1. Thinking is structured by the interaction of an organism with its environment.
2. Cognition develops via exploratory inference, which remains a core cognitive ability throughout the life cycle.
3. Inquiry/problem solving begins with genuinely irritating doubts that arise in a situation and is carried out by exploratory inference.

I begin with a discussion of the difference between a pragmatist and internalist approach to cognition and include a discussion of a Peircean approach to sign action. Thereafter, I provide an account of the first principle in terms of sensorimotor contingencies and Dewey's organism-environment relations. Finally, I introduce the second and third principles and show how they are related to contemporary work on active inference in the predictive coding framework, including a discussion of the relationship between Peirce's conception of abduction and Bayesian approaches to inference.

Pragmatism and Internalism

Thinking is an ongoing interactive process between the organism and the environment, where that interaction is at least partially constitutive of our thought processes. This is the key difference between pragmatists, who think that cognition is interactive, and internalists, who think that internal systems are sandwiched between environmental inputs and outputs. Internalists take thinking to be constituted only by processes which occur in an inner system with defined boundaries, where interaction with the environment is defined in terms of inputs to and outputs from the system. At least since Dewey (writing at the end of the nineteenth century), pragmatists have rejected an input-output picture of the mind.[1] Consequently, a genuinely pragmatic turn in cognitive science would see an end to the model of the mind as a system bounded at the periphery by environmental inputs and outputs and see a turn toward empirical studies of how we think by interacting with the world.

It is of real importance that the pragmatic turn gives the right framework for understanding the interactive nature of thought. Pluralism about styles of interaction is very much in the spirit of the classical pragmatists. So we should be wary of moving from one dominant view of cognition as only being about internal computations on informational states, to another which treats cognition as only a matter of sensorimotor contingencies (enactivism), or that all cognition is aimed at predicting (inferring) sensory inputs from an environment that is external and never directly experienced (predictive coding). Sensorimotor contingencies and predictive inferences will no doubt be core methods for understanding cognition as interactive,[2] but if there are differences in the interactions (call these styles of interaction), then it is likely to follow that there will be differences in how we model or explain those different styles.

In addition, pragmatists and internalists differ on the role and importance of representations in cognition. Pragmatists are usually aligned with enactivism and embodied cognition in reducing the importance of representational explanations of cognition, certainly in terms of internal representational states that cause behavior. This is often portrayed as a fundamental difference between whether cognition is primarily action oriented or concerns the acquisition of information and/or knowledge about the environment so as to be able to act upon it. Some pragmatists, however, deny that there has to be a complete break with the role of representation. For example, Clark (1998, 2008, 2013b) is a contemporary exponent of the pragmatic turn, who nevertheless embraces representation. In classical pragmatism, the great polymath and *enfant terrible* of classical pragmatism, Peirce, made the most significant strides toward a comprehensive theory of signs: from simple signaling systems in animals to complex linguistic interactions in humans. I briefly outline some of his central ideas here (with some modernization for a contemporary audience) to illustrate how pragmatists may maintain their pragmatic credentials while embracing the

---

[1] This is the sandwich model of the mind as described by Hurley (1998). I am referring to Dewey's famous (and now neglected) paper "the intentional arc," discussed further below.

[2] A residual problem exists concerning the relationship between an interactive model of cognition and a predictive processing one where external states of the environment are hidden from the internal states that make predictions on the basis of sensory inputs from the environment, especially if one of the leading ideas of the pragmatic turn is to reject the internalist input-output model. This issue will be addressed later in the chapter.

importance of signs in the mental lives of animals, even those strange primates *Homo sapiens*.

Peirce and Sign Action

Peirce is famous for his work on sign action and its relation to cognition. His work has been largely underexplored by contemporary cognitive science, with a few honorable exceptions (e.g., Barbara von Eckardt and Bill Ramsey). I have used a version of Peirce in several publications (Menary 2007, 2009, 2013). The primary difference between Peirce and most of the theories of representational content that have been formulated over the last forty years is that Peirce does not have a simple "vehicle as carrier of information" model of mental content, nor does he think that signs[3] (or representations) stand in a simple dyadic relationship to an object. His mature view is that signs develop in a process of continuous dynamical interpretation.[4]

What is essential to the sign relation? Gallie (1952:120) nicely illustrates the three essential features of sign action: "(i) A sign stands for (ii) an object by (iii) evoking some further sign of the original object." This conception of sign action fits with contemporary pragmatic cognition, precisely because it is a dynamical process: the evoked sign may determine a further sign and so on until some natural terminus is reached. A triadic relationship between sign, consumer, and object must be met for sign action to occur.

Peirce's system can be formulated in a modern context in terms of what I have elsewhere called the Peircean Principle (Menary 2007, 2009). The Peircean Principle maintains that any sign (including representations and intentionally directed traits) must involve the following components:

- The sign-vehicle has certain intrinsic or relational properties that make it salient to a consumer. It might have iconic properties, it might be indexical (causally connected), or it might be symbolic and subject to public convention. In more complex cases the sign may have combinations of these properties.
- The sign is exploited by a consumer in virtue of its salient properties, thereby establishing its function of signification (i.e., the function of signifying an object/environmental property, in the simplest case).

---

[3] I will stick with "signs" throughout the discussion, as "representations" imply determinate contents with truth or accuracy conditions. Signals and signs do not imply content and truth; they can be iconic in character, and thus quite indeterminate, and not imply any conditions for truth (since they make no determinate "statement" about the world). However, the conditions for when something can act as a sign are determinate and these are the conditions upon which I shall focus, leaving aside the vexed issue of content determination.

[4] Interpretation does not have to be thought of as requiring a mind with the ability to interpret a sign conceptually. Peirce uses the neologism "interpretant" to distinguish between minds that intentionally interpret signs and nonintentional consumers of signs (e.g., immediate behavioral responses or evolved producer-consumer systems of the kind familiar from Ruth Millikan's work).

However, an animal's vivid coloring may be a sign of its dangerous properties and may evoke a behavioral response).

- The triad of sign, consumer, and object is established only when the function of signification is recruited for some further end (e.g., food detection).

The recruitment of the sign in virtue of its function is established as a repeatable pattern. Millikan (1984, 1993) shows how such repeatable signs are established as proper biological functions, but the repeatability might very well be established by conventional means: as a social/public norm. Representations established biologically are teleonomic signs, and those established by convention are teleological signs (Liszka 1996). The conditions can be unpacked in the following way:

1. A token vehicle Φ is a sign when it has properties that can potentially be exploited by a consumer. For example: Φ is salient because it is reliably correlated with an object or environmental property X, or with objects/environmental properties X, Y, Z....

2. Φ functions as a sign when its salient features are exploited by some consumer Ψ. For example: Φ has the function of signifying X for consumer Ψ, because Φ is reliably correlated with an object/environmental property X.

3. Φ is a sign of X for consumer Ψ in the performance of some biological function (or for some conventional norm).

The conditions for sign action are simple: a vehicle has properties that are potentially exploitable by a consumer; these are its salient properties. It is consumed in virtue of its salient properties. However, for the repeatability of this sign triad, we need the coordination of producer and consumer mechanisms: a sign is produced which is consumed for some further end. This process is established as a teleonomic sign if it is adaptively successful. Consider the following example:

During cricket phonotaxis, the female cricket can locate and move toward the location of male cricket songs (Webb 1994). She does so through the activation of two dedicated interneurons, each of which is connected to one of her ears. The strongest activation determines the direction in which she will fly. If the interneuron connected to her left ear is more strongly activated than that connected to her right ear, then she flies to the left. There is an exquisite coupling between the iterated song of the male cricket and the activation threshold and decay time of the female's dedicated interneurons. Intuitively, we might think that the male's song functions as a sign that the female cricket consumes. The triangulation is produced by the coordination of the male cricket song with the female cricket interneuron activation and decay patterns. The male cricket song is the vehicle with salient properties, and the interneuron is the consumer of those properties, establishing its sign function. The iterated song is exploited by a consumer that recruits the sign to the production of some end, in this case directing the female toward the male to mate.

The very same conditions for sign action are the basis for teleological signs, and repeatability requires the coordination of producer and consumer. However, the process is established as a teleological norm by being part of a conventional system such as language or mathematics.

For our current purposes, the Peircean Principle demonstrates the complexity involved in establishing a sign/intentional/representational relation. It requires the coordination of producer and consumer mechanisms for some further end; therefore it requires either the coordination of mechanisms within the organism or the coordination of a mechanism in the organism with a mechanism in that organism's environment.

The Peircean Principle is valuable because it allows us to explain how sign action works in both natural and social environments, by giving the same structural conditions for teleonomic and teleological signs. It makes no commitment as to whether sign action must be internal, external, or distributed across brain, body, and world. It allows for all three possibilities.

Teleonomic signs and teleological signs should not be thought of as distinct categories. They are continuous with one another. There is a difference in that teleological signs will tend to be more flexible and open-ended in their range of interpretation. For example, teleological signs are subject to growth and development of meaning across populations of sign users and over time. However, their sign function is still established by the coordination of consumer and producer: a sign is produced which is consumed for some further end. Consequently, the Peircean Principle is an account of sign action that is based in a principle of continuity not discontinuity.

Now think of teleological sign use in the context of an inquiry or a problem-solving task, such as solving a mathematical problem using pen and paper. There are capacities for the creation and manipulation of external signs in the context of problem solving or inquiry. The act of problem solving takes on the sign-action cycle. Teleonomic sign action and teleological sign action are located on a continuous line of evolution and appear in the developmental trajectories of children. It is clear that pragmatic cognition will need to make use of both types of sign action.

Thinking Is Structured by the Interaction of an Organism with its Environment

How should we understand the relation between an organism and its environment? In biological terms, following Godfrey-Smith, we can give a symmetric or asymmetric account. The asymmetric externalist explanation of the relation between the organism and its environment denies that there is any significant level of feedback from the organic system onto its environment (Godfrey-Smith 1996:327).

By contrast Dewey (1929/1958) provided a symmetric externalist explanation which he called organism-environment transactions. In his epistemological work, Dewey denied the strict separation between mind and world, but this denial arose from his views on organism-environment transactions. These transactions allowed Dewey to bridge the gap between organism and environment because, although adaptations of organisms have evolved through the familiar process of natural selection, a strong distinction between the organism and its environment need not be made, as in the case of extended phenotypes (Dawkins 1982).

Godfrey-Smith identified a similarity between Dewey's view (1929/1958) and Lewontin's (1982, 1983): they both recognized a two-way interaction between organism and environment. Rather than the organism merely being the "passive" object of environmental selection pressures, the organism also reshapes its environment, thereby altering the "future course of the selection pressures to which they will have to respond" (Godfrey-Smith 1996:327). The organism and its environment are reciprocally coupled: the organism does not just passively reflect its environment; it affects that environment, through its responses and behaviors. Selective pressures upon the organism and environmental niche are built up by their reciprocal coupling, such that they coevolve as a single system.

The organism-environment system is a biological basis for cognitive science's turn to interaction. What would the interactive approach look like in a cognitive context?

Dewey and Sensorimotor Theories

Dewey's influential paper, "The Reflex Arc Concept in Psychology," was published in 1896, but it heralds many of the arguments made by the contemporary 4E movement (embedded, extended, embodied, enactive) in philosophy and cognitive science. Dewey begins by arguing that the existing model of the reflex arc, involving stimulus–nervous system–behavior, introduces a dualism between environmental input/output and central functions of the nervous system; this replaces the Cartesian dualism between mind and body with one between environmental inputs and outputs and neural functions. The distinction between the environment and the central nervous system reintroduces the problem of how to put mind, body, and world back together again. This model is anathema to Dewey's conception of organism-environment systems: "The idea of environment is a necessity to the idea of organisms, and with the conception of environment comes the impossibility of considering psychical life as an individual, isolated thing developing in a vacuum" (Dewey 2008:56).

Dewey's conception of, what we would call, the cognitive system is one that does not involve a self-contained inner system that processes inputs and produces outputs. His conception of the reflex arc is one that involves an "organic unity" rather than "disjointed parts" (Dewey 1981:97): [S]ensory stimulus, central connections and motor responses shall be viewed...as divisions of labor, functioning factors, within the single concrete whole, now designated the reflex arc." This "unity" of functions is best understood in terms of what Dewey terms "sensorimotor coordination" (Dewey 1981:97):

> Upon analysis, we find that we begin not with a sensory stimulus, but with a sensorimotor coordination, the optical-ocular, and that in a certain sense it is the movement of body, head and eye muscles determining the quality of what is experienced. In other words, the real beginning is with the act of seeing; it is looking, and not a Sensation of light.

It is remarkable that Dewey wrote this in 1896, for it captures much of the spirit of the contemporary sensorimotor contingency approach to perception and conscious experience (O'Regan and Noë 2001; Noë 2004; O'Regan 2011).

Conscious experience begins with a primary sense of embodied agency. The sensorimotor capacities of an infant are developing right from birth. Hence, it should come as no surprise that infants develop an exploratory and open-ended method of interacting with their environment. The interactive nature of cognition grows directly out of these sensorimotor explorations of the local environment.

Experiences are these interactions of organism and environment. Dewey's account of experience has strong affinities with recent developments in externalist accounts of cognition; particularly distributed and extended accounts of cognition (Hutchins 1995; Clark 2008) and enactive or sensorimotor accounts of perception (Noë 2004; O'Regan and Noë 2001). The importance of Dewey's and Peirce's work for embodied and extended approaches to cognition has recently been made explicit by some of those working in the field. For example, Gallagher (2009, 2014b) and Menary (2007) both discuss the importance of Dewey's conception of organism-environment relations as a grounding for contemporary discussions of embodied and extended mind. One important commonality between the pragmatic approach and the sensorimotor approach is that exploratory activity results in habits of action; or sensorimotor contingency. Rather than encoding representations of the environment which are then processed computationally, the system is set up to explore and sample the environment in the service of action.

Cognition Develops through Exploratory Inference

In this section I explore some of the key concepts of classical pragmatism and their importance for the pragmatic turn. The first of these is the idea of fallible cognitive agents who actively explore their environments. Thereafter

I examine this form of exploratory inference in relation to Bayesian accounts of cognition.

An embodied form of active or exploratory inference is evident early on in ontogeny. Exploratory inference and Bayesian or predictive accounts of active inference (often described as exploratory) are quite complementary. Indeed, fallible exploratory inference can help us to conceptually frame the role of active inference in models of predictive processing. In particular, a Peircean account of abduction supplements active inference. Finally I argue for a more externalist interpretation of the role of active inference, using the pragmatist conception of active inference as a springboard and thereby avoiding an internalist view of the Bayesian brain.

Fallibilism and Exploratory Inference

As a pragmatist philosopher and scientist, Peirce proposed that thinking is a form of self-corrective practice, much like that used by the experimental sciences. He provides an analysis of cognition that is thoroughly fallibilist (i.e., involving an epistemic agent who is capable of error and learning from error): humans begin their cognitive lives by fallibly exploring our environments. Through direct physical interactions with objects, we begin the process of gaining knowledge and an understanding of how to act effectively in our local, developmental, environment.[5] Fallible cognitive agents learn from their mistakes. Our early and formative explorations are corrected and constantly updated in real-time and over developmental time.

A consequence of Peirce's foundational work was to make fallible and active (exploratory) inference the fundamental form of inference in cognitive systems. Exploratory inference is key to pragmatist approaches to the mind and cognition. In addition, pragmatists take the development of cognition seriously. The pragmatist vision of the child is of a fallible agent who develops as a cognitive agent through exploratory interactions with the developmental environment. Consequently, the developing cognitive agent is open to the world through self-corrective actions (or practices). Cognition is thoroughly fallible, exploratory, open-ended, diachronic, and open to the local environment.

The Fallible Infant

As infants begin to explore their surroundings and interact with caregivers, their responses become more flexible and open-ended. A range of vocalizations,

---

[5] The classic articles where Peirce first begins to develop the fundamentals of pragmatism are: "Questions Concerning Certain Faculties Claimed for Man" (CP 5.213), "Some Consequences of Four Incapacities" (CP 5.264) and "The Fixation of Belief" (CP 5.358). All of his papers can be found in the Collected Papers of C. S. Peirce (1931), an eight volume compendium published by Belknap Press. In referring to his papers, I follow the convention of citing the paper by volume and paragraph numbers (e.g., CP 5.213).

gestures, and actions become generalized and habitual. In line with contemporary developmental psychology, pragmatists think that the infant begins to develop "biases and preferences" as well as a basic capacity to "form and test hypotheses" (Stern 1985:41–42; Gopnik et al. 1999). Hypotheses do not need to be conceived as propositions that need to be interpreted and tested; they might be entirely spatial or action based, with the testing itself an action or exploration. We can be pluralists about the nature of exploratory inference and hypothesis testing. The pragmatic view of development is one in which learning occurs in a richly structured niche, which produces regularities in the sensorimotor interactions of the neonate.

We might think about these early explorations as a way of adapting to the environment. Peirce, for example, held the view that the regularity of the infant's habitual activity parallels the regularity of the environment. This can be formulated in terms of a predictive coding account of perceptual learning: "Put simply, sustained exposure to environmental inputs causes the internal structure of the brain to recapitulate the causal structure of those inputs. In turn, this enables efficient perceptual inference" (Friston and Stephan 2007:433).

The learning environment should, therefore, be statistically regular with repeated gestures, sounds, facial expressions as well as stable objects, artifacts, and signs or representations. In Bayesian terms it facilitates the development of priors and stable predictions. "In summary, the free energy principle can be motivated, quite simply, by noting that systems that minimize their free-energy respond to environmental changes adaptively" (Friston and Stephan 2007:428).

Even so, the environment is not wholly predictable and stable; exploratory behavior is still required. Consequently, pragmatism offers only a very minimal nativism about cognitive systems with powerful learning mechanisms for interactively exploring the physical and social environment, rather than innate hierarchically organized modules with domain specific knowledge. The cognitive capacities of the fallible neonate include: associative inferences, imitation, and causal or Bayesian learning.

In evolutionary terms, pragmatism does not align well with evolutionary psychology in its standard formulation (Barkow et al. 1992). It aligns much better with a niche construction account of developmental biases that are both endogenous and exogenous (Menary 2014; Odling-Smee et al. 2003). The Deweyean organism-environment system sits well in the niche constructionist account of development, since organisms reciprocally influence their environmental niches. "The evolutionary significance of niche construction hangs primarily on the feedback it generates. Many organisms modify their own selection pressures, so that environment-altering traits coevolve with traits whose fitness depends on alterable sources of natural selection in environments" (Laland et al. 2000:134).

We are born with a high degree of developmental plasticity, and our brains exhibit high degrees of plasticity. It is evident that "development of neural circuits in the visual system and acquisition of visuomotor skills critically depend on sensorimotor interactions and active exploration of the environment (Held and Hein 1963). Even in the adult brain, there is considerable plasticity in cortical maps (e.g., in the somatosensory and motor system) that has been shown to depend on action context (Blake et al. 2002)" (Maye and Engel, this volume).

Modern humans are born into a highly structured cognitive niche that contains not only physical artifacts, but also representational systems that embody knowledge (writing systems, number systems, etc.) as well as skills and methods for training and teaching new skills (Menary and Kirchhoff 2014). Knowledge systems, skills, and practices are real and stable features of the sociocultural environment. Their early cognitive development is a process of exploring this environment. As their interactions become more stable and regular they begin to produce principled epistemic patterns of action, incorporating the skills, practices, and forms of knowledge that structure the niche in which they develop. Consequently, plastic brains that can learn in structured developmental niches are prerequisites for fallible agents.

Exploratory Inference, Abduction, and Bayesian Approaches to Cognition

The pragmatist conception of cognition as an exploratory and interactive form of active inference requires some kind of account. Is it inductive or deductive? According to Peirce it is abductive: "[a]bduction is the process of forming explanatory hypotheses. It is the only logical operation which introduces any new idea" (CP 5.172). It is important to be clear that abduction, in this sense, is not exactly the same thing as inference to the best explanation (IBE). It is a form of creative abduction (Schurz 2008); it introduces new concepts, models, and hypotheses to test. This is because abduction is part of the process of discovery and not of justification alone: we abductively generate hypotheses to test. However IBE is important for determining which hypotheses are most likely to be true; this is a form of selective abduction (Schurz 2008). We can see how this might be the case in the following two schemas. Abduction as IBE follows the schema:

> Given evidence E and hypothesis H1 and H2, infer the hypothesis which best explains E.

Criteria are then needed to decide what the "best" explanation is. Peirce's version of abduction provides a hint as to the nature of the criterion (CP 5.189):

Given the surprising evidence E,

If H1 were true then E would be a matter of course (explained by some systematic principle).

Therefore H1 is probably true.

A probabilistic version of Peirce's abductive schema can be given in the following way (Osei-Bryson and Ngwenyama 2011:412):

> If hypothesis H explains the evidence E better that H1 and H2 combined, H is the more general hypothesis. We can now satisfy the requirements for assessing the posterior probability of our hypotheses by defining the following constraint: IBE 1.2: Assuming that $P(H) > 0$ and $P(E) < 1$, if H entails E, then $P(H/E) > P(H)$.

Peirce introduced abductive inference as the third type of inference (other than inductive and deductive). Clearly, abductive inference, as a species of ampliative inference, is part of the logic of discovery in the philosophy of science. However, Peirce also thought of abductive inferences as part of everyday life and not solely at a personal level. So if we think of abduction as both creative and selective, then we can conclude that abductive inferences have a strategic as well as a justificatory function (Schurz 2008). They produce "promising conjectures" (Schurz 2008:203) that call for testing by experience and, as Hintikka (1998) puts it: "stimulate new questions." Schurz (2008:205), usefully encapsulates the abductive pattern of inference as "the crucial function of a pattern of abduction...consists in its function as a search strategy which leads us, for a given kind of scenario, in a reasonable time to a most promising explanatory conjecture which is then subject to further test."

Here I focus primarily on the selective role of abduction. Peirce hypothesizes that these patterns of abduction are instinctual (CP 5.47, fn. 12; 5.172; 5.212). I understand this claim as involving a minimal developmental bias for exploratory behavior, rather than as an innate module. Cognition is shot through with active exploratory inference that is abductive—a pattern of action that is a search strategy for a conjecture that can be further tested. As noted before, abductive inference can be an explicit hypothesis generation and test involving beliefs or theoretical posits; however, in early developmental, and at least some sensorimotor cases, the conjecture and test may be based on motor activity rather than on beliefs or representations. Therefore, it is possible to give a nonrepresentational account of active inference, and this would be entirely consistent with the likely evolutionary origin of those inferences in sensorimotor interactions with the environment. This interpretation is consistent with the reflex arc concept developed by Dewey (see previous section), as a matter of sensorimotor coordination.

The pragmatist abductive approach to cognition is complementary to the Bayesian approach to confirmation. Abduction is a process of selecting plausible conjectures that require testing (Psillos 2000). The test, then, conforms to Bayesian processes and may be neurally implemented along the lines suggested by Friston and the predictive coding framework (Friston et al. 2013).

Many examples of perceptual inference conform to this combination of abduction and probability. Take Schurz's (2008:7) example of the explanation of sandal prints on a beach:

If your evidence consists in the trace of the imprints of sandals on an elsewhere empty beach, then your immediate conjecture is that somebody was recently walking here. How did you arrive at this conjecture? Classical physics allows for myriads of ways of imprinting footprints into the sand of the beach, which reach from cows wearing sandals on their feet to foot-prints which are drawn into the sand, blown by the wind, or caused by radioactive decay of foot-shaped portions of the sand, etc. The majority of these physically possible abductive conjectures will never be considered by us because they are extremely improbable.

While there are many conjectures as to the causes of imprints in the sand, only some of them are probable and only probable against a set of prior expectations, assumptions, beliefs, or even physical laws. Consequently, we are unlikely to entertain the implausible conjectures and eliminate them, leaving only those that are most plausible. Plausibility will be dependent upon our priors (in the Bayesian sense). Given a set of priors, some hypotheses will be more "plausible" than others, and it is from this subset of plausible hypotheses that we will select one to test.[6]

Pragmatic cognition is defined by fallible and active explorations of the environment, which can generate hypotheses that can be tested by selective sampling of the environment. Hypothesis testing via active inference can be stimulated by an "irritating" experience and so has its origins in experience.

It looks plausible that active inference may play a role in abductive inference, in terms of reducing surprisal, but it also seems likely that certain principles or rules of thumb will already do that job.

To return to exploratory inference, a child forming and testing hypotheses (with actions), refining them and re-testing is a good example of everyday abductions (an example is given below in the section on The Manipulation Thesis). It is likely that her search space is not as restricted by priors as an adult's, but this is a process of learning and experience generated by exploring and epistemically interacting with the niche.

An Externalist Account of Bayesian Exploratory Inference

One question remains to be answered at this juncture: Is predictive coding compatible with pragmatism since pragmatism requires continuous interaction with the environment? At least one prominent formulation of predictive coding starts from a position where internal mental states are screened off from the states of the external world.

The answer depends on the flexibility of the boundaries determined by Markov blankets between predictive processing in the brain and the body and environment. If we think of them as limiting determinate boundaries between

---

6   Although we should note that "plausibility" should not constrain all conjectures given that some occasions require us to produce novel conjectures that might conflict with our existing background knowledge. Scientific breakthroughs are often like this.

mind and world, then this would contradict one of the leading principles of pragmatic cognitive science; namely, that cognition is to be studied in terms of organism-environment interactions. Following Hohwy (2013) I call the limiting role of Markov blankets the "isolationist" interpretation of predictive processing.

In the isolationist interpretation, there is a perceptual interface to an environment of hidden variables: the internal system creates internal models (representations) of those hidden environmental variables which then causally produce behavior. The internal states must predict the external variables via sensory input, but it has no direct access to the causal ancestry of the sensory input. This form of individualism is used as an explanation for why the models and predictions are required: "Because the brain is isolated behind the veil of sensory input, it is then advantageous for it to devise ways of optimizing the information channel from the world to the senses" (Hohwy 2013:238). Hohwy describes the mind-world relation as "fragile" because of the isolation of the brain, and this is why active inference is required.

For example, in Clark's version of predictive processing, active inference and cultural props help to minimize prediction errors (Clark 2013b) and, as such, there is a deep continuity between mind and world mediated by active inference and the cultural scaffolding of our local niche. This interpretation of active inference appears to be more in line with the pragmatic approach. However, Hohwy thinks that Clark's interpretation is consistent with his isolated brain interpretation. Hohwy agrees that active inference and the cultural scaffolding of the environment help to change sensory input so as to minimize prediction error, but also "by increasing the precision of the sensory input" (Hohwy 2013:238). The primary role of predictive processing is perceptual inference; as a matter of "second-order statistics," active inference helps to optimize sensory input so that perceptual inference is less error prone. "The key point…is that this is a picture that accentuates the indirect, skull-bound nature of the prediction error minimization mechanism" (Hohwy 2013:238).

Organizing and structuring our environments makes sense if the mind-world relation is fragile in the way that Hohwy presents it, and also because this structuring makes perceptual inference more reliable. Hohwy's position is radically at odds with the aim of the pragmatic turn. Remember that the pragmatic turn consists of "a paradigm that focuses on understanding cognition as "enactive," as skilful activity that involves ongoing interaction with the external world" (Engel et al. 2013:202). This is clearly at odds with a new paradigm founded on the notion that cognition is based on a skull-bound prediction error mechanism that is only indirectly connected to its environment.

Active inference can be saved for the pragmatic framework if it is put into the broader context of the exploratory inferences of a fallible agent. As I argue elsewhere (Menary 2015), predictive processing is a subpersonal account of neural processes that fits within a larger account of the brain-body-niche nexus. This is possible if we do not take the minimization of prediction error to be the only kind of cognitive processing. Predictive processing might be supportive of interactive processing, but then it is just part of the processing routines available to an organism for completing cognitive tasks.

With this issue resolved, we now turn to exploratory inference and inquiry.

Exploratory Inference as the Foundation of Inquiry

The pragmatist account of inquiry, primarily developed by Peirce and Dewey, begins from the irritation produced by a situation in which further thought and action are blocked. Inquiry consists in the responses of an organism that succeeds in overcoming the problem situation. Over time, patterns of response to a recurring problem situation can become habitual for the organism. The problem situation becomes more than simply physical stimuli; they take on a significant character in relation to the habituated organism. Consequently, the pragmatist account of cognition has an experiential origin in an "irritating" experience.

Take Scheffler's (1974) example of a cat placed for the first time in a box with a door and a latch on it, which, when struck, opens the door. With a saucer of cream placed beyond the door and in sight of the cat, it would not be surprising if the cat were to attempt to reach the cream. At first its actions might be random and spontaneous, being produced by the irritations presented by the circumstance (e.g., hunger, the inability to reach the cream, and so on). If the random movements of the cat were lucky enough to strike the latch and open the door, then the irritation would be appeased. Subsequent movements of the cat would become less random and more ordered, until they became directed at manipulating the latch and as such the cat would have acquired a habit with regards to the solution of the "puzzle box" situation. However, this is not a simple case of stimulus-response correlation; the cat has developed a habit directed at appeasing the irritation which gave rise to its behavior in the first place. The cat's movements are directed at manipulating the catch to reach a desired end. As summarized by Scheffler (1974:43): "given this situation S, with perceptual and motivational features p and m, it responds appropriately with response R to achieve the desired and perceptible consequence K." Scheffler stresses the importance of the meditational role of R; it mediates between the initial source of irritation and the state that the organism reaches in which the irritation is assuaged. This provides us with a model of belief and belief fixation, distinct from the traditional concept of belief as an intellectual state which is "removed" from the environment. Belief is tied to how we would intelligently act in a situation. (Although some beliefs may be far removed from possible actions, it is likely that beliefs evolved for action and that this remains their primary function.) The organism's actions become self-directed and even self-controlled; this is different from classical behaviorism where the external stimuli control the behavior of the organism. The aim of inquiry is to reach a settled state of belief, which is a settled habitual state that predisposes us to act intelligently.

The pragmatists were philosophical and psychological precursors to the current embodied and extended approaches to mind and cognition. Dewey and Peirce are explicit in their externalism about thinking. Thinking does not exclusively take place in some inner mental substance or in some inner cognitive system. Thinking is an activity involving the interaction between an organism and its environment. The fallible method of thinking is the direction of activity to achieve some desired end. Like the cat in the puzzle box, we learn to manipulate the environment to achieve our goals. These manipulations become habitual and we do not even notice them in the background of our cognitive lives. This is the view that cognition is extended by our bodily manipulation of the environment (Menary 2007; Rowlands 1999), to which we next turn.

Principles at Work in a Contemporary Pragmatic Approach to Cognition

So far I have argued that predictive coding and sensorimotor contingency approaches to cognition are entirely compatible with the three principles. In this section I outline an approach that helps give further detail to the nature of interactive cognition and exploratory inference. This approach comes from the cognitive integration framework (Menary 2007), which taxonomizes different ways in which cognitive agents interact with, explore, and manipulate their environments by articulating the different ways that we manipulate the environment to achieve cognitive goals. As such it is a contemporary pragmatist approach to cognition.

Cognitive integration is committed to the foundational pragmatist idea that cognition is fundamentally a matter of interaction with the environment (Menary 2007). This framework explains how we learn to be active cognitive agents who think by manipulating their environments and interacting with one another in social groups. The integrationist framework also draws on cultural inheritance (Boyd and Richerson 1985) and niche construction models (Odling-Smee et al. 2003), which explain the evolutionary conditions under which richly structured cultural, and cognitive, niches are inherited. Ultimately, the integrationist framework explains how our minds are transformed while learning the cognitive practices by which we carry out many of our routine cognitive and epistemic tasks. The core of the argument is that our cognitive capacities endowed by evolution are not sufficient, on their own, to explain how we develop higher-order cognitive capacities (e.g., those that require mastery over public representational systems). The capacities we acquire through our learning and training histories during the extended developmental period in human ontogeny are layered over, but continuous with, those basic evolutionary endowments.

The primary focus of cognitive integration is on how we create, maintain, and manipulate cognitive niches. We can give a causal or coordination dynamics style explanation of these manipulations. A coordinated process allows the organism to perform cognitive tasks which it otherwise would be unable to, or to perform tasks in a way that is distinctively different and an improvement over the way that the organism performs those tasks via neural processes alone. However, we can also think of these coordinated processes as normative patterned practices spread out over a population or group. Some of these practices will be cognitive in nature (Menary 2012, 2014, 2015); in cognitive integration they are referred to as cognitive practices.

Cognitive practices are enacted by creating and manipulating informational structures in public space, for example, by creating shared linguistic content and developing it through dialogue, inference, and narrative or by bodily creating and manipulating environmental structures, which might be tools or public and shared representations (or a combination of both). Examples of linguistically mediated actions (or sign action in Peirce's sense) include (a) self-correction by use of spoken (or written) instructions, or by coordinating actions among a group, and (b) solving a problem in a group by means of linguistic interaction. Examples of creating and manipulating public and shared representations include using a graph to represent quantitative relationships; a diagram to represent the layout of a circuit or building; or a list to remember a sequence of actions, to solve an equation, to model a domain mathematically, to make logical or causal connections between ideas, and so on. Practices can be combined into complex sequences of actions where the physical manipulation of tools is guided by spoken instructions, which are being updated across group members.

The Manipulation Thesis

Task-driven manipulation of the environment constitutes the contribution of bodily and environmental processes. Mark Rowlands (1999:23) describes the idea as:

> [C]ognitive processes are not located exclusively in the skin of cognising organisms because such processes are, in part, made up of physical or bodily manipulation of structures in the environments of such organisms.

The manipulation thesis concerns our embodied engagements with the world, but it is not simply a causal relation. Bodily manipulations are also normative; they are embodied practices developed through learning and training (in ontogeny). Below I outline six different classes of bodily manipulation of the environment, with the general label of cognitive practices:

1. *Biological interactions* or direct sensorimotor interactions with the environment: An obvious example is sensorimotor contingencies (O'Regan and Noë 2001). A direct example of the first principle in action and anticipated by Dewey (see above).

2. *Corrective practices* are a form of exploratory inference and are clearly present early in cognitive development. A classic example from Vygotsky helps to illustrate: A four-and-a-half-year-old girl was asked to get candy from a cupboard with a stool and a stick as tools. The experiment was described by Levina in the following way (his descriptions are in parentheses, the girls speech is in quotation marks):

> (Stands on a stool, quietly looking, feeling along a shelf with stick.) "On the stool." (Glances at experimenter. Puts stick in other hand.) "Is that really the candy?" (Hesitates.) "I can get it from that other stool, stand and get it." (Gets second stool.) "No that doesn't get it. I could use the stick." (Takes stick, knocks at the candy.) "It will move now." (Knocks candy.) "It moved, I couldn't get it with the stool, but the, but the stick worked" (Vygotsky 1978:25).

The child uses speech as a corrective tool: "that didn't work, so I'll try this." Speech as a corrective tool is normative, because it is a medium through which the child can correct her activity in the process of achieving the desired result.

3. *Epistemic practices*: A classic example is Kirsch and Maglio's example of epistemic action in expert Tetris players (Kirsch and Maglio 1994). Experts would often perform actions that did not directly result in a pragmatic goal. The actions were designed to simplify cognitive processing. Other examples include, the epistemic probing of an environment and epistemic diligence, maintaining the quality of information stored in the environment (Menary 2012).

4. *Epistemic tools*: Many tools aid in the completion of cognitive tasks, from rulers to calculators, pen and paper to computers. Manipulating the tools as part of our completion of cognitive tasks is something that we learn, often as part of a problem-solving task.

5. *Representational systems*: Behaviorally modern humans display an incredible facility for innovating new forms of representational systems. (Remember that according to Peirce teleological sign systems are open-ended and flexible.) Humans also display a general capacity for learning how to create, maintain, and deploy representations. Alphabets, numerals, diagrams, and many other forms of representation are often deployed as part of the processing cycle that leads directly to the completion of a cognitive task (Menary 2015).

6. *Blended interactions* are complex cognitive tasks that may involve combinations of practices in cycles of cognitive processing.

Conclusion

A pragmatist approach to cognition entails three core principles. First, thinking is structured by the interaction of an organism with its environment. Second, cognition develops via exploratory inference, which remains a core cognitive ability throughout the life cycle. Cognitive agents are fallible: they start out by exploring their environments, develop inferential techniques for active exploration, and maintain those techniques (scaffolded in development) throughout the life span. Exploratory inferences should be thought of as a combination of abductive search and Bayesian constraint. Third, genuinely irritating doubts arise out of a particular situation to initiate problem solving, thus prompting the organism to search actively for concrete solutions. Exploratory inference also serves to affix belief.

If the "pragmatic turn" in cognitive science is considered to be a matter of explaining all cognition by sensorimotor interactions, I argue that it will likely be found lacking. Although the pragmatists developed a view of cognitive inference by actively exploring the environment, they did not think that all of cognition could be reduced to sensorimotor exploration, due to the role that representation, norms, and practices play in guiding exploration.

The pragmatic turn can make a real difference to the methodology and theory of cognitive science if it concentrates on the different styles of interaction. The turn should be away from inner mechanisms crunching information toward engaged cognitive agents who explore and interact with their environments and who think in action.

Acknowledgments

The research for this article was supported by the Australian Research Council Future Fellowship Project: FT130100960. Thanks to two anonymous referees for their comments.

14

Consciousness in Action

The Unconscious Parallel Present Optimized by the Conscious Sequential Projected Future

Paul F. M. J. Verschure

Abstract

This chapter outlines the historical cycle of the dominant views in the study of mind, brain, and behavior and the resulting trajectories taken in science. It discusses the grounded enactive predictive experience (GePe) framework, capturing contemporary science and philosophy of consciousness. It advances the hypothesis that consciousness is a necessary ingredient in a behavioral control architecture that has to solve action in a multi-agent world (the H5W problem). Using the distributed adaptive control theory, it shows how apparent heterogeneous approaches can be synthesized to gain greater understanding of a broad range of properties of mind and brain. As the "pragmatic turn" in cognitive science continues to be analyzed, it advocates a reorientation to the study of mind by returning to fundamental issues involved in consciousness.

Introduction

Understanding the nature of consciousness is one of the grand scientific challenges still confronting science today. Fundamentally, the problem involves how to account for phenomenal first-person experience in a third-person verifiable form. Also referred to as the hard problem or the explanatory gap (Levine 1983; Chalmers 1995), it is rooted in the rejection of structuralism by behavioralists about 100 years ago. Interestingly, the scientific study of consciousness has led to two extreme views: (a) as an epiphenomenon (Dennett 1992) or (b) as a fundamental property of matter on a par with mass, charge or space-time (Chalmers 2010; Koch 2012; Tononi 2012). The former perspective deems the phenomenon irrelevant while the latter resorts to panpsychism. This sets up somewhat of a paradox as the ontology of this phenomenon is placed beyond the scientific method: something to assume rather than to explain. Obviously, a number of alternative proposals fall in between these extremes and aim at finding a link between consciousness and its neuronal substrate (Crick and Koch 1990; for a review, see Dehaene and Changeux 2011). The paradox in the scientific study of consciousness signals a deep conceptual crisis at the heart of phenomenology and psychology: Have we reached the end of the science of mind because we are unable to get past an unsolvable riddle (Horgan 1997)? Or, as summarized by a recent popular newspaper article: "Why can't the world's greatest minds solve the mystery of consciousness?" (Burkeman 2015).

The distributed adaptive control (DAC) theory has been widely tested for over twenty years in the domain of both H4W and H5W. DAC has explained a broad range of properties of mind and brain, made predictions that have been corroborated and further elaborated, and allowed for the control of real-world systems ranging from interactive installations and robots to virtual reality-based neurorehabilitation interventions (Verschure 2012b; Verschure et al. 2014). In doing so it has taken an inclusive approach incorporating the core positive values expressed in the mind-brain cycle (Figure 14.1). In short, it links to behaviorism in its focus on embodied action, the core behavioral paradigms of classical and operant conditioning, and the insistence on empirical grounding of knowledge while allowing explicitly defined intervening variables to enter the explanatory framework and avoiding scientism. It incorporates key values

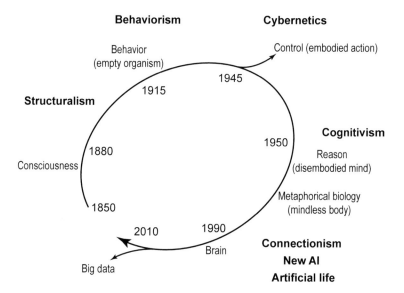

Figure 14.1   The mind-brain-behavior loop starting with experience and performance in the mid-nineteenth century and currently focusing on the brain and its derived data. See text for explanation.

from cybernetics by looking at the mind-brain as an embodied control system that can be studied through synthesis and explains key aspects of reasoning and problem solving as pursued by artificial intelligence while solving the symbol-grounding problem. In addition, it is consistent with the objectives of more recent approaches to anchor the science of mind in that of the brain, staying away from a metaphor. Lastly, DAC does not follow the mirage of big data but rather sees the challenge to get back to the fundamental question of consciousness.

I begin with a small historical detour to demonstrate that the situation in which we find ourselves is no accident but rather a logical consequence of the trajectories that the study of mind has followed over the last 150 years of scientific enquiry. I will then use this to propose a reorientation of the study of mind, in particular consciousness, toward a synthetic and action-oriented paradigm. I argue that consciousness is a transient memory system that functions specifically to mediate between the self and the world, providing valuation of parallel control systems and being causal with respect to future action.

The Mind-Brain-Behavior Loop

The scientific[1] study of mind and brain has followed a very specific development of concepts and methods. The main phases of this process have impacted a range of disciplines including psychology, neuroscience, computer science, linguistics, and philosophy. Our current situation in terms of the study of mind and the renewed focus on consciousness and action can best be understood if it is grounded historically. This will allow us to identify with greater clarity the novel contributions to the discussion of the pragmatic turn in the study of mind and brain and distinguish it from repetitions, redundancies, and noise (for reviews, see Koch and Leary 1985; Gardner 1987; Pfeifer and Scheier 1999).

Structuralism and the Primacy of Subjective Experience

The scientific study of mind began in the second half of the nineteenth century with the continental psychology of Fechner, Helmholtz, Donders, and Wundt. Wundt's structuralism, seen as the first school in psychology, was characterized by the systematic scientific study of phenomenology and human performance. Structuralism tried to combine rigorous empirical methods with the study of instantaneous experience through introspection. Wundt gave precedence to free will over reason, or voluntarism, and advocated a combined passive associative and active interpretative or "apperceptive" process in the construction of experience. This approach, among other things laid the foundation of

---

[1] I adhere to the restricted definition of science as the attempt to ground knowledge in direct observation.

modern psychophysics and sequential-processing models of the mind (Miller in Koch and Leary 1985). Hence, the initial scientific study of psychology took conscious experience as its explanandum.

Behaviorism and the Empty Organism

Behaviorism emerged in the first half of the twentieth century, driven by the confluence of (a) the comparative study of behavior set forth by the Darwinian revolution, (b) the pragmatism of Peirce, James, and Dewey, which grounds knowledge in its practical outcomes, and (c) what we can call "physics envy" or scientism. Behaviorism was a direct reaction to structuralism, negating its core dogmas and advancing three fundamental ideas (Kendler in Koch and Leary 1985): translate the methods of the natural sciences to the study of behavior; rely on the study of psychology on behavior as the dependent variable; and reject mental states on methodological grounds. In adopting Bridgman's physics-inspired philosophy of operationalism, the behaviorism of Watson and Skinner advanced the notion of an empty organism: a passive object fully controlled by the external forces of its world and explained in terms of the reflex atom.

Although Bridgman's model turned out to be a caricature of late nineteenth century physics, the drive to force the conceptualization of the phenomena under study into the methods employed became a prime example of *scientism*, or the excessive need for quantification even in domains that do not satisfy its specific conditions (Hayek 1943). This empiricist perspective on knowledge, formalized by the logical positivism of the time (which sought a link between logic and observation), also served the *unity of science* agenda; the psychology of adaptive behavior could be reduced to the biology of the brain, which would in turn give way to explanations at lower levels of description (i.e., chemistry and physics).

After about half a century of trying, behaviorism failed to deliver on its promise of identifying universal principles of adaptive behavior. Behaviorism also did not manage to scale up to more advanced forms of behavior beyond salivating, twitching, freezing, pushing levers, or pecking. Most importantly, organisms were not enslaved by the reinforcement received from the environment, as the empty organism dogma prescribed. Instead, autonomously structured learning and behavior was dramatically demonstrated in the experiments of Tolman, for example, leading to the notion of the cognitive map (Tolman 1948) and the Rescorla and Wagner laws of classical conditioning: animals only learn when events violate their expectations (Rescorla and Wagner 1972). Behavior could just not be explained by ignoring the agency and self-structuring of adaptive behavior by the organism itself.

Behaviorism negated structuralism and its explanandum, consciousness, thus giving precedence to the empirical methods deployed in the study of mind, which gravitated to the ultimate dependent variable: behavior. By placing method before concept, however, behaviorism went down the rabbit hole of scientism.

The Disembodied Mind of Cognitivism

The crisis that ensued through the collapse of behaviorism was resolved by switching to the computer metaphor of mind that gave rise to artificial intelligence (AI) and the cognitive science of the second half of the twentieth century, spearheaded by McCarthy, Newell, Simon, and Minsky, and the linguistics work of Chomsky (for a review, see Gardner 1987). AI won out over cybernetics, the latter placed emphasis on control and real-world action, whereas the former advanced a study of the logical operations performed by a disembodied rational mind. The neo-functionalism that followed (i.e., mind as explained in terms of rules and representations rather than its substrate or operant expression) argued for a unique level of explanation akin to that of a Turing machine (Putnam 1960). Hence, the move toward the computer metaphor negated the dogmas of behaviorism and its link to the rigorous empirical investigation at the level of brain and behavior—logically advancing the notion of a special level of explanation, yet isolated from implementation by virtue of multi-instantiation. As a result, the notion of an algorithmic computational reasoning system replaced that of mind as part of an embodied acting system, as demonstrated by the General Problem Solver of Newell and Simon (1963) and the SOAR cognitive architecture (Newell 1990).

Unfortunately, cognitivism and AI, in turn, stumbled over its claims of being able to both explain and synthesize intelligence. This intellectual failure combined with a lack of impact in the real world led to the so-called AI winter, during which funding and interest evaporated. As a research program, AI got bogged down in symbolic grounding (Harnad 1990) and the frame problem (McCarthy and Hayes 1969), both of which relied critically on the prior specification of the rules and representations that purportedly should have explained "intelligence" (Verschure 1998). In other words, AI's successes were largely due to designers defining appropriate priors into their systems as opposed to these artificial intelligences autonomously acquiring them, also referred to as the *problem of priors* (Verschure 1998).

Hence, the cognitive revolution, and its AI spearhead, sacrificed the empirical methods of behaviorism to develop a *new* science of reason decoupled from the natural science approaches, capitalizing on synthetic methods afforded by the emergence of the computer. This disembodied mind and its associated theoretical framework, however, was more a reflection of the designers' fancy coupled with technical capabilities than of any kind of natural system.

A Metaphorical Biology of Mind and Brain

Although a historical account of the recent past concerning the study of mind and brain requires a perspective that only time will provide, clear trends can already be distinguished (for an optimistic initial rendering, see Pfeifer and Scheier 1999). The disembodied mind of AI was followed in the late 1980s by a period of research in which biological metaphors guided the study of mind and brain, based partially on the construction of artificial systems. Examples include behavior-based AI, new AI, artificial life, genetic algorithms, neural networks and connectionism (Pfeifer and Scheier 1999) combined with a philosophy of "eliminative materialism," where the whole of human experience would be described in "brain speak" (Churchland 1986) that harked back to the philosophical behaviorism of Ryle (1949). "New" AI directly negated its predecessor (symbolic AI) by proposing a nonrepresentational, behavior-based, and embodied explanation of mind, whereas connectionism sought out the "subsymbols" that would link mental states to the neuronal substrate. Neither approach had a lasting conceptual impact on the study of mind and brain beyond signaling a shift toward new methods, such as those used in computational neuroscience, embodied cognition, or biomimetic robotics. However, this methodological advance, which opened up a universe of *in silico* experimentation, was realized at the expense of an organizing theoretical framework and a clear coupling back to empirical science.

Big Data and Avoiding Conceptual Defeat

The emergence of big data at the beginning of the twenty-first century exemplifies the regression of the study of mind and brain: the idea of advancing a theory was explicitly abandoned in favor of collecting large amounts of data and is well exemplified by the large-scale brain-oriented projects currently being pursued in the United States and Europe (Verschure 2016). From data-free theory, we fall into the antithesis of theory-free data. Placed in historical context, big data appears to be the logical result of the preceding period: failing to crack the conceptual problem of mind and brain over the previous two centuries, we resort to the last viable vestige of the scientific method—the collection of data. Big data emerges at a critical point in the study of mind and brain and can be seen as signaling a scientific crisis (Horgan 1997; Kuhn 1962/1970).

This crisis has been further amplified through the contemporary trend of studying consciousness outside the scope of science, due to the exceptional status attributed to consciousness: it was essentially removed from the scientific agenda by Dennett (1992), who holds that it is epiphenomenal (for a discussion, see Robinson 2010); by Rosenthal (2008), who advocates that it has no function; and by Chalmers, Tononi and Koch, who want us to believe that it is neo-panpsychistic, part of the fundamental fabric of nature (Chalmers 2010; Tononi and Koch 2014). It is indeed ironic that after distinguishing the easy from the hard problem, the latter was solved by assumption and declared not to be a problem at all! This can be viewed as a form of explanatory nihilism (Price 2002):

> What shall we do? Many would find relief at this point in celebrating the mystery of the Unknowable and the "awe," which we should feel at having such a principle to take final charge of our perplexities. Others would rejoice that the finite and separatist view of things with which we started had at last developed its contradictions, and was about to lead us dialectically upwards to some "higher synthesis" in which inconsistencies cease from troubling and logic is at rest. It may be a constitutional infirmity, but I can take no comfort in such devices for making a luxury of intellectual defeat. They are but spiritual chloroform. Better live on the ragged edge, better gnaw the file forever! (James 1890/1950).

This short analysis shows that in five easy steps the study of mind and brain has sacrificed its explanandum, its methods and theories and an intellectual limbo has resulted wherein the questions have shifted from the explanation of psychological constructs to the description of neuronal correlates and its underlying data. These steps are easy because each paradigm followed as a negation of the premises underlying the preceding one, terminating in a science that is about data as opposed to ideas. What shall we do?

Following James's recommendation, I argue that an alternative is to embark on a new study of mind and brain; let us use the unpopular notion of psychology that addresses the fundamental question of consciousness and combines the strength of preceding approaches, as opposed to their overstated promises, while answering their weaknesses. This explains the idea of a mind-brain cycle depicted in Figure 14.1: back to experience as our explanandum! Essentially we need to focus on explaining consciousness, linking it to overt behavior and reasoning based on rigorous empirical, formal, and synthetic methods, and grounding this explanation in the biological principles that govern bodies and brains. Such a program is not necessarily incompatible in its realization with aspects of the approaches listed above. The big difference, however, is that it steps away from the brink of nihilism and declares consciousness, yet again, a phenomenon to be explained, hypothesizing a distinct function and an augmented method to investigate it. This approach is grounded in the *distributed adaptive control* (DAC) theory of mind and brain that has been advanced using embodied biologically grounded models linking the neuronal substrate to action (Verschure et al. 2003; Verschure 2012b). The DAC program realizes what this Forum seeks: linking mind to action.

Instead of assuming that consciousness is a fundamental property of the physical world, an alternative and more straightforward hypothesis (which has not yet been exhausted) is that consciousness is a unique feature of a subset of living systems: it is the product of biology rather than physics as advocated, for instance, by Searle (1998). This means that we have to place its study in the context of evolution to follow Dobzhansky (1973) and consider its function in terms of function and fitness. From this perspective, two features stand out and seem paradoxical: consciousness is defined in terms of one coherent unitary scene (James 1890; Bayne 2010), yet experimental evidence shows that this conscious scene is experienced with a significant delay, relative to the real-time action of the agent (Libet 1985; Haggard et al. 2002; Soon et al. 2008), that is not necessarily the cause of action and thought (Wegner 2003; Custers and Aarts 2010). The resulting paradox is that in optimizing fitness, evolution appears to have rendered solutions to the challenge of survival which include putatively epiphenomenal processes like consciousness.

I propose a solution to this paradox and advance the hypothesis that consciousness is a necessary ingredient of a behavioral control architecture that has to solve action in a multi-agent world, or the, so-called, H5W problem. Before turning to it, I outline the most dominant views on consciousness and show how these can be integrated into one coherent framework, which serves as a context from which we will launch the DAC theory of consciousness and H5W.

The GePe Framework

It has become standard to acknowledge that there are addressable and non-addressable problems in the study of consciousness, or easy and hard problems. With respect to the "easy" problems, a number of core principles underlying consciousness and qualia have emerged. These can be summarized in the *grounded enactive predictive experience* (GePe) model of consciousness (Verschure 2012b, 2013) that will guide model construction and validation. The GePe model utilizes five principles:

GePe 1: Consciousness Is Grounded in the Experiencing
of the Physically and Socially Instantiated Self

Experience requires a self that does the experiencing (Nagel 1974; Metzinger 2003; Edelman 1989; Craig 2009). For instance, Edelman (1989) proposed primary and secondary forms of consciousness that relate to the expanding temporal horizon of the self, from the instantaneous physical experience (primary) to the imagined future and remembered past (secondary). Metzinger (2003) refined this notion further: the self progresses from a globalized identification, with the body or first-person perspective, to a transparent spatiotemporal self-localization in the world or minimally phenomenal self based on a form of representation of the self, to a fully fledged phenomenal first-person perspective or strong first-person perspective. The first-person perspective begins as a point of convergence of sensory (but also proprioceptive and interoceptive) experience; then it coalesces into a strong form where the self is internally represented as reflecting the organization of the body and its sensorimotor coupling to the world (see GePe principle 2); this is followed by the representation of the object- and action-directedness of the self (i.e., intentionality) found in the strong first-person perspective. As interactive and social dynamics are rich sources of sensory experience and feelings, they become part and parcel of the representation of the self, which is not only physically but also socially instantiated (Frith 2008). In a sense, this view on self and consciousness also reflects a trend in cognitive science to ground knowledge and experience in embodiment, situatedness, and interaction dynamics (Verschure et al. 1992; Pfeifer and Bongard 2006; Barsalou 2008). Damasio (2012) has recently advanced a similar proposal, retracting his version of the James-Lange theory of emotional experience to suggest now that consciousness requires representations of self to enter into memory.

GePe 2: Consciousness Is Defined in the Sensorimotor
Contingencies of the Agent in the World

In neuroscience, cognitive sciences, and robotics there is a shift from a representation-centered framework toward a paradigm that focuses on the intimate relation between perception, cognition, and action (see above). Although many proponents have supported such an "action-oriented" paradigm over the years, starting with Pavlov (Pavlov 1927; Verschure 1992) and his mentor Sechenov, it has only recently started to regain traction. In this view, cognition is not isolated from action and a database-serving planning in a strict sense-think-act cycle, as already advanced by Donders in the nineteenth century. Rather, cognitive processes are closely intertwined with action and can best be understood as "enactive," as a form of practice itself (Pulvermüller and Fadiga 2010; Verschure et al. 1992). The intrinsic action-relatedness of cognition is the core consideration of the *sensorimotor contingency theory* put forward by O'Regan and Noë (2001) that addresses the fundamental role of action for perception and awareness. Accordingly, the agent's sensorimotor contingencies are law-like relations between movements and sensory inputs which provide the foundations for knowledge and experience. O'Regan (2011) has proposed that these laws of sensorimotor contingencies define the qualia of conscious experience. A challenge for this framework, as for its behaviorist ancestors, is to scale-up to cognition, affect, and rich experiences which might appear non-motor.

GePe 3: Consciousness Is Maintained in the Coherence
between Sensorimotor Predictions of the Agent and the
Dynamics of the Interaction with the World

The idea that perception is defined by predictive models of environmental causes of sensory input enjoys a rich pedigree that extends back at least as far as Helmholtz and the seminal work of Tolman (1932). Indeed, sensorimotor contingencies not only exist instantaneously, they can also be predicted by virtue of their invariance (Bar 2007). It has been proposed that cognition and consciousness are based on such internal simulations of the possible scenarios of interaction with the world using *forward models* (e.g., Hesslow 2002; Cisek 2007). Indeed, it has been proposed that concepts themselves can be seen as simulations (Barsalou 2008). It is through simulation that an "internal" world can appear in consciousness, freeing the organism from its immediate physical environment (Hesslow 2002; Revonsuo 2006). Merker (2005) has argued that the simulated internal world compensates for the uncertainties generated by the dynamics of sensory states due to self-induced motion and that it can be seen as a self-generated virtual reality (Revonsuo 1995).

That the brain is organized around prediction has reached recent prominence in the Bayesian Brain and "predictive coding" frameworks (Bar 2007; Clark 2013b; Verschure et al. 1992; Rao and Ballard 1999; Friston 2005; Barsalou 2008), and was anticipated by Massaro (1997) in his analysis of speech perception. In these views, core structures of the brain (including the thalamocortical and corticobasal ganglia systems as well as the cerebellum) are engaged in hierarchical Bayesian inference, extracting generative models of both sensory inputs and the consequences of action across multiple timescales and modalities (Hesslow 2002; Lau and Rosenthal 2011). In neurophysiological terms, "top-down" connections are suggested to convey the content of these generative (predictive) models, whereas "bottom-up" signals convey prediction errors (Bar 2007; Mathews and Verschure 2011). The resulting models have received growing support (Friston 2005). However, the exact relation between predictive processing and biological consciousness remains poorly understood, although some correlate of this view has been reported in coma patients (Boly et al. 2011). For example, there is no consensus on which sorts of predictive model give rise to conscious contents, and which do not. Furthermore, it is unclear what the relations are between probabilistic representations postulated by the Bayesian brain and the fact that (apparently) we do not perceive our conscious states as being probabilistic. Merker (2005) argues that information, while in cortex, is generally maintained in the form of probability distributions yet the content of consciousness is linked to the "collapsing" of probability functions into a simpler format (hence the reason why our conscious percepts do not appear to be probabilistic). This format is proposed to be required for subcortical processing (Ward 2011) to provide global best estimates of variables of interest within narrow time windows. However, this hypothesis remains to be tested and compared with other mainstream theories that view cortex as the locus of consciousness.

A variation on the prediction-based theories on consciousness is the attention schema theory proposed by Graziano (2013). In this proposal, underlying consciousness is the process of attention; its role of identifying subsets of sensory information of relevance to an agent is attributed to it by an observer. Consciousness is thus seen as the ascription of such attentional states to others and the self.

GePe 4: Consciousness Combines High Levels of
Differentiation with High Levels of Integration

More progress has been made with respect to another structural property, namely *complexity*. Following Edelman and Tononi, it is a deeply significant fact that each and every conscious scene is both *integrated* (or unitary) and massively *differentiated*, such that it provides for a highly informative discrimination among a very large repertoire of possible experiences (Edelman 1989; Tononi and Edelman 1998; Tononi 2008, 2012). Viewing consciousness from the perspective of such integrated information suggests theoretically grounded and empirically applicable quantifications of consciousness, such as information theoretic measures (Tononi and Edelman 1998) or multivariate autoregressive modeling (*causal density*) (Seth 2009). Tononi's *integrated information theory* (IIT) provides a precise definition of information integration given a number of assumptions on how to segment informational spaces. IIT introduces a fundamental quantity, integrated information (Φ), expressed in bits, which measures to what extent a system integrates information as a whole via its causal dynamics, over and above that of its subparts. For systems composed of largely independent modules, Φ is low as it is for nonmodular systems that are connected in a homogeneous or random manner. IIT is high only for systems that are both functionally specialized and integrated. This measure has been used to distinguish between different levels of consciousness, where sleep states show a lower complexity than awake and alert states (Massimini et al. 2009). However, this in itself is also a possible drawback because one can confuse the measure with the ontology of consciousness. For instance, what is the bound on Φ, and what are its discriminative capabilities? The notion that complexity reflects consciousness can lead to a misunderstanding of the ontological significance of these measures such that any complex system (e.g., the Internet) must be conscious by definition, leading to the aforementioned panpsychism (Koch 2012). Within this perspective, consciousness does not have a declared function.

GePe 5: Consciousness Depends on both Highly Parallel, Distributed
Implicit Factors and Metastable, Continuous, Unified Explicit Factors

Theories of consciousness are tightly constrained and informed by evidence regarding unconscious processing (Baars 1988). In Baars's "global workspace" architecture, specialized unconscious processors compete for access to a central resource: the conscious global workspace. Accordingly, consciousness is ascribed to content that is received from and broadcast back to a broad network of unconscious modules or processors. In this way consciousness provides a serial and integrated stream of qualia that are produced by many subconscious "processors." The key parameter that defines whether content becomes conscious is the ability to penetrate many of these processors. In this respect the global workspace is an example of *access consciousness* (Block 2007). The integration and serialization provided by the global workspace provides for behavioral flexibility by allowing unconscious processors to generate fast responses in familiar situations, while in novel situations the integrated qualia that are broadcast from the global workspace can facilitate the production of new responses (Baars 1988). The global neuronal workspace hypothesis proposes that the workspace comprises perceptual, motor, attention, memory, and value areas which form a common higher-level unified information space that serves a similar role as the global workspace largely dependent on the specific anatomy of cortiocortical projections (Dehaene et al. 1998; Dehaene and Changeux 2011). The main function ascribed to the global neuronal workspace is that of assisting in problem solving and executive control (Dehaene 2014).

Epiphenomenalism and the Case against Free Will

Whereas Descartes places phenomenal subjective states at the center of mental existence, a number of converging lines of evidence show that humans are largely unaware of the causes of their own thoughts and actions (Wegner 2003). This observation, corroborated by a large set of experiments, has fueled the interpretation that consciousness is an epiphenomenon; that is, it is an evolutionary leftover with no operational relevance (Dennett 1992). A large amount of cognitive processes can be performed without reportable awareness of the relevant stimuli or contingencies, and some processes (e.g., overlearned motor responses) are supposedly even more effective when implemented by unconscious systems (Baars 1988; Milner and Goodale 1995). The latter claim of an unconscious thought advantage has been put in doubt in a recent meta-analysis (Nieuwenstein et al. 2015). Less appreciated but equally fundamental is the notion that motor actions and intentions can be unconscious as well as conscious (Dijksterhuis and Bargh 2001; Frith et al. 2000a), that unconscious intentions are known to reliably precede conscious awareness of motor actions (Desmurget and Sirigu 2009; Libet 1985), and that behavioral goals can be set by unconscious factors (Custers and Aarts 2010). Another category of implicit factors in experience and action are emotions. Indeed, emotion and consciousness are tightly coupled, and conscious experiences generally involve affective (emotional) components, both transiently (e.g., delight, surprise) and as a background mood (e.g., sadness, contentment, anxiety) (Tsuchiya and Adolphs 2007). Since James-Lange it has been suggested that emotions arise as perceptions of bodily states (Critchley et al. 2004) and that autonomic signals can reflect implicit reactions to salient stimuli, including prediction errors (Uhlhaas et al. 2009). It has further been argued that the processes underlying volitional behavior (e.g., implicit learning, evaluative conditioning, unconscious thought) are intrinsically goal dependent, requiring forms of attention while operating outside of awareness (Dijksterhuis and Aarts 2010). In all cases, conscious and unconscious processes are closely coupled and interact strongly in generating the stream of consciousness and adaptive behavior (Baumeister et al. 2011). They can be seen as complementary since unconscious processing can be sensitive to patterns, regularities, and other structures within signals prior to conscious awareness, suggesting that the content of consciousness is biased and based on unconscious factors (Baars 1988; Haggard and Eimer 1999). Such dual-process theories (Evans 2008)—such as fast and slow processes in decision making (Kahneman 2011) and the distinction between reasoning, planning, and monitoring processes (Gazzaniga 2011)—face the fundamental question of how these processes are maintained in isolation and interfaced as well as how the exchange of information between them is regulated. In particular, we need to know whether these multiple processes are coherent or are descriptions of a further heterogeneous set of subsystems, possibly leading to an infinite regress (Evans 2008).

Distributed Adaptive Control: A Theory of the Mind, Brain, Body Nexus

The perspectives on consciousness and its putative functions outlined above can come across as rather heterogeneous. However, when each are viewed as highlighting specific and complementary aspects of consciousness and its function, they can be brought together and reconciled. I have synthesized this from the perspective of the distributed adaptive control, illustrated in Figure 14.2, and begin with a brief explanation of the DAC principles.

A highly abstract representation of the DAC architecture is depicted on the left-hand side of Figure 14.2, wherein the brain is organized as a layered control structure with tight coupling within and between the somatic, reactive, adaptive, and contextual layers. Across these layers a columnar organization exists to process the states of the "world" or exteroception (left column), "self" or interoception (middle column), as well as "action" (right column), which mediates the previous two. The somatic layer equips the body with its sensors, organs, and actuators. The reactive layer is made up of dedicated behavior systems which combine predefined sensorimotor mappings with drive reduction mechanisms predicated on the needs of the body (somatic layer).

Depicted in the right lower panel of Figure 14.2 (allostatic control) we see that each behavior system follows homeostatic principles supporting the self essential functions (SEFs) of the body (somatic layer). To map needs onto behaviors, the essential variables served by the behavior system have a specific distribution in space called an affordance gradient. In this example, we consider the (internally represented) "attractive force" of the home position supporting the *security* SEFs or of open space defining an *exploration* SEFs. The values of the respective SEFs are defined by the difference between the sensed value of the affordance gradient (red) and its desired value given the prevailing needs (blue). The regulator of each behavior system defines the next action so

248

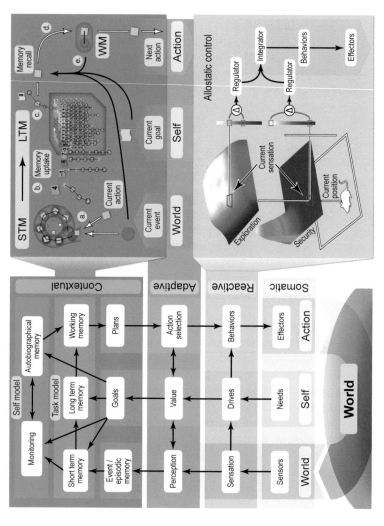

Figure 14.2 The distributed adaptive control (DAC) theory of mind and brain. See text for details. Figure adapted from Verschure et al. (2014); see text for explanation. For an overview of the mapping of DAC principles to the brain, see Verschure (2012b).

as to perform a gradient ascent on SEFs. An integration and action selection process across the different behavior systems forces a strict winner-take-all decision, which defines the specific behavior emitted. The allostatic controller of the reactive layer regulates the internal homeostatic dynamics of the behavior systems to set priorities defined by needs and environmental opportunities through the modulation of the affordance gradients, desired values of SEFs, and/or the integration process. The adaptive layer acquires a state space of the agent-environment interaction and shapes action. The learning dynamics of the adaptive layer is constrained by the SEFs in the reactive layer which define value. Crucially, the adaptive layer contributes to exosensing by allowing the processing of states of distal sensors (e.g., vision and audition). These are not predefined; instead they are tuned in somatic time to properties of the interaction with the environment. In turn, acquired sensor and motor states are associated through the valence states signaled by the reactive layer. The adaptive layer is modeled after the paradigm of classical conditioning (Pavlov 1927), and the acquisition of the sensorimotor state space is based on predictive mechanisms to optimize encoding and counteract biases due to behavioral feedback. The adaptive layer has been mapped to the cerebellum, amygdala, cortex, and hippocampus. The contextual layer is divided between a "world" and a "self" model. It expands the time horizon in which the agent can operate through the use of sequential short- and long-term memory (STM and LTM, respectively) systems. These memory systems operate on the integrated sensorimotor representations generated by the adaptive layer and acquire, retain, and express goal-oriented action regulated by the reactive layer. The contextual layer comprises a number of interlocked processes (right upper panel):

a. When the error between predicted and encountered sensory states falls below an STM acquisition threshold, perceptual predictions (red circle) and motor activity (green rectangle) generated by the adaptive layer are stored in STM as a *segment*. The STM acquisition threshold is defined by the time-averaged reconstruction error of the perceptual learning system of the adaptive layer.

b. If a goal state (blue flag) is reached (e.g., reward or punishment), the content of STM is retained in LTM as a sequence conserving its order, goal state, and valence marker (e.g., aversive or appetitive), and STM is reset so that new sequences can be acquired. Every sequence is thus defined through sensorimotor states and labeled with respect to the specific goal it pertains to and its valence marker.

c. If the outputs generated by the reactive and adaptive layers to action selection are below the threshold, the contextual layer realizes its executive control and perceptual predictions generated by the adaptive layer are matched against those stored in LTM.

d. Action selected in the contextual layer is defined as a weighted sum over LTM segments.

e. The contribution of LTM segments to decision making depends on four
    factors: perceptual evidence, memory chaining, the distance to the goal
    state, and valence. The working memory (WM) of the contextual layer
    is defined by the memory dynamics that represents these factors. Active
    segments in WM that contributed to the selected action are associated
    with those which were previously active in establishing rules for future
    chaining.

The core features of the contextual layer have been mapped to the prefrontal
cortex. The self-model component of the contextual layer monitors task performance and develops (re)descriptions of task dynamics anchored in the self.
In this way, the system generates meta-representational knowledge to form
autobiographical memory.

To create structure in the tangle of neuronal processes and subprocesses that
make up the brain and its multilevel organization, we need to define unambigu-
ously what the overall function of this system is. DAC follows Claude Bernard
and Ivan Pavlov in defining the brain as a control system that maintains a metastable balance between the internal world of the body and the external world
through action. This pertains both to its physical and informational needs. The
question thus becomes: What does it take to act?

The DAC theory proposes that to act in the physical world, the brain needs
to optimize a specific set of objectives which are captured in answering the
questions: *Why* do I need to act? *What* do I need? *Where* and *when* can this
be obtained, and *how* do I get it? Embedded within these questions is a complex set of computational challenges that has been termed the *H4W problem*
(Verschure 2012b). In short, an agent needs to determine a behavioral procedure to achieve a goal state (the *how* of action). This, in turn, requires defining
the motivation for action in terms of needs, drives, and goals (the *why*); the
objects and their affordances in the world that pertain to these goals (the *what*);
the location of objects in the world, the spatial configuration of the task domain
and the location and confirmation of the self (the *where*); and the sequencing
and timing of action relative to the dynamics of the world and self (the *when*).
DAC theory proposes that goal-oriented action in the physical world emerges
from the interplay of these different processes subserving H4W.

Each of the *W*s can be seen as a specific objective that the brain must satisfy.
In turn, each can be decomposed into a large set of sub-objectives of varying
complexity organized across different levels and scales of organization of the
central nervous system. At a first level, the brain must assess the motivational
states derived from homeostatic self-essential variables defined at the level
of the soma and reactive control. These motivational states, in turn, need to
be prioritized so that goals can be set: this is the *why* problem, requiring the
modulation of associated behavior systems. Next, a second layer of control is
called upon to classify, categorize, and valuate states of the world, to identify
the spatial layout of the task, including the agent itself, and the dynamics of

the task and its affordances: *what*, *where*, and *when* also engages the learning systems of the adaptive layer. These labeled multimodal states are grouped in sequences around prioritized goals at the level of contextual control; for example, in a rodent navigation set-up, to go toward and push a lever placed at the northeast corner of the environment, given that the cue signal has appeared. At this stage the *how* has been generated and expressed. Using the accumulated spatiotemporal knowledge of the task and the self in which goal pursuit is framed, a procedural motor strategy (*how*) can be composed and its elements selected from the set of available options to achieve a goal state (Verschure et al. 2014).

The H4W framework is an exclusive set of processes that directly maps onto the functions of the different layers of DAC, capturing core brain mechanisms that mediate and control instrumental interaction with the physical world as in the adaptation to an open field (Figure 14.3) or to foraging tasks including neocortex, hippocampus, basal ganglia and the cerebellum (for a review, see Verschure et al. 2014). To solve H4W, we have constructed an architecture that comprises all components of GePe: an embodied self, generating and acquiring sensorimotor contingencies, relying on forward models, displaying the integration of information and maintaining a global workspace in its memory systems (Figure 14.3). This DAC H4W realization has been tested on a range of robots and shows all signatures of GePe; however, it is not conscious, contrary to claims that even simpler models can be called conscious (Tononi and Koch 2014). The reason for this, I propose, is that a fundamental aspect is missing; namely, the ability to simulate *hidden* states of the external world.

H4W solely addresses the interaction of an agent with its physical world. DAC theory proposes that the Cambrian explosion (ca. 550 million years ago) created environments dominated by one more critical factor, which demanded a specific objective function: *Who* is acting? The resulting move from the H4W to the H5W problem leads to a fundamental change in information processing: reciprocity and hidden states.

Reciprocity results from a behavioral dynamic: the agent is now acting on a world that is, in turn, acting upon it. The states of other agents, which are predictive of their actions, are however, hidden. At best they can be inferred from incomplete sensor data, such as location, posture, vocalizations, or social salience (Inderbitzin et al. 2013). As a result, the agent must unequivocally assess, in a deluge of sensor data, those extero- and interoceptive states that are relevant to ongoing and future action. In addition, the agent must deal with the ensuing credit assignment problem to optimize its own actions. In this partially observable intentional world, the solution to survival entails assessing (a) the relevant (hidden) states of the world and its agents, (b) the relevant states of self, and (c) the specific action which gave rise to relevant outcomes. I propose that consciousness is a necessary component of the control system that solves this H5W problem.

(a)

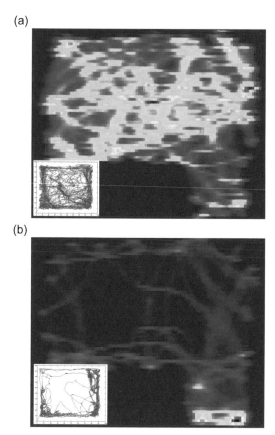

(b)

Figure 14.3 H4W solved by DAC in the real world. Density plots are shown of the positions visited by a robot controlled by the DAC architecture while (a) familiar or (b) novel environments are explored. Behavioral trajectories are modulated by the behavioral subsystems of exploration and security respectively. Insets show example trajectories from rats which performed under similar conditions. Adapted from Sánchez-Fibla et al. (2010).

The control system generated by DAC-based consciousness incorporates all elements of the GePe principles and adds a few new elements. Let us first map GePe to DAC:

- Grounded in the experiencing of physically and socially instantiated self: the somatic layer constitutes the foundation of the embodied hierarchy.
- Co-defined in the sensorimotor coupling of the agent to the world: both the reactive and adaptive layers establish immediate sensorimotor loops with the world (the former predefined, the latter acquired). Acquired sensorimotor states form the representational building blocks of DAC's cognitive processes.

- Maintained in the coherence between sensorimotor predictions of the agent and the dynamics of the interaction with the world: the adaptive level relies on prediction-based systems for both perceptual and behavioral learning (Duff and Verschure 2010). The memory systems of the contextual layer operate on a combination of forward and feedback models.
- DAC combines high levels of differentiation (each conscious scene is unique) with high levels of integration: the contextual layer integrates across all sensory modalities and memory systems and provides selection mechanisms to define a unique interpretation of the state of the world and the agent.
- Consciousness depends on highly parallel, distributed implicit factors with metastable, continuous, unified explicit factors: the contextual layer integrates memory-dependent implicit biases in decision making and interpretation of states of the world with explicit perceptual states. Task relevant states are "ignited" by the confluence of perceptual and memory evidence to form the dominant state of the contextual layer memory system.

If DAC resonates so well with GePe, why has it not reported conscious states? The answer is simple: GePe is incomplete. The DAC-based theory of consciousness, however, adds new considerations to the GePe framework:

1. *Simulation and its virtualization memory*: The hidden states of the world (i.e., other agents) are resolved through simulations which allow predictions on hidden states to be generated and maintained through forward models. As a result, action takes place in an augmented reality where sensor data (reflecting physical sources of stimulation and projected intentional states) are merged and tested against the world. This augmentation cannot take place in the physical world and thus requires a dedicated memory system which supports the virtualization of the world model. Some have characterized such a feature as a brain-based virtual reality (Revonsuo 1995; Merker 2005; Metzinger 2003).

2. *The intentionality prior*: To bootstrap the semantics of the simulations of hidden states of other agents, they are anchored in an intentionality prior, or pervasive intentionality, where novel states are automatically treated as being caused by other agents (Verschure 2012a). This implies that intentionality detection is operating at the level of the reactive layer. A further interpretation of intentional cues detected in the world or ascribed to it capitalizes on a self as other process (Merleau-Ponty and Edie 1964), which implies that the self and world columns of the architecture (see Figure 14.2) are tightly coupled, and that the self model continuously serves as an anchor of intentional cues detected in or projected onto the world.

3. *Parallel multi-scale operations*: Given the number of variables to be considered in a complex multiagent world and the finite operation powers of physical systems (i.e., brains), there is strong pressure on implementing components 1 and 2 through parallel operations. In addition, all real or imagined agents in the environment must be tracked in real time, thus defining a further functional need for parallelization. Indeed, parallel processing is one of the characterizing features of social brains, from the mushroom bodies of bees to the cerebellum of vertebrates.

4. *Serialization and unification*: The agent and its physical instantiation by necessity can only commit to a single action realized through its singular body at each point in time. These actions are informed by massively parallel simulations of possible world states that support real-time inference in an intention-laden world, thus creating a fundamental credit assignment problem: Given the outcome of a singular act, which value or action should be assigned to which property of the real or imagined world?

5. *Consciousness solves credit assignment in a parallel world model*: DAC allows us to rephrase the challenge of finding alignment between the singular and serial self model with a parallel and probabilistic world model. Real-time control of action requires parallel processing. For instance, the human cerebellum (credited with controlling real-time action) comprises about 15 million parallel segregated loops, constituting about 70% of the neuronal volume of the brain. Learning in this system is regulated through an error signal generated by the inferior olive, which matches reactive and adaptive modes of control (Herreros and Verschure 2013). Consciousness is a necessary counterpart to such a real-time parallel control system: a highly integrated sequential process that runs adjacent to the many parallel unconscious processes, integrating across many parallel states, valuating performance and projecting back error signals. In this way, cooperation between parallel unconscious and serial conscious control assures operational coherence through the reinterpretation and optimization of unconscious parallel loops. To realize this function, the process of consciousness requires a transient memory system that maintains the serialized and unified description of the world model in terms of the self-model. Unified intentionality is subsequently ascribed to the world and interpreted based on sequential conscious processing, in which self-generated actions are (re) interpreted, valued, and reorganized for future use (Verschure 2012b). Hence, consciousness serves goal-oriented performance in the future in a world filled with intentionality, while real-time action is under the control of the parallel unconscious systems that it optimizes (Figure 14.4). The problem of unifying the optimization of subconscious control is thus solved by shifting the representational frame from signal-based to intention-based, i.e., an intentional stance (Dennett 1988)

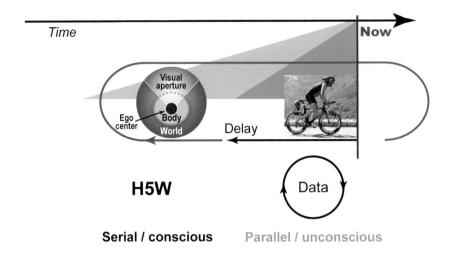

Figure 14.4 The interplay of parallel, real-time unconscious processing and sequential conscious monitoring. The green-filled circle represents the delayed serial conscious experience derived from the parallel control that generates real-time action.

framework, with respect to both the world and the self. The latter is achieved through *"ap*-presentation*"* (Merleau-Ponty and Edie 1964)— that is, the interpretation of the other in terms of the self—as well as by relying on the *a priori* ascription of intentionality to the environment, or the pervasive intentionality prior (Verschure 2012a). We can take this pervasive intentionality as another Kantian prior. Indeed, as Dan Sperber puts it: "the attribution of mental states is to humans what echolocation is to bats" (Gallagher 2005:207). Humans make social judgments based on simple geometric shapes (Heider 1944) or moving point models of the human body (Scholl 2001). Hence, the assumption of an intentionality prior as acting already at the level of reactive control seems to be defendable and a small price to pay to save consciousness from explanatory nihilism.

Addressing the "Hard" Problem: A Methodological Proposal

How can we make these predictions measurable or, more generally, how can we overcome the so-called hard problem? First, we do not need to ask physics to explain the form of a chair or even to provide a full explanation of the properties of materials. In fact, the dream of a unity of science is still stuck at crossing that bridge. Second, it can be argued that the hard problem is equally difficult in memory research as it is for consciousness: What is it like to remember some episode experienced in the past, and how is this information stored and retrieved? However, this might avoid essential issues, so let us consider a third option. The explanatory gap can be crossed when we focus on the *process* of consciousness rather than insisting that each specific quale be deciphered. DAC theory has progressed by insisting on a methodology of convergent validation (Verschure 1997): constraining models through simultaneously addressing anatomy, physiology, and behavior. As a result, all DAC models are validated using real-world behaving systems (i.e., robots). This approach is grounded in the philosophy of the eighteenth century Neapolitan philosopher Giambattista Vico, who famously proposed that we can only understand that which we create: *Verum et factum reciprocantur seu convertuntur* (Vico 1730). We can use this same approach to address the hard problem and parse quale, which are essentially expressing memory and which, in turn, reflect a specific prior experience at a specific point in time by a specific agent with its specific embodiment and history. We merely have to follow the machine in time and record its states to be able to parse its mental states, including conscious ones, on future occasions. We have already performed such semantic parsing of memory states successfully with complex autonomous DAC-based artifacts, such as the interactive sentient space Ada visited by over 500,000 humans (Eng et al. 2005) and different behaving robots (Verschure et al. 2003). Hence, testing a theory of consciousness requires us to build a conscious machine and there is no *a priori* reason why this should be impossible.

Empirical Consequences

The H5W DAC theory of consciousness provides an explanation of consciousness that focuses on the notion of the unitary nature of conscious experience and its delayed realization relative to real-time performance. The model behind it is advanced through controlling real-world systems, including buildings and robots. To complete all criteria of a scientific theory, the question is: What testable predictions can we derive from H5W DAC?

The first candidate is the assumption of pervasive intentionality or the intentionality prior. This suggests that infants would ascribe excessive intentionality to the world and that as a result of maturation, this intention reflex is suppressed. Indeed, it has been shown that both seven-month-old infants and adults model the intentional states of others in a form similar to their self models (Kovács et al. 2010). In addition, there is a negative correlation between age and the propensity to favor teleological explanations of social behavior, biological properties, artifacts, and life events (e.g., Banerjee and Bloom 2015). The fact that infants, starting at five months of age, show slowly maturing neurophysiological signatures of consciousness perception (Kouider et al. 2013), opens up a wide range of experimental questions on the relation between consciousness, pervasive intentionality, and maturation (e.g., how and when the intentional reflex is suppressed and how this is reflected in the signatures of conscious experience).

We have looked directly at the question of the hierarchical structuring of conscious experience in an effort to disentangle subconscious from conscious processing in the context of goal-oriented psychophysical tasks (Mathews et al. 2015). Using a displacement detection task combined with reverse correlation, it was shown that bottom-up fast saccades, top-down driven slow saccades, and conscious decisions follow distinct regions of the sensory space, or validation gates, modulated by the conscious task the subject performs. This experiment demonstrated that conscious decision making can be largely dissociated from subconscious parallel processing; it also provides support for the DAC notion of parallel layered control and for the view that consciousness provides a time-delayed description of an effective task that comprises a subset of the world in which the subject is acting. In addition, the idea of a continuous perceptual hierarchy linking sensation to perception and experience, popular in current Bayesian brain accounts (see Friston, this volume) does not hold up under these conditions; such hierarchical relations are dynamically formed dependent on the task conditions faced by the self.

Conclusion

In this chapter, I have presented the DAC H5W theory of consciousness in the context of a historical cycle in the study of mind, brain, and behavior and discussed the GePe framework, capturing contemporary science and philosophy of consciousness studies. I propose that returning to the question of the function of consciousness and its implementation by the brain is a historical prerogative, if we want to avoid sliding down the sinkhole of big data and the scientism of explanatory nihilism. It is important to emphasize that we should avoid the fallacy of mistaking our measures for the phenomena they are designed to measure, as behaviorism discovered at its peril. One can interpret the move into neo-panpsychism of the proponents of IIT as an artifact of such a scientism fallacy. Indeed, IIT and the Bayesian brain frameworks illustrate a trend to collapse the complexity of mind and brain into relatively simple quantitative measures, uncoupled in any relevant way to the neuronal substrate or action (i.e., the levels of observation that are accessible for a science of mind). The global workspace framework faces a similar problem of multi-instantiation; it could be realized in arbitrary hardware systems and is not constrained by any fundamental property of the brain. Hence, the challenge is to derive a convergent science of consciousness that is able to show how the brain generates and expresses consciousness in action.

I propose that consciousness is critically related to action in an intentional world or the transition from an agent that solves H4W to solving H5W.

Consciousness provides the interface between the singular self and the parallel world. In this proposal, conscious is by necessity intentional because it pertains to a single agent engaged with an intentional world. Grounded in the physical existence of the agent over time, the self-constructed conscious narrative defines its subjectivity and quale and assures the coherence of its operation. Thus consciousness is the coherent experience that results from the large-scale integration of perception, affect, memory, cognition, and action along the neuroaxis in a dedicated memory system. It is a form of memory that unifies and interprets the states of the agent to facilitate the optimization of its parallel real-time control loops that are driving action. This memory is only active when the agent is. This H5W hypothesis predicts that the conscious scene is a transient memory implemented in the thalamocortical system, which provides a unitary description and valuation of real-time performance and is able to project this valuation onto the parallel control loops of the brain (e.g., those found in the cerebellum). In this way the solution to the H5W problem reinstates free will as the ability to *will* improvement of performance in the future, as opposed to stopping at the contemporary interpretation that we lack the will to control our performance in the "now." Thus, the H5W hypothesis aims at explaining consciousness as a natural phenomenon—a property of specific biological systems that emerged during the Cambrian to act in an intentional world to survive. It would be premature to say that DAC has *explained* consciousness, but we can observe that it does capture the main components of the GePe framework, while advancing a concrete research agenda that poses specific questions about perception, emotion, cognition, and actions structured along H5W. With this in hand, we can turn to the more specific question of the functional role of consciousness and what this would imply for future extensions of the DAC theory.

Others have also advanced hypotheses which emphasize the social origins of consciousness (Mead 1934; Humphrey 2006; Baumeister and Masicampo 2010; Graziano 2013). These proposals have emphasized the contribution of consciousness to specific aspects of social interaction (e.g., rational thought, language, attention). The DAC H5W hypothesis emphasizes the role of consciousness in optimizing the control structures that social interaction and its underlying intentional stance requires.

As we continue to analyze the pragmatic turn in cognitive science, we need to be mindful of the damage previous forms of pragmatism have caused: behaviorism was the primary cause behind the disappearance of consciousness from the scientific landscape, because its methods and philosophy could not address it. A similarly dogmatic narrow view must be avoided at all costs. We need to return to a science that insists on gnawing the file of consciousness until it gives way to a deeper understanding of nature and ourselves.

Acknowledgments

I am grateful for the feedback I have received to an earlier draft of this chapter by the two reviewers appointed by the Forum. The work reported here is supported by the EU Integrated Project CEEDS funded under the Seventh Framework Programme (ICT-258749) and ERC grant cDAC (ERC-341196).

First column: (top to bottom): Anil Seth, Kevin O'Regan, Richard Menary, Marek McGann, Pierre Jacob, Judy Ford, Pierre Jacob
Second column: Paul Verschure, Judy Ford, Miriam Kyselo, Anil Seth, Chris Frith, Richard Menary, Martin Butz
Third column: Ezequiel Morsella, Chris Frith, Peter König and Olaf Blanke, Martin Butz, Miriam Kyselo, Paul Verschure, Kevin O'Regan

15

Action-Oriented Understanding of Consciousness and the Structure of Experience

Anil K. Seth, Paul F. M. J. Verschure, Olaf Blanke,
Martin V. Butz, Judith M. Ford, Chris D. Frith,
Pierre Jacob, Miriam Kyselo, Marek McGann,
Richard Menary, Ezequiel Morsella, and J. Kevin O'Regan

Abstract

The action-oriented approach in cognitive science emphasizes the role of action in shaping, or constituting, perception, cognition, and consciousness. This chapter summarizes a week-long discussion on how the action-oriented approach changes our understanding of consciousness and the structure of experience, combining the viewpoints of philosophers, neuroscientists, psychologists, and clinicians. This is exciting territory, since much of the resurgent activity in consciousness science has so far focused on the neural, cognitive, and behavioral correlates of perception, independent of action. Our wide-ranging discussions included questions such as how actions shape consciousness, and what determines consciousness of actions. The specific context of self-experience, from its bodily aspects to its social expression were considered. The discussions were related to specific theoretical frameworks, which emphasize the role of action in cognition, and identified an emerging empirical agenda including action-based experiments in both normal subjects and clinical populations. An intensive consideration of action is likely to have a lasting impact on how we conceive of the phenomenology and mechanisms of consciousness, as well as on the ways in which consciousness science will unfold in the years ahead.

Introduction

There is a renewed emphasis within cognitive science on the role of *action*. Set in contrast to classical paradigms which emphasize computation involving mental representations, action-oriented perspectives emphasize the enactive, embodied, and embedded nature of cognitive systems. While there are many variations of the action-oriented approach (for summaries, see Thompson and Varela 2001; Wilson 2002; Engel et al. 2013; Dominey et al., this volume), they have in common the important idea that actions are not just the outputs of a cognitive system; rather, cognitive processes are shaped and may even be partly constituted by the actions they subserve. Accordingly, cognition is "for" action, not for the generation of abstract world models subserving planning and problem solving. In this chapter, we examine how this pragmatic turn in cognitive science (Engel et al. 2013) impacts our understanding of consciousness and the structure of experience. This is challenging and exciting territory, especially when set in contrast to prevailing approaches in consciousness science, which generally studies the neurocognitive correlates of perceptual scenes (e.g., Dehaene and Changeux 2011) independently of action.

There is a trivial sense in which action impacts consciousness through the selection of sensory samples. The action-oriented approach suggests much deeper influences, which are explored in this chapter. We start by offering some working definitions of consciousness and action, noting that not only can action shape consciousness, but also that we are (sometimes) conscious of actions (our own, and those of others). Thereafter we outline candidate theoretical frameworks, which turn out to be useful in organizing a discussion of action and consciousness: the Bayesian brain (Friston 2009; Clark 2013b; Seth 2014), the sensorimotor contingency (SMC) theory (O'Regan and Noë 2001; O'Regan 2011), the distributed adaptive control (DAC) theory (Verschure et al. 2003), and enactive autonomy approaches (Varela et al. 1992; Di Paolo et al. 2010). Two key questions are addressed: How does action shape consciousness, and what determines our conscious awareness of action?[1] Notions of *goal-directedness* and *hierarchical organization* (of functional architectures) turn out to be critical in these discussions.

We next examine possible neuronal substrates and functional roles relating action and consciousness, again capitalizing on the four candidate theoretical frameworks. This leads to an analysis of action and consciousness in the specific setting of the *self*, where "selfhood" can be understood to operate at multiple levels: from physiological homeostasis and interoception, to social and cultural structures and norms. We consider the notion of *joint action* (in particular, the mother-infant dyad) to be an especially illuminating example of how also social influences structure conscious experience through action.

Considerations of selfhood provide a useful context to identify specific empirical challenges for an action-oriented view on consciousness. We discuss whether such a view could shed new light on pathologies involving disordered conscious experience (e.g., schizophrenia) and whether pathologies of motor control (e.g., like locked-in syndrome, and amyotrophic lateral sclerosis) might illuminate the action-consciousness relation. These considerations highlight

---

[1] "Consciousness" and "awareness" are used synonymously here.

the experimental opportunities and limitations in studying consciousness from an action-oriented perspective, and we close with a brief survey of future and emerging directions.

Consciousness and Action: What Are They?

What Is Consciousness?

Discussions of consciousness always confront issues of definition. Watertight definitions are not needed in advance of scientific progress; they advance in lockstep and eventually emerge from a mature understanding of mechanisms. Here we offer instead some basic "identifications," recognizing that these cannot do justice to the diversity of views on what consciousness "is."

A first distinction can be made between *creature consciousness* (whether X is conscious at all), which can be titrated into distinct conscious *levels* (e.g., from dreamless sleep to vivid conscious wakefulness), and *state* or *content* consciousness (what X is conscious of; e.g., the components of a conscious visual scene at a given time) (Rosenthal 2005; Seth et al. 2008). Importantly, conscious contents are, at least for humans, remarkably diverse. Very broadly they include experiences of the world (including other selves), of one's own body (from the inside as well as from the outside), of action, of emotion, and even of abstract cognitive operations (cognitive phenomenology). Having conscious contents about one's own mental states is sometimes referred to as "higher-order" or "reflexive" consciousness (Rosenthal 2005). In what follows, we focus primarily on conscious contents, as opposed to creature consciousness.

A second and more controversial distinction is between *phenomenal consciousness* and *access consciousness* (Block 2005). Roughly, the former means "what it is like" to have a particular experience, whereas the latter refers to the information in conscious experiences that is accessible or available to consumer cognitive mechanisms, which include those mechanisms that can supply explicit behavioral report. "Report" here means an action (verbal or otherwise) that conveys what one is conscious of to an experimenter. Importantly, since report itself is an action, it may shape or even constitute conscious contents.

Other definitions of consciousness are more closely tied to particular theoretical frameworks. For instance, in SMC theory, consciousness is defined as a set of abilities to interact with the world (O'Regan and Noë 2001), whereby we are (phenomenally) conscious of X when X is a quality defined by a SMC and when we are poised to use X for flexible behavior. SMC theory is an example of an approach according to which consciousness constitutively rests on action.

What Is Action?

Action is simpler to define, though there are still interesting boundary conditions. A straightforward approach is to say that *an action is any goal-oriented manipulation of an external or internal situation* (Jeannerod 2006). Not all movements are actions, and not all actions are movements. "Covert" actions that do not involve movements include top-down attention switching, manipulation of mental states, and autonomic control including, for example, glands and smooth muscle control. Covert actions could also include planned but unexecuted bodily movements. In addition, the same movement can be an action in some situations but not in others (contrast the patellar reflex with an attempt to kick a ball), and the same movement can participate in different actions depending on the goal (waving a hand to scare a fly or to say goodbye). Thus, the key feature of actions is the association with a *goal* or *intention* (Dretske 1988).

Actions can also be classified as *instrumental*, *epistemic*, or *communicative*, according to the goal (Kirsh and Maglio 1994; Gergely and Jacob 2013; Seth 2015). The goal of an instrumental action is to effect a change of a particular kind. The goal of an epistemic action is to generate new information. The goal of a communicative action is to effect a change in somebody else. Note that communicative actions can be either epistemic (to indicate to somebody else a property of the world or of a mental state) or instrumental (to get somebody to do something).

From a neurobiological perspective we can distinguish multiple levels of the control of the skeletal-muscle system starting with the spinal cord reflex circuits, which directly control the skeletal-muscle system defining movement primitives (Mussa-Ivaldi and Bizzi 2000), and brainstem-dependent discrete behaviors such as eye blinks, grasp, and posture associated with reticular formation and the red nucleus regions, and stereotyped patterns such as those involved in feeding, defense, and reproduction regulated by the periaqueductal gray (Panksepp 2005). These latter systems interface with learning systems that can render discrete experience-dependent action such as those observed in classical conditioning, which depend on the amygdala and the cerebellum. Lastly, forebrain structures (e.g., cerebral cortex, basal ganglia) are centrally involved in goal-directed voluntary actions which can comprise complex sequences of movements. These different layers of movement, behavior, and action are all tightly coupled in the interaction between the organism and the environment.

Frameworks for Cognition and Action

Considering the importance of action within consciousness can be helped by declaring (though not necessarily endorsing) *frameworks* or *architectures* which express particular theories or provide structures by which action and experience can be related. Here we consider four candidate frameworks which put specific emphasis on action (see Figure 15.1):

1. Bayesian brain (Friston 2009), when equipped with concepts of active inference (Friston, Samothrakis et al. 2012; Seth 2014; Friston, this volume)
2. SMC theory (O'Regan 2011; O'Regan and Noë 2001)
3. Distributed adaptive control (Verschure et al. 2003; Verschure, this volume)
4. Enactive autonomy and autopoiesis (Varela et al. 1992; Di Paolo et al. 2010)

The Bayesian Brain

Within the Bayesian brain approach, perception is understood as a process of inference on the (hidden) causes of sensory signals. Although its origin in the work of Helmholtz emphasizes that the mechanisms and processes of inference can be (and usually are) *un*conscious, the outcomes from this process may shape or constitute conscious contents. An emerging consensus suggests that conscious phenomenology is shaped more by (Bayesian) priors or top-down expectations than by (bottom-up) prediction errors (Melloni et al. 2011; Hohwy 2013; Chang et al. 2015; Mathews et al. 2015), a position which fits nicely with evidence for the importance of top-down signal flow for consciousness (Lamme and Roelfsema 2000; Pascual-Leone and Walsh 2001; for a review, see Lamme 2010).

Three aspects of the Bayesian brain approach deserve emphasis in the context of the current discussion. First, the Bayesian brain is hierarchical, so that posteriors at one level can form priors in the level below, instantiating a process of "empirical Bayes." This means that high-level goals or intentions can percolate throughout the hierarchy to shape priors at levels descending all the way to the sensory epithelia or spinal cord.

Second is the concept of *active inference*, which says that prediction errors can be minimized not only by updating prior predictions but also by performing actions to change sensory samples. Accordingly, the active inference view underlies both perception and action: actions are generated through the minimization of proprioceptive prediction errors through engagement of classical reflex arcs (Friston, Samothrakis et al. 2012). Importantly, active inference emphasizes the deployment of predictive models for control rather than representation, calling on parallels with theories of predictive homeostasis in cybernetics (Conant and Ashby 1970; Seth 2015).

Third, priors, predictions, and prediction errors are always associated with *precisions* (inverse variances), which determine how strongly they affect inference. Attention corresponds to optimization of precision weighting, which corresponds to modulating the gain of prediction errors at specific hierarchical

266

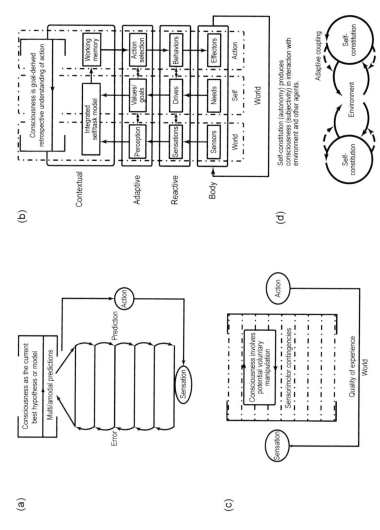

Figure 15.1 Schematics of four theoretical frameworks underlying discussions of action-consciousness relations: (a) Bayesian brain or predictive processing, (b) distributed adaptive control (DAC); (c) sensorimotor contingency (SMC) theory, and (d) enactive autonomy. Reprinted with kind permission from Ezequiel Di Paolo.

levels (Feldman and Friston 2010). This modulation impacts the updating of the forward (generative) models underlying inference.

Although the "standard" Bayesian brain approach does not offer an explicit account of how goals are generated, broader formulations under the rubric of the free energy principle link goal states to a fundamental imperative to the maintenance of homeostatic or allostatic physiological integrity (Friston 2010; Seth 2015) through active inference. Explicit mechanistic models of goal-emergence in this framework remain, however, to be elaborated (see, however, Friston et al. 2014).

Sensorimotor Contingency Theory

SMC theory holds that perception is an activity involving skillful engagement with the world. Inheriting from the Gibsonian notion of "affordance" (Gibson 1979), conscious perceptual content is given by mastery of the SMCs, which are regularities governing how sensory stimulation depends on the activity of the perceiver. For instance, the conscious perceptual quality of redness is given by an implicit knowledge (or mastery) of the way red things behave given specific actions. Thus, in SMC theory, action plays a constitutive role in conscious phenomenology.

SMC theory also gives primacy to intentional actions, which effect manipulations of the objects of perception. Accordingly, conscious perception requires that the actions underpinning SMCs be exercised intentionally. Provocatively, there can be no conscious phenomenology without (potential) voluntary action.

Distributed Adaptive Control

DAC theory explains the mind in terms of the embodied brain interacting with its environment (Verschure et al. 2003; Verschure, this volume) and postulates that brains evolved to generate action to maintain the agent by means of a multilayered control structure (see Figure 14.2 in Verschure, this volume) involving somatic, reactive, adaptive, and contextual layers. Across these layers, three columns address states of the *world*, the *self* and *action*. DAC thus proposes a highly specific architecture that has been matched to specific brain regions and realized in real-world systems. It assigns explicit roles for goal representations in the structuring of action, as well as for Bayesian perception as inference. DAC also proposes that internal representations of SMCs lie at the heart of goal-oriented action, providing a link to SMC theory (although DAC, but not SMC theory, endorses a representationalist functional architecture). DAC also expresses central features of the enactive autonomy view (see below), by taking coupled mind, brain, body, and environment systems as its explanatory target.

The DAC architecture proposes a specific role for consciousness: the retroactive reconstruction of the sequentialized course of action(s) that led to goal achievement, to enable the derivation of norms and values for the optimization of the unconscious parallel control systems that generated the course of action in real time. Put simply, consciousness serializes and interprets the present to define future goal-oriented action (Verschure, this volume).

Enactive Autonomy

This framework emphasizes the importance of autonomy and self-organization in cognition (Varela et al. 1992; Di Paolo et al. 2010) and shares with SMC theory an emphasis on skillful engagement between the agent and its environment. It is distinguished, however, by an emphasis on how this engagement supports the autonomous identity of the agent (or, more generally, system) (Thompson 2007). Accordingly, an autonomous system is one where component processes that continually self-produce the system's identity have a mutual dependence, creating an implicit organizational identity that the system maintains. This ongoing need for self-maintenance forms the basis of inherent values and goals for the system, and where such an agent can adaptively modulate the way it is coupled to the environment, these values and goals form the basis of cognitive activity (i.e., how it makes sense of, or adaptively copes with, the world around it). This emphasis echoes both the Bayesian brain approach, when seen in the wider context of the free energy principle, which grounds inference on the maintenance of organismic homeostasis (Friston 2010, 2013) and the DAC theory, which is predicated on the interaction between self-regulatory allostatic-embodied agents and their environment (Verschure et al. 2003). In contrast to the Bayesian brain and DAC approaches, proponents of enactive autonomy tend to place a very strong emphasis on phenomenology, but say less about the specifics of underlying mechanisms.

Summary

These frameworks are just four among a range of possibilities, chosen for their broad representation within the field of cognitive science and their explicit consideration of either action or consciousness (or both). The frameworks are not necessarily exclusive. SMC theory and enactive autonomy both underline the importance of skillful engagement with the world, though neither makes claims about mechanistic implementation. It is possible to provide SMC theory with a mechanistic foundation in Bayesian terms by conceiving of SMCs as (active and counterfactually informed) inferences about sensorimotor regularities (Seth 2014) or via the principles of DAC theory (Verschure et al. 2003). Free-energy principle readings of the Bayesian brain also connect with the enactive emphasis on organismal homeostasis and autopoeisis (Friston 2013). Finally, the DAC architecture is more concrete than SMC theory (and enactive autonomy) or Bayesian brain while arguably encompassing concepts from both.

Action and the Structure of Experience

Equipped with these working definitions and candidate frameworks, we turn to the core questions regarding the relationships between consciousness and action, broadly grouped into two categories: the influence of action on consciousness and awareness of action.

Influence of Action on Consciousness

There is a trivial sense in which action determines conscious contents, by specifying the configuration of sensory receptors in relation to the environment (e.g., through eye movements). In our discussions of the action-oriented approach, we asked whether there are deeper, nontrivial influences that arise from the tight coupling of perception and action.

Prefiguring this question, Sperry (1952) noted that the outputs of a system are often more informative about its working than are its inputs. For instance, one can learn much more about the inner workings of a blender by observing its output (e.g., juices) than its input (e.g., fruits). Sperry thus concluded—consistent with Gibson—that conscious percepts are more isomorphic (i.e., similar in form) with potential action plans than with the proximal patterns of sensory inputs.

A global perspective on this issue comes from Merker, who argues that the organization of the entire phenomenal or conscious field is defined to support flexible action. For adaptive influence over the motor domain, most conscious contents should appear *as if* they transpired from a first-person perspective (Merker 2013) (perhaps with the exception of contents underlying communicative actions). In addition, accurate spatial representation is essential for adaptive action selection. It would not be adaptive for a nearby object to be represented as if positioned far away. Insofar as the action selection process in the motor domain must take into account *spatial distance from the organism* as one of its factors in the selection process, then all perceptual contents about the external world (including the body) must have a common, egocentric reference (Merker 2013). An interesting corollary of this idea is that different action possibilities (e.g., from different body morphologies) would necessarily give rise to different organizations of the global conscious scene.

A second relatively global aspect reflecting the influence of action on consciousness concerns the influence of goals. Several issues are involved: The existence of a goal distinguishes an action from a (mere) movement (a kick from a patellar reflex). In addition, goals distinguish between different types of action (instrumental, epistemic, communicative). These distinctions may operate at the level of the unfolding movements (internal or external) or at the level of how resulting changes in sensory samples are used to update conscious perceptions. For example, it seems plausible (although evidence is lacking) that a movement, if deployed as an epistemic action, may preferentially modulate conscious contents related to a (target) object of the world, but when deployed as a communicative action may modulate specific conscious contents related to others' mental states. Another way to say this, again using vision as an example, is that we never simply "see"; rather, we "see socially" or "see walkingly" (Gibson 1979).

There are several examples where action and action preparation have been shown to impact directly on conscious contents. In one study, distance perception was found to be altered when participants were holding a baton for use in reaching movements (Witt et al. 2005). Vishton et al. (2007) extended this finding in the context of the Ebbinghaus illusion, showing that planning a reaching movement affected the perceived size of a visual object. Interestingly, Fleming et al. (2015) recently demonstrated action-specific disruption of perceptual metacognition. In a simple visual discrimination task, they used transcranial magnetic stimulation to disrupt motor responses underlying the response not chosen (targeting premotor cortex). Although objective discrimination performance was unaffected, confidence in correct responses was reduced. This selective reduction in metacognitive capacity implies an effect of motor representations of unexecuted actions on conscious perception.

In a different context, Hayhoe et al. (2003) used eye-tracking data to show how visual memory is affected by motor planning in natural tasks (e.g., making a sandwich). Their results suggest that object- and hand-relative spatial structures as well as object identities are determined and constructed on-the-fly. Thus, current working memory content appears to be modified by current intentional actions, and this is relevant to conscious content inasmuch as current conscious contents are closely associated with working memory (Bor and Seth 2012; see also Hagura et al. 2012, who show that motor planning of ballistic reaching movements induces subjective time dilation).

Other examples of modulation of conscious content by action include experiments which show that directional instructions or directional intentions can prime—or bias—the perception of bistable, ambiguous visual motion displays in the direction of the intended action (Wohlschlager 2000). Similarly, Butz et al. (2010) have shown that rotating tactile stimuli can bias the conscious perception of similar, bistable ambiguous visual motion displays. In this case, tactile bias depended on the orientation of the hand in space relative to a head-centered frame of reference (see also Salomon et al. 2013, who show that proprioceptive signals can bias conscious access, in a masking paradigm).

More dramatic effects of action on consciousness can be found in the domain of self- and body-related consciousness (discussed later) and in the phenomenon of sensory suppression. It is well known that the performance of actions causes the *attenuation of sensory experiences*, which explains why self-tickling is largely ineffective (Blakemore, Wolpert et al. 2000). Recent interesting evidence suggests that actions may suppress *auditory* sensory experience across a broad range of actions, mediated by direct ipsilateral projections from secondary motor cortex to auditory cortex (Schneider et al. 2014).

Behavioral report (i.e., actions which convey to an experimenter information about current conscious contents) provides an intriguing context in which to consider the influence of action on consciousness. Methodologically, it is (on most accounts) imperative for investigating consciousness, as explicit reports (verbal or otherwise) are the primary means for obtaining information about the conscious contents of another person. Yet report, by definition, involves a communicative action. The question therefore arises as to whether report actions change or shape the reported experience. As yet, evidence is thin. Recent work has shown that frontal brain activations, often associated with (access) consciousness, are absent when report is not required (Frassle et al. 2014), but this does not demonstrate directly any change in the experience as a consequence of report. This question also pertains to early work by Marcel (1993) which demonstrated that different perceptual judgments were made when different reporting channels were employed (e.g., button presses vs. eye blinks).

Awareness of Actions

The relationship between consciousness and action is perhaps most directly expressed in *when* and *how* we experience actions themselves. To underline this point, awareness of actions should be understood as awareness of the actions themselves (at different levels of abstraction), not as awareness of the consequences of actions. A general way to think of action awareness is that a focus on action execution and on the required motor (or mental) control (i.e., a focus on the control of the unfolding SMCs) might constitute the continuously experienced phenomenology of action.

An examination of action awareness first requires, however, a phenomenology of action. Compared to perceptual phenomenology, the terrain is only dimly lit (Pacherie 2008). The phenomenology of action is "thin and evasive" (Metzinger 2006). It is useful to characterize two distinct dimensions along which action phenomenology can be organized. The first reflects a scale from high-level goals and intentions to low-level individuated muscle contractions prescribing movements (or their equivalents for autonomic control or mental acts). The second distinguishes experiences of intention from experiences of agency or authorship.

An extensive line of work, originating with Benjamin Libet, has studied experiences of intention and agency (for a helpful review, see Haggard 2008). Put simply, the experience of intention has to do with awareness of a goal and is sometimes associated with the phenomenology of an "urge," whereas the experience of agency depends on action-outcome association, often reflecting goal fulfillment. A now classic finding in this area is that of *intentional binding*: if an event is experienced as the effect of an intended action, the time between the action and the event is perceived as being shorter in duration (Haggard et al. 2002). This not only provides an additional example of an effect of action on consciousness (in this case, of perceived time), it also shows that the conscious experience of agency depends on both the prior expectations about outcome of the action and the observed outcome of the action. This implies that a role for action in consciousness is to create the experience of agency; that is, I am in control of my actions, in the sense that it was I who chose the option (action selection) and caused the outcome, and that I could have chosen another option, giving rise to the possibility of regret (Frith 2014). It is interesting to note that voluntary actions appear to have a specific kinematic signature, as compared to stimulus-driven actions (Becchio et al. 2014).

Goals remain important in action awareness. Our awareness of our actions tends to follow the scale of the goal driving the action in question. Habitual and well-practiced actions are performed relatively automatically, with little deliberate awareness. Generally, as our expertise increases so does the abstractness of associated goals, and this changes our awareness of the situation (Speelman and Kirsner 2005). Expert musicians are frequently as focused on the emotional tone of their performance as on the playing of the music, much more so than the movements of their fingers required to produce notes. Similarly, your consciousness of your actions in a conversation has much more to do with the message you wish to convey, or the effect you wish your words to have, than on the physical movements involved in producing the sounds (and sometimes even on the choice of words themselves). The execution of practiced action is frequently robust to minor perturbations (as in the case of dynamic stability of the articulation system for jaw movement interruptions; Kelso et al. 1984). However, where an action sequence meets failure, when goals and reality sharply diverge, we become keenly aware of the more fine-grained resolution of the movements involved.

Architectures, Mechanisms, and Implementations

At this point it is useful to return to the four candidate frameworks (Bayesian brain, SMC, DAC, enactive autonomy) to try to connect action-consciousness relations to specific mechanistic properties. The four frameworks specify these relations to different degrees, as will be seen.

Under the Bayesian brain account, conscious content can be associated with the Bayesian model or "hypothesis"—spanning multiple hierarchical levels—that best suppresses prediction error (Hohwy 2013). When these hypotheses have to do with proprioceptive (and possibly vestibular[2]) predictions, they reflect awareness of actions. As discussed above, a key feature of the Bayesian brain approach is the deployment of precision weighting (attention) to optimize inference. Accordingly, the specific phenomenology of action may

---

[2]   The "broken escalator" phenomenon is a wonderful example of vestibular mis-prediction. It is the sensation of losing balance reported by most people when they step onto an escalator that is not working. Strikingly, the illusion is not diminished by knowledge that the escalator is broken (Reynolds and Bronstein 2003).

depend on the hierarchical assignment of precision to proprioceptive predictions. Action awareness may shift from low-level movements to high-level goals during the acquisition of expertise. The converse shift may occur when an action is blocked, frustrated, or otherwise goes wrong: the resulting cascade of prediction errors will lead to reassignment of precision weighting targeting lower hierarchical levels. (If I try to pick up a glass and almost knock it over, I suddenly become aware of the specific movements of my fingers.) Sensory attenuation is also naturally accommodated within the Bayesian brain framework: the precision of proprioceptive prediction errors needs to be underweighted so that predictions are fulfilled rather than updated. Under the SMC theory account, actions (including "mental" actions) are constitutive of consciousness. SMC theory makes no claims about specific brain mechanisms and thus is compatible with both Bayesian brain and DAC accounts. Enactive autonomy approaches, however, tend to be incompatible with representational accounts (Hutto 2012).

With respect to goals, priors at one level of a hierarchy of control operate as goals for the level below. The character of our awareness of an action may in this view be governed by the level of either highest prediction error (indicating where more attention and control is most needed) or highest prediction confidence (where action is at the most abstract level of description consistent with the amount of practice or habituation). DAC theory makes a similar prediction, pointing out that consciousness involves a maintenance of coherence between an agent's predictions (including forward models of its own actions) and the interaction with the world (Verschure, this volume). Within the DAC perspective, consciousness serves to derive value from goal-oriented action and its outcomes through retrospective reconstruction, thus marking a distinction with generic Bayesian brain approaches in which conscious experiences are temporally aligned with the dynamics of perception and action. Both SMC and autonomy-based approaches assume that consciousness always involves action or potential action, so that our awareness of our own actions is complementary to our awareness of properties of the world. Insofar as we are frequently engaged with multiple goals simultaneously—or at least hierarchies of goals at different granularities—conscious experience of our actions and the world will be variable and textured by the various forms of goals and the associated skills being deployed (O'Regan and Noë 2001; Di Paolo et al. 2010; McGann 2010; O'Regan 2011).

Multiple Levels of Self-Experience

Experiences of being a self, having a body, and perceiving the world from a first-person perspective are intimately tied to action, perhaps more so than experiences of the external world. Some aspects of self-experience have already been discussed (namely, experiences of intention and agency). Other

more basic experiences of selfhood have to do with *being* and *having* a body
as well as the first-person perspective (Merker 2013). At the other extreme,
self-experience—at least in humans—is co-constituted by social interactions
and the elaboration of a self-narrative which evolves over time. Importantly, all
aspects of the self involve actions, but the actions are different, ranging from
real bodily actions to mental and ultimately social actions.

The Bodily Self

Considering the body brings into focus processes of *interoception* (sense
of internal physiological condition) and *autonomic control* (Craig 2003), as
well as multisensory integration (Blanke and Metzinger 2009; Tsakiris 2010).
Importantly, autonomic control signals can be thought of as internalized actions: they are associated with goals and have discrete somatic effects, following a hierarchical structuring akin to that associated with external action
(see above). At the foundation of this stand notions of homeostatic control and
allostasis (i.e., maintenance of multiple homeostatic loops) linked to concepts
of drive reduction (Sánchez-Fibla et al. 2010). A long tradition extending at
least as far back in time as James and Lange has associated conscious emotional or feeling states with the perception of changes of physiological condition
(James 1894). Recently, this tradition has been extended via the Bayesian brain
approach to suggest that emotional states arise through a process of *active
interoceptive inference* (Seth 2013, 2015).[3] Accordingly, autonomic control
can be understood as implementing active interoceptive inference to suppress
interoceptive prediction errors and as a consequence, maintain physiological
homeostasis.[4] Thus, internal actions may be fundamental to emotional experience. In this light, it is fascinating to note recent evidence for direct projections
from motor cortex to internal organs (Levinthal and Strick 2012).

Interoceptive signals also influence the experience of body ownership and
first-person perspective, another key constituent of conscious selfhood. This is
shown by the fact that individual interoceptive sensitivity predicts susceptibility to the "rubber hand illusion" (Tsakiris et al. 2011), and that cardiac feedback,
presented visually on augmented-reality representations of body parts and ava-
tars, can modulate experienced ownership and first-person perspective depending on the accuracy of this feedback (Aspell et al. 2013; Suzuki et al. 2013).

Action seems especially critical for shaping the experience of body ownership and first-person perspective. This might be because actions induce specific correlations between proprioceptive, visual, and perhaps other modalities,

[3] SMC theory claims that only those internal actions that are voluntarily controlled and executed
with an intent to sense an individual's emotional state can be experienced consciously.

[4] Alternative views associate emotion with a prediction error between allostatic targets and
states of the world (Verschure 2012b), rates-of-change of prediction error or free energy (Joffily and Coricelli 2013), or embodied predictions shaping action selection (Barrett 2012).

which can be used to infer a distinction between one's own body and the rest of the world. These actions, therefore, are epistemic actions and might be related to motor babbling during infancy (Bullock et al. 1993). Notably, proprioceptive motor tactile correlations, along with interoceptive signals, are available before birth and thus might play an extremely important role in the development of the experience of body ownership before vision comes into play (Gori et al. 2008; Rochat 2010).

The Stable Self

A Bayesian brain perspective may shed light on the phenomenology of self-stability: though the experienced contents of the world continually change, our experience of "being a self" is highly continuous (as pointed out by William James, long ago). The reasoning is as follows. As we move through the world (and continuously move our eyes), we are bombarded with an ever-changing sequence of sensory signals and experiences which, in their raw form, are extremely difficult to predict (by analogy with pre-Copernican attempts to predict the motion of the planets). The solution is to develop the prior that there is a self, moving through a relative stationary and stable three-dimensional world (Butz 2008). Given this prior, we can then predict the changing pattern of stimulation received by our senses. The same process applies in the social sphere. Predicting the various interactions we have with others is greatly simplified given the prior that there is population of individuals, each with their own (and this includes the self) relatively stable preferences and styles (Robalino and Robson 2012).

The Social Self and Joint Action

An important dimension of the experience of selfhood and the structure of the sense of self involves intersubjectivity and our social and cultural relations with the environment. Several, not necessarily compatible, perspectives in current embodied cognition help shed light on this issue. Here we broadly distinguish three: (a) the neural implementation of the social self, (b) the particularly rich set of evolved social abilities in humans, and (c) the intersubjectivity as constitutive for minimal sense of self and the dynamic construction of embodied identity.

Neural Implementation

One way to think about the (experience of the) social self is in terms of its neurological implementation and the question of how the social self can be functionally represented and processed in the brain. "Mirror" neurons and activations in biological motion areas when an individual performs and observes actions (Rizzolatti and Sinigaglia 2010) suggest that actions, intentions, and goals of others are partially represented by means of self-representations. The resulting inference-making capabilities about others imply that we understand others somewhat in the way we understand and control ourselves. Importantly, once such representations and simulations of others are active and overlap with our self representation and self-control mechanisms, the representations of others need to be separated from those pertaining to self (Gallese et al. 2004; Butz 2008). Moreover, by simulating actions of others through self-grounded representations, perspective taking becomes possible in a social context (including self-reflection) by monitoring one's self from the perspective of another person (Frith and Frith 2006; Hassabis et al. 2014).

Social Abilities and Joint Action

Distinguishing self-representations from representations of others is particularly important in the performance of cooperative and communicative actions, awareness of communicative goals, and awareness of others' psychological states. Indeed, much comparative evidence suggests that compared to nonhuman primates, humans are uniquely cooperative (Tomasello 2014) in at least three fundamental ways: through *trust*, a basic ability for *perspective taking*, and a basic receptivity to *cultural learning*, each of which contributes to shaping the human social self. First, cooperative breeding, which is uniquely human among great apes, requires an unusually high level of trust in one's in-group members (Hrdy 2009). Second, humans engage in joint actions in an unprecedented way compared to nonhuman primates. In joint action, two (or more) agents must represent a common goal and fulfill two (or more) complementary roles, and switch roles according to context. Arguably, role taking and role switching require some perspective taking (see above). Third, developmental evidence suggests that human infants are uniquely tuned to the ostensive signals that reflect assumed intentions to convey relevant information. These ostensive signals include speech in "motherese," contingent responses, and direct gaze, which is facilitated by the fact that, compared to the eye of nonhuman primates, the human eye involves a uniquely dark iris on a white background (Kobayashi and Kohshima 1997). In addition to this three-tiered basic social self, the social self is further enriched by membership to linguistic, religious, nationalistic, and other groups of various sorts.

Autonomous Dynamics in the Social Domain

An important question emphasized by the enactive autonomy perspective is whether social interactions are constitutively necessary for individual agency and self-identity. Social contexts can be incorporated into Bayesian brain, SMC, and DAC through specification of hierarchically high-level predictions that transcend the individual. The enactive autonomy account (at least in line with most readings) goes further in this regard: the social is not merely the context in which cognitive systems are embedded, but figures in the constitution of agency and self-identity. In other words, an agent's mechanism of self-organization is not fully determined in terms of *individual* sensorimotor activity but is continuously *co*-enacted with others. Arguably, this entails a sense of openness and connectivity associated with individual agency and selfhood (Kyselo 2014).

Empirical Challenges

The issues discussed so far in this chapter point to a number of important empirical challenges, which we outline below.

Disorders of Motor Control

If action is important in shaping, or even constitutive in, conscious contents, then disorders of motor control (where the capacity for movement is severely diminished or even lost altogether, as in amyotrophic lateral sclerosis or locked-in syndrome) should have measurable effects on consciousness. In addition to investigating basic levels of consciousness in these patients, it would be interesting on a more fine-grained level to use functional magnetic resonance imaging to examine a range of action-related brain responses. These include whether the mirror system still responds to the perception of goal-directed actions (e.g., reaching and grasping) executed by others, whether the superior temporal sulcus still responds to the perception of biological motion, and whether one finds responses in motor systems, as in healthy subjects, during perception of action sentences or action verbs (e.g., "kick," "lick," "pick"). Finally, one might examine whether their responses are in accordance with the somatotopic representation of the respective effectors with which the action described by the verb is normally executed, as has been found by Pulvermüller and colleagues in healthy subjects (see Pulvermüller 2005; Pulvermüller, Hauk et al. 2005).

Another interesting context arises in neuroprosthetics, for instance in cases where a robotic arm is controlled directly by brain signals obtained from motor cortex (see Dominey et al., this volume). Here, in some cases it is possible to modulate explicitly the degree to which the robotic arm is controlled by the brain signals, or by "helping" signals provided by an external computer (given an identifiable goal, like reaching toward a target) (e.g., Miele et al. 2011). This raises the intriguing possibility of calculating a psychometric curve to relate the experience of agency to the degree of control the subject's brain has over the robotic limb.[5]

---

[5] In a recent study with stroke patients, in which the movements of the paretic limb were amplified beyond its physical characteristics, the point of subjective equality of the upper extremities could be shifted without loss of sense of agency (Ballester et al. 2015).

Regret

The experience of regret provides intriguing possibilities for the empirical study of the importance of action in consciousness (Frith and Metzinger, this volume). Like disappointment, regret is a decision-related emotion, closely tied to action (or inaction). Disappointment is elicited if the outcome of a chosen action is not as good as expected on the basis of previous experience. In contrast, regret occurs when we realize that an action not chosen (including doing nothing) would have achieved a better outcome. Thus, regret involves counterfactual cognition. Regret is an unpleasant experience and thus behavior is altered by counterfactual thinking about possible future regret. For example, humans can take steps to avoid finding out about the outcomes associated with alternative actions (Reb and Connolly 2009) or we can adjust our choices to minimize anticipated regret (Filiz-Ozbay and Ozbay 2007). It has also been argued that rats experience regret (Steiner and Redish 2014).

Several open questions concerning regret require further study: How does the phenomenology of regret relate to other aspects of awareness of action such as agency, responsibility, and selfhood? How does the phenomenology of regret relate to that of other emotions, and are there specific interoceptive and autonomic signals associated with regret? How does the conscious monitoring of regret have its impact on planning and future action? What is the relation of regret to social emotions such as shame and guilt?

Empirical Possibilities in the Further Study of Action and Consciousness

One reason why action-consciousness relations have not been studied extensively, as compared to perceptual consciousness, is that it can be difficult to design experimental protocols involving rich movement. Most psychophysical and especially neuroimaging environments impose severe restrictions on movement, such that action repertoires are often restricted to simple eye movements and button presses.

Emerging technologies like virtual and augmented reality may ameliorate some of these limitations, by providing "virtual" analogs of rich action repertoires in physically restrictive situations. Even these technologies, though, face important limitations, since virtual actions are necessarily not grounded in the subject's own physiology. Nonetheless, a great many interesting experimental opportunities can be identified. Many of these have to do with experiences of body ownership, where virtual reality can induce systematic and highly controllable manipulations. Prospectively, these techniques could be used to examine social impacts on body experience and misperception of self-consciousness as somebody else's consciousness; for instance, the clinical condition of feeling "a presence" (i.e., the feeling of being in the presence of another person, although there is nobody there) may be of interest for study (see Blanke et al. 2014).

Experimental opportunities can also be identified with clinical populations, other than those with extreme deficits in motor control (the locked-in syndrome and amyotrophic lateral sclerosis patients mentioned earlier). These are predicated on the notion that inaccurate signaling related to actions may cause a variety of symptoms. In schizophrenia, inaccurate signaling may underpin delusions of control (the belief that one's actions are being controlled by another agent) (C. D. Frith 2012). The normal attenuation of sensations resulting from our own movement depends on relating the intention to move (or speak) with the anticipated sensory changes (the forward model) (Frith et al. 2000a; Ford et al. 2007). This explains why we cannot tickle ourselves. Schizophrenic patients with delusions of control *can* tickle themselves, presumably because something has gone wrong with this action-perception loop (Blakemore, Smith et al. 2000). Abnormalities in neural activity that precede simple, self-paced, noncontingent button presses have been related to a lack of motivation and general apathy (Ford et al. 2008). Perhaps the mismatch between predicted sensations and the resulting experiences ultimately diminishes the motivation for action. In terms of the Bayesian brain framework, failure to suppress proprioceptive feedback would create problems for active inference and thus for action. This might contribute to general motor awkwardness and neurological soft signs in schizophrenia (Bachmann et al. 2005) as well as the motor symptoms that often precede the onset of the illness (Walker and Lewine 1990; Cannon et al. 1999). The implications of this system for the social self and its dysfunction in schizophrenia are also evident. Patients with schizophrenia have reduced activity in the posterior superior temporal sulcus during imitation of actions and during action observation, possibly reflecting a breakdown in internalized mimicking, one route to understanding the minds of others and increasing social facility (Thakkar et al. 2014).

Open Issues

The topic of action and consciousness is unusually broad and any single review is necessarily incomplete. Several important areas have not been discussed in any detail, including the *role of language* (speech acts are canonical examples of actions which, like any other, can be instrumental, epistemic, or communicative). Another neglected area has been the role of *synthetic modeling*. Synthetic models are extremely valuable in connecting theoretical frameworks to specific empirical predictions. Here, concrete architectures like DAC might play an important role (Verschure, this volume). Another aspect of experience emphasized by an enactive autonomy approach is its *temporally extended* nature. Here, relations to different timescales in action need further attention. Finally, the scope of this treatment has not extended very far into an examination of the brain regions and circuits involved in action awareness and consciousness. This, too, demands close attention.

There is also a need for more work on action phenomenology, following Pacherie (2008). It is worth noting that there is also phenomenology to *reflex* movements (which are not actions) or to the movement of my hand when you move my hand, but I don't, and which is therefore *your* action, but *not mine*.

Concluding Remarks

Bringing an action-oriented perspective to the study of consciousness clarifies several important issues. A first clear outcome is that *action actively shapes and structures conscious experiences* in ways that extend beyond the trivial case of selecting sensory samples. Action emphasizes the *openness of consciousness* to extrapersonal influences. More controversial is the suggestion, emerging in particular from SMC theory and enactive autonomy approaches, that actions (possibly social actions) are constitutive of (some) conscious experiences. Although this seems highly plausible for nonhallucinatory experiences of action, it is less clear that all conscious experiences constitutively rest on action. Interestingly, actions may shape conscious experiences in specific ways depending on whether they are instrumental, epistemic, or communicative. Data on this question would be very valuable.

A second outcome is that actions both shape conscious scenes and engage a specific phenomenology of action awareness, which depends on goals and emerges at multiple hierarchical levels and spatiotemporal grains. Action awareness, in the specific guises of experiences of intention and agency, is also a key determinant of experiences of selfhood.

A third outcome is that it makes sense to speak about action and awareness at multiple scales, spatial and temporal. Actions can take place within the body (autonomic control), at the body-world interface, and within larger social spheres. Timescales can range from fine finger movements and rapid autonomic contractions to long-term intentions to, for example, become an academic and write book chapters on action and consciousness. Actions may shape experiences at each of these scales and may give rise to specific action-related phenomenology at each of these scales. Social actions, including joint actions as an illuminating example, have emerged as especially important domains in which action determines self-experience, carrying important implications for psychological well-being.

A fourth outcome is a general (though not universal) appreciation that a Bayesian framework seems valuable when considering the effects of action and embodiment on awareness. Importantly, this framework does not exclude (and may benefit from) insights gained from alternative frameworks like SMC theory and enactive autonomy as well as specific operational architectures like DAC. Indeed, suggesting specific mechanisms brings to light interesting new questions, such as whether action-related experiences are constructed prospectively (as suggested by Bayesian brain) or retrospectively (as suggested by DAC). These points emphasize the importance of systems-level embodied modeling.

A final cautionary remark is that the degree to which action affects or constitutes consciousness remains unclear and thus it is not possible to conclude, at this stage, whether the "pragmatic turn" will constitute a revolution in our understanding of consciousness. Action is important, but it may not be everything, when it comes to experience.

Acknowledgments

The authors gratefully acknowledge the following. AKS: the Dr. Mortimer and Theresa Sackler Foundation, which supports the work of the Sackler Centre for Consciousness Science. MVB: Emmy Noether grant BU1335/3-1 from the German Research Foundation (DFG). JF: Department of Veterans Affairs (I01CX000497) and National Institute of Mental Health (MH-58262). CDF: Wellcome Trust and Aarhus University. MK: Marie-Curie Network TESIS: FP7-PEOPLE-2010-ITN, 264828. MM: Convergent Science Network of Biomimetics and Neurotechnology. JKO'R: ERC Advanced Grant "FEEL" No. 323674. All authors are grateful for illuminating input from other participants at this Ernst Strüngmann Forum.

Implications of Action-Oriented Paradigm Shifts in Cognitive Science

16

Do We (or Our Brains) Actively Represent or Enactively Engage with the World?

Shaun Gallagher

Abstract

This chapter reviews scientific discussions of several problems (free will, social cognition, perception) that reflect a representational approach to cognitive science, and contrasts them with embodied, enactive approaches. It asks whether predictive coding models can adjudicate between these different views and suggests that predictive coding models can go either way. An interpretation is then offered that is designed to push such models toward the enactive camp. It concludes by considering the suggestion that enactivism is a philosophy of nature (as defined by Godfrey-Smith), rather than a scientific research program, and suggests that enactivism's attempt to rethink the nature of mind and brain also involves rethinking the concept of nature.

Introduction

Recent developments in embodied cognition motivate the following question: Is cognition *in the head* or *in the world*, or in some mix of brainy and worldly processes? There is a long tradition that takes cognition to be a fully in-the-head event. I will discuss a few of the more recent versions of this view and then offer the contrasting view of enactive cognition—an embodied approach that has roots in phenomenology and pragmatism. Enactive approaches to cognition suggest that, at least in basic (perception- and action-related) cases, cognitive processes are not just in the head. In trying to weigh the balance of these ideas, I address the challenge of predictive coding models and suggest where they might fit into this debate. In addition, I will discuss issues that concern time and dynamics, since these are important issues and are treated differently in different approaches.

To provide a framework in regard to temporality I borrow a threefold division from the work of Varela (1999). Different aspects of cognition involve processes on three different scales of duration:

1. The *elementary* scale (varying between 10–100 milliseconds)
2. The *integration* scale (varying from 0.5 to 3 seconds)
3. The *narrative* scale involving memory (above 3 seconds)

In terms of neurophysiology, the elementary scale corresponds to the intrinsic cellular rhythms of neuronal discharges, roughly within the range of 10 ms (the rhythms of bursting interneurons) to 100 ms (the duration of an excitatory-inhibitory sequence of postsynaptic potential in a cortical pyramidal neuron). Neuronal processes on this scale are integrated into the second scale, which, at the neurophysiological level, involves the integration of cell assemblies. Phenomenologically, the integration scale corresponds to the experienced living present, the level of a fully constituted cognitive operation; motorically, it corresponds to a basic action (e.g., reaching, grasping). On a dynamic systems interpretation, neuronal events on the elementary scale synchronize (by phase-locking) and form aggregates that manifest themselves as incompressible but complete acts on the integration scale.[1] The narrative scale is meant to capture longer time periods. Further distinctions could be made, and other more rhythmic time patterns could be explicated (as suggested at this Forum by Marek McGann, pers. comm.). For purposes of this article, the threefold distinction should, however, be sufficient.

Cognition in the Head: Some Standard Approaches

Even if we define cognitive processes broadly to include not just beliefs and desires but also states that refer to bodily action and to interactions with other people, we still find that mainstream cognitive science offers narrow accounts which place all the action required for full explanation in the mind or in brain processes. The term "narrow" is a technical one in philosophy of mind, referring to internal mental representational processing or content. Standard explanations in cognitive science define cognition as constituted by mental or neural representations. A few examples will provide a good sense of this approach.

The first example concerns action, specifically issues that involve action planning and intention formation. The processes involved can be characterized at subpersonal and personal levels of explanation, but all of them remain narrowly within the traditional boundary of the mind-brain. Concerning the very basic elementary scale, consider the well-known Libet experiments. Libet et al. (1983) asked about neural dynamics involved in the readiness potential (*Bereitschaftspotential*) and its relation to our immediate sense of deciding to act (see also Soon et al. 2008). These experiments are not only well known, they are also controversial. I will not go into details about the experiments or controversies pertaining to methodology, but I will briefly summarize the basic idea and say something about the philosophical controversy.

---

[1] This currently has the status of a working hypothesis in neuroscience (Thompson 2007:332).

The question Libet tries to answer is whether consciousness plays a role in the initiation of action, and he interprets this to be a question about free will. Libet's results indicated that on average, 350 ms before the subject is conscious of deciding (or having an urge) to move, the subject's brain is already working on the motor processes that will result in the movement. That is, the readiness potential is already underway, and the brain is preparing to move before the subject makes the decision to move. The conclusion is that voluntary acts are "initiated by unconscious cerebral processes before conscious intention appears" (Libet 1985:529).

There are different interpretations of what these results mean. Most of them focus on the question of free will. Libet himself finds room for free will in the approximately 150 ms of brain activity remaining after we become conscious of our decision, and before we move. He suggests that we have time consciously to veto the movement. Others, however, think that the brain decides and then enacts its decisions; consciousness is epiphenomenal in this regard. The brain inventively tricks us into thinking that we consciously decide to act and that our actions are controlled at a personal level. On this view, free will is nothing more than a false sense or illusion.

Rather than enter this debate, I will simply point to a central assumption made about the kind of cognitive processes that are supposed to be involved in free will expressed by Haggard and Libet (2001:47), who frame the question and refer to it as the traditional question of free will: "How can a mental state (my conscious intention) initiate the neural events in the motor areas of the brain that lead to my body movement?" They are right in that this is the traditional way to ask the question: it is precisely the way that Descartes and many thinkers in the modern philosophical tradition would frame the question. It is a question of mental causation, which places the cognitive processes of free will in the head where brain and mind meet up.

To assume that this is the right way to ask the question overlooks the possibility that free will is not something that can be explained simply by looking where Libet experiments look. For example, one can argue that these experiments have nothing to do with free will (Dennett 2004; Gallagher 2006). This challenges the assumption that free will can be characterized in terms of the short, elementary scale of 150 ms, or even on the integrative scale which might involve 3–5 s. This type of response can go one of two ways. The first simply leads us back into the head, into discussions of intention formation where cognitive deliberations generate prior intentions that have a later effect on intentions-in-action. Since the Libet experiments address only intentions-in-action or motor intentions, they miss the mark since free will is more about deliberation and prior intention formation. Such explanations are worked out in representational terms of beliefs and desires in processes best characterized in terms of a space of reasons and on the narrative timescale, but still very much in the head. The second way leads to the idea that free will is not a property of one individual brain, mind, or organism, but is relational, and that social and environmental factors contribute to or detract from our ability to act freely (discussed in more detail in the next section).

To say that something like social relations are involved in free will, however, does not necessarily lead beyond traditional concepts of the mind. This is clear in ongoing debates about social cognition that are dominated by methodological individualism; that is, the idea that theory of mind is explained by a causal mechanism (a theory-of-mind mechanism or a mirror system) or process inside the individual's head. Today the growing consensus is that there are two networks in the brain responsible for our ability to understand others: a theory-of-mind network that includes the temporo-parietal junction, medial parietal cortex, and medial prefrontal cortex (e.g., Saxe et al. 2009), and mirror areas in premotor and parietal cortexes. Taken together, the neuroscientific findings may justify a hybrid of theory theory and simulation theory, or suggest a two-system approach of online perspective taking and offline social reasoning (Apperly and Butterfill 2009).

Complicating such views, mainstream theories of social cognition have started to take note of objections coming from embodied cognition and action-oriented approaches. This and more general concerns about the claims made by embodied cognition theorists have motivated a way of thinking about the role of the body consistent with standard representationalism—so-called "weak" or "minimal" embodiment (Alsmith and de Vignemont 2012). For example, Goldman and de Vignemont (2009) suggest that none of the things that embodied cognition theorists usually count as important contributors to cognitive processes—anatomy and body activity (actions and postures), autonomic and peripheral systems, relations with the environment—really do count. Rather, the only "bodily" things relevant to an account of cognition in general, or social cognition in particular, are body-formatted (or B-formatted) representations in the brain. As they put it, B-formatted representations offer a "sanitized" way of talking about the body and "the most promising" way to promote embodied cognition (Goldman and de Vignemont 2009:155).

B-formatted representations are not propositional or conceptual in format; their content may include the body or body parts, but they may also include action goals, represented in terms of how to achieve such goals by means of bodily action. Somatosensory, affective, and interoceptive representations may also be B-formatted, "associated with the physiological conditions of the body, such as pain, temperature, itch, muscular and visceral sensations, vasomotor activity, hunger and thirst" (Goldman and de Vignemont 2009:156). Goldman (2012:74) argues that one can develop an overall embodied cognition approach simply by generalizing the use of B-formatted representations.

Social cognition, on this view, is embodied to the extent that B-formatted representations involved in perceptual mirroring are used to represent the actions or states of others. This is precisely the sense in which Gallese talks of *embodied* simulation. Gallese (2014) endorses the idea of B-formatted representations in contrast to more enactive views of embodied cognition. Mirror neurons "can constitutively account for the representation of the motor goals of others' actions by reusing one's own bodily formatted motor representations, as well as of others' emotions and sensations by reusing one's own visceromotor and sensorimotor representations" (Gallese 2014:7).

Similar strategies aiming to "sanitize" embodied cognition can be found in accounts of other aspects of cognition. As one example, several theorists point to body-related simulations (representations) as important for language and concept processing (e.g., Glenberg 2010; Meteyard et al. 2012; Pezzulo et al. 2011). All of this is consistent with the standard representationalist "mentalistic enterprise" of reconstructing the world (Jackendoff 2002), of "pushing the world inside the mind" (Meteyard et al. 2012), and a very narrow conception of embodiment.

Cognition in the World: Phenomenologically Inspired Enactive Approaches

Enactive approaches to cognition are inspired and informed by phenomenological philosophy. Varela et al. (1992), who first defined the enactive approach, found significant resources in the phenomenological tradition. For example, Husserl's (1989) concept of the "I can" (the idea that I perceive things in my environment in terms of what *I can* do with them); Heidegger's (1962) concept of the pragmatic ready-to-hand (*Zuhanden*) attitude (we experience the world in terms of pre-reflective pragmatic, action-oriented use, rather than in reflective intellectual contemplation or scientific observation); and especially Merleau-Ponty's (2012) focus on embodied practice, which influenced both Gibson's (1977) notion of affordances and Dreyfus's (1992) critique of classical cognitivism (see also Di Paolo 2005; Gallagher 2005; Noë 2004; Thompson 2007). Less noted are relevant resources in the American pragmatist tradition; many of the ideas of Peirce, Dewey and Mead can be considered forerunners of enactivism (see Gallagher 2014; Menary, this volume).

Enactive versions of embodied cognition emphasize the idea that perception is *for action*, and that action orientation shapes most cognitive processes. Most enactivists call for a radical change in the way we think about the mind and brain, with implications for methodology and for rethinking how we do cognitive science (see below). Clark (1999) provides a succinct three-point summary of the enactive view, endorsed by Thompson and Varela (2001):

1. Understanding the complex interplay of brain, body, and world requires the tools and methods of nonlinear dynamic systems theory.
2. Traditional notions of representation and computation are inadequate.
3. Traditional decompositions of the cognitive system into inner functional subsystems or modules ("boxology") are misleading and blind us to arguably better decompositions into dynamic systems that cut across the brain-body-world divisions (Thompson and Varela 2001:418).

Enactive approaches, similar to the idea of extended mind or distributed cognition, argue that cognition is not entirely "in the head," but is distributed across brain, body, and environment. Enactivists, however, reject functionalism and claim that the specific nature of (human) bodily processes shape and contribute to the constitution of consciousness and cognition in a way that is irreducible to representations, even B-formatted representations. In contrast to Clark (1998), who argues that specific differences in body type or shape can be transduced and neutralized via the right mix of representational processing to deliver similar experiences or similar cognitive results, enactivists insist that biological aspects of bodily life, including organismic and emotion regulation of the entire body, have a permeating effect on cognition, as do processes of sensorimotor coupling between organism and environment. Noë (2004), for example, developed a detailed account of enactive perception where sensorimotor contingencies and environmental affordances take over the work that had been attributed to neural computations and mental representations (see also O'Regan and Noë 2001).

To be clear, enactivists do not deny the importance of the brain, but they understand the brain to be an integrated part of a larger dynamic system that includes body and (both physical and social) environment. The explanatory unit of perception (or cognition, action, etc.) is not just the brain, or even two (or more) brains in the case of social cognition, but dynamic relations between organism and environment, or between two or more organisms, which include brains, but also their own structural features that enable specific perception-action loops, which in turn effect statistical regularities that shape the structure and function of the nervous system (Gallagher 2005; Thompson 2007).

If I reach out to grasp something (or someone), my hand is involved, as is my arm, my shoulder and back muscles, my peripheral nervous system as well as my vestibular system, no less than my brain, which in all of its complexity is making its own dynamic adjustments on the elementary timescale as part of this process of reaching out to grasp. A full account of the kinematics of this movement does not add up to an explanation of the action, nor does a full account of the neural activity involved. Likewise, if I reach a decision about how to act, the neural components of this activity are a necessary part of it, but also where I happen to be located, who I'm with, my past practices, current physical skills, and health status, not to mention my mood, will to some degree play contributory roles in the decision formation. Some of these elements enter into the process on a narrative timescale and are not under my current control. In this respect, my body is not just a sensorimotor mechanism. Affect plays an important role—things like hunger, fatigue, physical discomfort or pain, as well as emotion and mood. Such things are not well behaved in terms of timescale—they involve all three scales. With respect to the discussion of free will, whatever agentive action is, it is both constrained and enabled by all of these different factors. As Clark and Chalmers (1998:9) suggest, if one of the extra-neural components is taken away, "the system's behavioral competence will drop, just as it would if we removed part of its brain." At the very least, a removal (or an addition) of any component will entail compensatory adjustments across the system.

Evan Thompson (2014) provides a nice analogy. Saying that cognition is in the brain is like saying that flight is inside the wings of a bird. Just as flight does not exist if there is just a wing, without the rest of the bird, and without an atmosphere to support the process, and without the precise mode of organism-environment coupling to make it possible (indeed, who would disagree with this?), cognition does not exist if there is just a brain without bodily and worldly factors. "The mind is relational. It's a way of being in relation to the world" (Thompson 2014:1). For some, these claims may seem obvious or even trivial, and yet we often find ourselves doing science as if the only things that counted as explanatory were neural representations.

Processes of social interaction are also not reducible to neuronal processes (or B-formatted representations) within the individual, since they include physical engagement with another person and/or a socially defined environment, processes of "primary intersubjectivity," affective processes where distinct forms of sensorimotor couplings are generated by the perception and response to facial expression, posture, movement, gestures, etc. in rich pragmatic and social contexts. Again, this is not to say that all the essential processes of social cognition are extra-neural. Mirror neurons may indeed make a contribution, not by simulating actions of others, repeating a small version of them inside one's head, but by being part of larger sensorimotor processes that respond to different interaction affordances (e.g., Caggiano et al. 2009). On the enactive view, social cognition is an attunement process that allows me to perceive the other as someone to whom I can respond or with whom I can interact. In the intersubjective context, perception is often *for inter-action* with others. In some cases, a relational understanding is accomplished in the social interaction between two people where some novel shared meaning (or some decision or even some misunderstanding) is instituted in a way that could not be instituted within the single brain of either one of them (De Jaegher et al. 2010).

Embodied Prediction: How to Be an Embodied Theorist without Losing Your Head

Take any example of cognition and one can run two different explanations: a standard representationalist one and an enactivist one. Sometimes it seems to be simply a vocabulary substitution; sometimes the enactivist description seems to work better, especially if we think of examples that involve problem solving rather than belief, whereas at other times the representationalist description seems to have the upper hand. Even when the representations involved are action-oriented, minimal, or B-formatted there are clear differences in explanation.

Consider the example of fielding (trying to catch) a ball. We can run the account in both ways, where running it in one case means representing various aspects of speed and trajectory, and in the other case literally running rather than representing.

In the classical representational account, the problem is first solved in the fielder's head. Speed and trajectory of the ball are calculated and represented by the brain, which, having solved the problem offline, then simply sends instructive signals to the limbs to move in the most efficient way to catch the ball. It is unlikely that anyone believes this story, and there is evidence against it since it does not predict the actual pattern of movement that the fielder makes to catch the ball. In a more likely, weak-embodied, *action-oriented representation* (AOR) account, calculations are made online as we move, but part of the process involves quick (on the elemental timescale) offline AORs formed in forward models that contribute to motor control. Sensory feedback is too slow to update the system in a timely fashion; the forward model generates a simulation or representation (an internal model) that anticipates sensory feedback from intended body positions on the run and allows for a fine-tuning of motor control. The AOR stands in for a future state of some extra-neural aspect of the movement—a body position (or proprioceptive feedback connected with a body position) that is just about to be accomplished in the action of catching the ball. Since the model represents a state of the system that does not yet exist—a predicted motor state—it is said to be offline, or decoupled from the ongoing action (Clark and Grush 1999), and to occur in the self-contained brain. Such representations can then be taken further offline and reused (e.g., in memory systems), scaling up to enable additional cognitive states. The brain can run such offline models to accompany states in which no running and catching is involved at all—when, for example, I imagine or remember catching a ball. No need for the body itself or for "a constant physical linkage" (Clark and Grush 1999:7).

On the enactive account, we solve the problem by vision and movement. We run on a curved line so as to keep the ball's trajectory pointed straight. This reliably gets the fielder to the catching spot (McBeath et al. 1995). There is no need to compute in-the-head mental representations of the ball, its speed, its trajectory, and so on. Rather, the cognitive component of this action depends to a significant extent on how we directly act in the world. The processes involved are dynamic sensorimotor processes that are fully online. Indeed, it is unclear in what sense the AOR account should describe anticipatory motor control processes as offline or decoupled. On the enactive account, this kind of forward anticipatory aspect of neural processing is a constitutive part of the action itself, understood in dynamic terms, rather than something decoupled from it. The anticipation of a future state or position (of the ball, or of the body grabbing the ball in the next second) requires ongoing reference or "constant physical linkage" to current state or position. To think that such processes are decoupled (or in some sense off-line) is to think that such anticipations are in someway detached from perceptual and proprioceptive input, which they clearly are not. Such processes may be one step ahead of real-world proprioceptive feedback, but they are also at the same time one step behind the previous moment of feedback, integrated with ongoing movement and perception.

On some views, decouplability is part of the very definition of representation (Clark and Grush 1999). On the enactive account, however, to scale up to cognitive states such as imagining or remembering, the brain does not decouple or recreate a process that was not representational in the first place (since the process had not been decoupled from the action itself); rather, the system (using the same motor control or forward control mechanism) enacts (or reenacts) a process that is now coupled to a new cognitive action. In remembering, for example, there may be reactivation of perceptual areas that had been activated during the original experience. We do not know to what extent other nonneural bodily factors may be (re-)activated (e.g., subliminal tensing of muscles, facial expressions, gestures).

Here, however, the line between accounts of AORs and the idea of enactive cognition gets blurred, and some may suspect that the difference is merely one of preferred vocabulary. Thus, defenders of AORs, like Wheeler (2005:219), give up the criterion of decouplability as part of the concept of an AOR, and both Wheeler as well as Rowlands (2006:224) suggest that AORs involve aspects of a system that includes brain, body, and environment: "The vehicles of representation do not stop at the skin; they extend all the way out into the world." What enactivists refer to as affordances, proponents of weak embodiment call AORs (Clark 1998:50).

Can predictive coding models somehow adjudicate between representationalist and enactivist accounts? One might think that predictive coding has already settled on the representationalist side, since much of the predictive coding literature assumes or adopts the representationalist vocabulary (e.g., Hohwy 2013; Hohwy, this volume). An alternative interpretation could push predictive coding a bit toward the enactivist account.

On one reading of predictive coding, the brain is pictured as having no direct access to the outside world; accordingly, it needs to represent that world by some internal model that it constructs by decoding sensory input (Hohwy 2013). On this basis, the brain makes probabilistic inferences about the world and corrects those inferences by addressing prediction errors. This process involves synaptic inhibition based on an empirical prior. Predictions are matched against ongoing sensory input. Mismatches generate prediction errors which are sent back up the line, and the system adjusts dynamically back and forth until there is a relatively good fit.

Do we have to think that the outcome of this process is the creation of a representation in the brain? Why should we not rather think of this process as a kind of ongoing dynamic adjustment in which the brain, as part of and along with the larger organism, settles into the right kind of attunement with the environment—an environment that is physical but also social and cultural.

We know that one's beliefs and values as well as one's affective states and cultural perspectives (phenomena defined for the most part on the narrative scale) can shape the way that one quite literally sees the world. How such cognitive and affective states and perspectives enter into (elementary-scale) subpersonal processes can be explained in terms of predictive coding models. With respect to affect, for example, Barrett and Bar's *affective prediction hypothesis* "implies that responses signaling an object's salience, relevance or value do not occur as a separate step after the object is identified. Instead, affective responses support vision from the very moment that visual stimulation begins" (Barrett and Bar 2009:1325). Along with the earliest visual processing, the medial orbital frontal cortex is activated initiating a train of muscular and hormonal changes throughout the body, "interoceptive sensations" from organs, muscles, and joints associated with prior experience, and integrated with current exteroceptive sensory information that help to guide response and subsequent actions. Accordingly, along with the perception of the environment, we also undergo certain bodily affective changes that accompany this integrated processing. In other words, before we fully recognize an object or other person, for what it or he or she is, our bodies are already configured into overall peripheral and autonomic patterns based on prior associations. In terms of the predictive coding model used by Barrett and Bar (2009), priors, which include affect, are not just in the brain, but involve a whole body adjustment.

On the enactivist view, brains play an important part in the dynamic attunement of organism to environment. Social interaction, for example, involves the integration of brain processes into a complex mix of transactions that involve moving, gesturing, and engaging with the expressive bodies of others; bodies that incorporate artifacts, tools, and technologies are situated in various physical environments, and defined by diverse social roles and institutional practices. Brains participate in a system, along with all these other factors, and it would work differently, because the priors would be different, and therefore the surprisals would be different, if these other factors were different.

Changes or adjustments to neural processing will accompany any changes in these other factors, not because the brain represents such changes and responds to them in central command mode, but because the brain is part of the larger embodied system that is coping with its changing environment. Just as the hand adjusts to the shape of the object to be grasped, so the brain adjusts to the circumstances of organism-environment. It is not clear that we gain anything by saying that the shape of the grasp represents the object to be grasped (Rowlands 2006). At the very least, it remains an open question about how the neural processes described by predictive coding models are most usefully characterized, whether as inferential and representational or as part of a dynamic attunement of organism to environment.

Concluding Remarks

Enactive embodied cognition approaches present a challenge for science. By focusing on not just the brain, not just the environment, not just behavior, but on the rich dynamics of brain-body-environment, enactivists offer a holistic conception of cognition. To put it succinctly, however, it is difficult to operationalize holism. Neither experimental control nor the division of labor in science allows for all factors to be taken into consideration at once. It is also unclear whether there could be one critical experiment that might decide the issue between the representationalist and the enactivist.

This motivates serious consideration of a suggestion made made by Cecilia Heyes at this Forum, drawing on work by Peter Godfrey-Smith (2001). Godfrey-Smith, discussing developmental systems theory, distinguishes between a "scientific research program" and a "philosophy of nature." Enactivism, Heyes suggests, has elements of both, but may be more successful as the latter.

On one hand, enactivism makes empirical claims, for example, about the work of sensorimotor contingencies, and in this sense it resembles a research program that can suggest new experiments and new ways of interpreting data. On the other, its emphasis on holism presents problems for empirical investigations. One does not get far in experimental science without controlling for variables. With respect to its holistic approach, enactivism resembles a philosophy of nature. As Godfrey-Smith makes clear, a philosophy of nature is a different kind of intellectual project from science, and although science may be its critical object, the two enterprises do not have to share the same vocabulary. A philosophy of nature "can use its own categories and concepts, concepts developed for the task of describing the world as accurately as possible when a range of scientific descriptions are to be taken into account, and when a philosophical concern with the underlying structure of theories is appropriate" (Godfrey-Smith 2001:284). A philosophy of nature takes seriously the results of science, and its claims remain consistent with them, but it can reframe those results to integrate them with results from many areas of knowledge. The requirements of such a reframing may indeed call for a vocabulary that is different from one that serves the needs of any particular science. To work out a philosophy of nature is not to do science, although it can offer clarifications relevant to doing science, and it can inform empirical investigations. In this sense, a philosophy of nature is neither natural philosophy nor the kind of naturalistic philosophy that is necessarily continuous with science. It offers critical distance and practical suggestions at the same time. In some cases it may make doing science more difficult.

Is enactivism a philosophy of nature? Indeed, from the very start enactivism involved not only a rethinking of the nature of mind and brain, but also a rethinking of the concept of nature itself (see Di Paolo 2005; Thompson 2007:78ff). If enactivism is a form of naturalism, it does not endorse the mechanistic definition of nature presupposed by science, but contends that nature cannot be understood apart from the cognitive capacity that we have to investigate it. As Heyes suggests, in the context of a philosophy of nature meant to offer an encompassing view, holism is a strength rather than a practical complication.

Does this make enactivism irrelevant to the actual doing of science? Enactivism may still motivate experimental science in very specific ways. Even if in some cases it is difficult to apply a holistic view to a given question, in many cases there may not be any special complication in designing experiments that can test enactive ideas. It is not that in every case we must include absolutely everything when addressing a particular concrete question, but in the end it may be easier to include than to ignore a factor that is crucial. For example, including embodied interactions in explanations of social cognition might actually involve less complexity if keeping them out of the picture requires the elaboration of more convoluted explanations in terms of theory or simulation mechanisms (De Jaegher et al. 2010). Athough in this, and other cases, much will depend on circumstances like the availability of the right lab technology, the whole may sometimes lead to simpler explanations. In short, even if enactivism is to be considered a philosophy of nature, it would not be right to conclude that it cannot offer or test concrete hypotheses or raise novel scientific questions.

Acknowledgments

The author acknowledges support received from the Marie-Curie Initial Training Network, "TESIS: Toward an Embodied Science of InterSubjectivity" (FP7-PEOPLE-2010-ITN, 264828), European Commission Research, and the Humboldt Foundation's Anneliese Maier Research Award.

17

Ways of Action Science

Wolfgang Prinz

Abstract

This chapter argues that an action-oriented view of cognition is nice to have, but not enough. The study of cognition most certainly needs to be extended to include action. However, the study of action requires more than simply understanding its cognitive foundations. This chapter discusses two functional features of action that a cognitive approach fails to capture: *top-down control* and *action alignment*. Top-down control operates within individuals and requires a framework that addresses the formation of motives, goals, and intentions as precursors of action selection and execution. Action alignment operates between individuals, necessitating a framework that addresses the common representational basis of perception and production. To accommodate these features we need to proceed from including action in cognitive science to including cognition in action science.

Introduction

The aim of this Forum was "to examine the key concepts of an emerging action-oriented view of cognition and the consequences of such a paradigm shift." Although I certainly share this aim, I address a more ambitious aim in this article; namely, to examine key concepts of an emerging action science and the consequences of such a paradigm shift for cognition. My aim is to find a place for cognition in action science, not just for action in cognitive science.

As has been elegantly pointed out in Neisser's classical foundation of modern cognitive psychology, the task of research in cognition is to trace the fate of the input and study stimulus information and its vicissitudes in attention, perception, memory, imagery, thought, etc. (Neisser 1967). This input-oriented way of viewing cognition is not surprising, since it reflects the roots of cognitive science in endeavors such as sensory physiology, psychophysics, and philosophical epistemology. Accordingly, action was not included in Neisser's list, in spite of the fact that much of the research he reported relied on reactions and reaction times. However, reactions were not regarded as targets of study in themselves. They served as mere indicators of cognitive states.

Since Neisser, the *zeitgeist* has changed and action has come to the fore. Today it is broadly acknowledged that cognitive processing is intimately intertwined with action processing. As a result, researchers claim that the study of cognition needs to be extended to also include the study of action. This is what we may call the action extension view of cognition.

While I certainly want to support an extension along these lines, my proposal here is to suggest a still more radical move. Numerous authors claim that the chief adaptive function for which minds/brains evolved and have been optimized pertains to smart action rather than true cognition. To put it briefly, smart actions let the animal do right things at right times, aiming at altering conditions in accord with currently given inner and outer circumstances. If this is true, cognition is a secondary, subsidiary function: true cognition helps with smart action but it is not the proper function for which mind/brain systems are optimized. If one takes this view seriously, the science that one needs to strive for is *action science*: How is smart action possible, and what can cognition contribute to it? A move in this direction takes us from action extensions of cognition to cognitive contributions to action. Moreover it allows us to view action as a target of study in its own right, including both its cognitive and noncognitive foundations.

The claim that action builds on noncognitive foundations can be understood in two different ways. One reading implies that action builds on noncognitive foundations that cannot be captured by the theoretical language of representationalism. This is what various brands of enactivism claim. The other reading invokes the notion that action goes beyond cognition in a descriptive, but not theoretical sense. This reading maintains that actions exhibit features that do not come into view from the perspective of an input-oriented approach, without, however, implying that these features cannot be captured by the theoretical language of representationalism. This view, which I am following here, considers representationalism a useful framework for all action science, powerful enough to capture both the cognitive and noncognitive foundations of action.

In what follows I sketch two basic ways of action science: individual and social. The individual way studies actions as tools of control (i.e., of altering events in accord with demands, needs, and desires). The social way studies actions as tools of interindividual alignment (i.e., of coordinating own and foreign actions).

The Individual Perspective: Action for Control

The notion of adaptation provides a convenient starting point for understanding what it means for animals to control their actions. Adaptation can be studied at two levels: structural dispositions and functional interactions. At the structural level, we may study how long-term dispositions match global, long-term conditions in the environment. Here we speak of a species being adapted to its environment. At the functional level, we may study local matches between fluctuating inner states and fluctuating outer conditions in individuals. Here we speak of their capacity to adapt, or adapt themselves, to changes in bodily states or environmental conditions.

When we speak about action control, we focus on particular kinds of such local adaptations—those that involve bodily movements as a means for altering environmental conditions or bodily states. The adaptive value associated with these movements does not reside in themselves but in the alterations of inner and outer states achieved through them. Accordingly, we think and talk about these movements in terms of the goals toward which they are directed. In other words, we consider them actions, not just movements. Actions are segments of bodily activity that converge on some goal state. When a lion chases a zebra, that action terminates when the lion eventually catches it. Likewise, when someone hammers a nail into the wall, that action terminates when the nail is eventually embedded in the wall.

Modes of Control

How are means and ends related to each other? Does control proceed from means to ends or from ends to means? In addition, how are means and ends related to circumstances under which the action is performed? Answers to these questions can be divided into two major camps: bottom-up and top-down control. Both camps agree that actions have the potential to achieve desirable outcomes in terms of the current needs and interests of individuals, but they disagree on the machinery involved. Bottom-up control posits that ends follow from means; that is, goals are attained as outcomes of given actions. Top-down control posits that means follow from ends: actions get selected to achieve given goals.

Bottom-Up

The notion of bottom-up control captures the idea that goal-directedness is an emergent property of the workings of control systems whose operation does itself not draw on goals or goal representations at all. Bottom-up control posits that goal-directed behavior can be explained as a consequence of currently given states of affairs (or representations thereof), with no role being played by future intended states of affairs (or anticipatory representations thereof). Explanatory strategies along this line have been advanced by several classical approaches (e.g., Skinner 1953; Thorndike 1911; Tolman 1959). These approaches have devoted much effort to explaining purposeful behavior without purposes and goal-directed action without goals.

How can this happen? A useful framework is provided by the technical metaphor of control. Engineers furnish technical systems with controllers:

computational devices that determine, for each configuration of current circumstances, which action to take in order to establish or maintain certain inner and outer circumstances. Likewise, we may think of animals as being furnished with controllers. In this case, controllers determine for the animal, under each configuration of current inner and outer circumstances, which actions to take to establish or maintain satisfactory or desired inner and outer circumstances.

These devices can be characterized in terms of the output they provide, the input they require, and the algorithms on which they rely. On the *output* side, controllers steer bodily movements suited to modulate inner or outer conditions. These movements may, for instance, act to alter environmental conditions (e.g., a frog catching a fly) or to move the body relative to the environment (e.g., navigating around an obstacle). To perform these computations properly, controllers need to be informed, on the *input* side, about the configuration of current outer and inner circumstances (i.e., the position of the fly or the obstacle). Third, controllers need to dispose of *algorithms* for input interpretation and output generation. The operation of these algorithms depends on event knowledge and action knowledge. Event knowledge reflects what the controller "knows" about the to-be-controlled events. Conversely, action knowledge reflects what the controller "knows" about possible actions and interactions with these events.

At first glance, this scheme looks like a linear sequence that leads from event interpretation to action generation (and, hence, from stimuli to responses). Yet, to complete the picture, we need to realize that these sequences are embedded in cyclical interactions between controllers and target events: actions alter events which in turn give rise to new cycles of interpretation and generation, and so forth. Still, within each given cycle, action generation always depends on event interpretation.

How is event and action knowledge acquired? Different kinds of learning mechanisms have been proposed, both on the ontogenetic and the phylogenetic scale. They all share the same functional logic—the logic of trial, error, and success. On the ontogenetic scale, learning applies to individual animals. Here one of the most powerful mechanisms for the acquisition of mappings between stimulus events and responding actions relies on the production of behavioral variation plus subsequent selection of those variants which prove to be successful. Eventually, the animal will home in on stimulus-response mappings that turn out to be successful in terms of its current needs under current conditions. On the phylogenetic scale, learning applies not to individuals but to populations and their gene pools. Here, through production of genetic variation and selection, populations may "learn" that some stimulus-response mappings are more advantageous than others, and eventually they will come to incorporate those mappings into their gene pools. Such phylogenetic learning may require hundreds or perhaps thousands of generations, based on variation, selection, and survival of the fittest mapping.

In sum, bottom-up control explains the occurrence of goal-directed actions as a causal consequence of two kinds of factors: currently given circumstances and outcomes of previous learning. Outcomes of learning are stored in the controller's machinery for event interpretation and action generation. This generates actions previously proven to be successful under similar conditions. Naturally, such actions are often meaningful and may even, in many cases, look as if they were goal-directed in a literal sense (i.e., guided by explicit intentions).

Top-Down

Bottom-up control is, in essence, a stimulus-triggered, stimulus-guided affair. Since cognitive scientists like stimuli and their consequences, this explains why generations of them have tried to push the limits of this control mode as far as possible. Still, there is no reason to believe that bottom-up control is the only game in town. Top-down control may perhaps be a late arrival in the evolution of action control, but in humans it is certainly in place and plays a strong, if not dominant, role. Top-down control posits that means follow from ends; that is, movements derive from goals. Put somewhat paradoxically, top-down control implies that movements are selected on the basis of desired or intended circumstances—circumstances which can only be attained through (and hence after) performing those movements.

Whether this sounds like magic depends on what we are considering: circumstances in the world or a controller's representations thereof. When we talk about circumstances in the world, we are in fact invoking a teleological explanation which claims that certain movements are performed at t_1 to achieve certain states of affairs at t_2. Yet, when we express the same relationship in terms of representations, the paradox goes away. At that level, movements performed at t_1 go back to desires and intentions at t_0. What we thus require at that level are goal-related representational states such as desires and intentions, which act as temporal and causal antecedents of movements.

What does top-down control require? To build a top-down controller, we need to take two major steps beyond the scheme for bottom-up control. First, we need to furnish our device with the ability to create action goals; that is, to form representations of events that are independent from the configuration of currently given circumstances. Second, we need to furnish goal representations with the power to make an impact on action selection and generation. The first step invokes dual representation, the second ideomotor control (cf. Prinz 2012, chap. 7).

The notion of *dual representation* refers to the ability to maintain two parallel streams of event representations and keep them apart: one for events that are currently present and another for events not present in the current situation. Whereas the first stream is fed from external sources (stimulus information pertaining to the current situation), the second emanates from internal sources (memory information pertaining to events beyond the current situation).

When applied to action control, the implementation of dual representation provides a first step toward top-down control. Top-down controllers need to be capable of tracing perception-based and intention-based episodes. Whereas perception-based episodes address ongoing and upcoming events, intention-based episodes address to-be-attained events. One of the two streams thus refers to factual events that are actually happening, the other to fictitious events which the agent would like to make happen. Strict separation between these streams is required since mistaking fact for fiction is no less maladaptive than mistaking fiction for fact.

The notion of *ideomotor control* explains how goal representations contained in intentional episodes become functional for action control. Put in a somewhat old-fashioned language, ideomotor theory views actions as creations of the will (James 1890; Lotze 1852). Thus phrased, two conditions must be met to carry out a voluntary action: there must be a mental image of what is being willed, and conflicting ideas or images must be removed. When these two conditions are met, the mental image acquires the power to guide the movements required to realize the intention.

Ideomotor theory claims that the links between intentions and movements arise from learning. Whenever a movement is performed, it is accompanied by perceivable effects. Some are directly linked to carrying out the movement itself, such as the kinesthetic sensations that accompany each movement (resident effects). Others are linked to the movement in a more indirect way since they occur in the agent's environment at a spatial and/or temporal distance from the actual movement (remote effects).

In any case, the regularities between actual movements and their resident and remote effects are captured in associations. Thus, representations of movement outcomes become associated with representations of the movements leading to them. Once established, these associations work in two directions. One allows the anticipation of movement outcomes (i.e., to predict perceivable consequences from given movements). This is the case of forward-directed computation in the service of bottom-up control. The other allows for backward-directed computation in the service of top-down control (i.e., to select and generate movements required to achieve given intentions). Such backward computations guarantee that events which have been learned to go along with, or follow from, a particular action will hereafter exhibit the power to call that action forth.

In sum, ideomotor control relies on two principles: one for learning and one for performance. The learning principle claims that the system is capable of establishing associations between actions and their outcomes (both resident and remote). The performance principle claims that, once established, these associations can also be used in the reverse direction (i.e., from outcomes to actions effectuating them).

Implications for Cognition and Action

What does it mean for cognition to be tailored to the needs of action and for action to be grounded in cognition? These issues have been broadly discussed over the past decades (cf. Braitenberg 1986; Hommel et al. 2001; Jeannerod 1997, 2006; Morsella et al. 2009; Neumann and Prinz 1990; Noë 2004; O'Regan 2011; Prinz 2012; Prinz et al. 2013; Prinz and Hommel 2002).

The first implication is that *cognition is for action*. While competing approaches make use of different theoretical frameworks for addressing these issues, they all share the basic idea that cognitive functions are embedded in an architecture for control. This idea can be read genetically as well as functionally. The genetic reading maintains that cognitive functions evolve for the sake of action control, on both phylogenetic and ontogenetic scales. However, the fact that evolutionary history has shaped the cognitive machinery to serve the needs of control does not necessarily imply that its online operation must always include elements of control. Accordingly, this reading does not imply that each and every act of cognition must be associated with an act of control. This is what the functional reading maintains. The genetic reading claims that acts of cognition always entail elements of control—be it in the weak role of associated extensions or the strong role of constitutive ingredients.

Both readings have important implications for research agendas in the respective fields of study. On the phylogenetic scale, the genetic reading has inspired neuroethological programs in the study of the natural history of mental functions as they emerge from architectures for sensorimotor interaction (e.g., Dean 1990; Gallistel 1980). The same applies to the ontogenetic scale, where the genetic reading has inspired psychological programs to study the construction of an architecture for cognition from basic interactions between perception and action (Bertenthal and Longo 2008; Piaget 1954; Thelen and Smith 1996; Vygotsky 1979).

The functional reading is associated with research programs which aim at demonstrating the secret workings of action in cognition (i.e., the implicit involvement of action control in perception, memory, and thought). In recent years, these programs have made much progress, as new technologies for online recording of brain activity have provided new tools for assessing the latent involvement of action in cognition. As a result, there is now convincing support for the functional reading of the cognition-for-action claim. As has been shown in a variety of experimental paradigms, input-related cognitive processing is intimately intertwined with output-related action processing (Barsalou 2008; Hommel et al. 2001; O'Regan 2011; Prinz et al. 2009; Viviani 2002). We may therefore conclude that action makes essential contributions to cognition—not only at the level of architecture construction but at the level of information processing as well.

A further implication is, of course, that *action relies on cognition*: If cognition is for action, then action must rely on cognition. This implication is not trivial vis-à-vis the classical view that action is a thing that commences only after cognition has terminated. While this view maintains that the two draw on disjunct representational resources, the new view maintains that event interpretation and action generation draw on the same pool of knowledge resources. Actions are, accordingly, represented like any other types of events, so that representations are entirely commensurate (cf., e.g., Hommel et al. 2001; Jeannerod 1997; Rosenbaum 2009, 2013).

A final implication is that we need an *architecture for voluntary action*. This architecture must capture three basic segments of voluntary action: motivation, volition, and execution. Several proposals have been made concerning underlying mechanisms and dividing lines as well as transitions (de Wit and Dickinson 2009; Gollwitzer and Moskowitz 1996; Hassin et al. 2009; Heckhausen and Heckhausen 2008). One of the crucial features that the architecture must capture pertains to the origin of motives, goals, intentions, and their underlying representations. Importantly, these representations originate in the agent, not the environment, and the architecture must take means to keep them apart from representations of ongoing external events. A further important feature that needs to be covered by the architecture pertains to the dynamic nature of volition. Unlike percepts, memories, or thoughts which act as "cool" placeholders for things that are there, motives, goals, and intentions serve as "hot" placeholders for things that are wanted and desired and ways of getting at them. Importantly, both of these features cannot be captured by an extended architecture for bottom-up control.

At the same time, it should be clear that a machinery for top-down control requires a machinery that is nested within for bottom-up control. This is because goals and intentions in the mind can only be realized if agents are in a position to shape their actions according to current circumstances in the world. Accordingly, an architecture for voluntary action must build on communications between representations of wanted and given events. Top-down controllers can only work with built-in bottom-up controllers.

The Social Perspective: Action for Alignment

A peculiar functional condition arises when we turn to control scenarios in which individuals address other individuals' actions as events of control. When two (or more) people interact, each person can be seen to become involved in controlling the other's actions through his/her own actions. Here, event interpretation turns into action interpretation, based on event knowledge that pertains to foreign action. As a result, control draws on action knowledge on either side. Action generation relies, as usual, on knowledge pertaining to one's own action, but event interpretation now relies on action knowledge as well; namely, knowledge pertaining to foreign action.

This peculiar condition opens a unique opportunity for action alignment across individuals. Action alignment requires that knowledge resources for own and foreign action are integrated and combined, so that perception of foreign action and production of own action draw on common resources. For social animals, like humans, this functional condition provides an invaluable asset since it offers a direct, effortless way to align own with foreign and foreign with own action.

Common Coding

The notion of common coding invokes the idea that production of one's own action and perception of foreign action draw on common representational resources. In other words, tokens of one's own action get entries in the same representational domain and on the same dimensions as tokens of foreign action. Common coding thus makes it possible to assess similarity relationships between one's own and foreign action. For instance, as concerns production, own action may replicate or continue foregoing foreign action. Likewise, as concerns perception, foreign action may be understood to replicate or continue foregoing own action. In the first case, perceived action primes production of own corresponding action, whereas in the second case, production of own action primes subsequent perception of corresponding foreign action.

There is now ample evidence that this principle is instantiated in human minds and brains. Evidence comes from various fields of study. Results from behavioral experiments on action imitation, action induction, and perception/action interference have lent support to a strong role of similarity between own and foreign action (Prinz 2012, chap. 5). Parallel to this, numerous electrophysiological studies on mirror neurons and mirror systems in the monkey brain provide what may be considered an existence proof of shared representational resources for action production and perception (Rizzolatti and Sinigaglia 2008). Similarly, brain imaging studies on humans have shown that shared brain circuits may not only be involved in processing own and foreign actions but own and foreign sensations as well as emotions (Keysers and Gazzola 2009; Schütz-Bosbach and Prinz 2015).

In spite of the overwhelming evidence, common coding is often considered a strange and somewhat mysterious notion. Two brief remarks may help to demystify this perception. First, as has been pointed out in several places, the emergence of shared resources for own production and foreign perception can easily be explained in terms of classical principles of association and connectionist models instantiating them (Cook et al. 2014; Keysers and Gazzola 2009; Keysers and Perrett 2004; Pulvermüller et al. 2014). Second, common coding is, of course, not everything. The claim that production and perception draw on shared representational resources does not imply that *all* resources on which they draw are shared. Claiming shared resources at one level is entirely compatible with acknowledging unshared resources at other processing levels.

Social Mirroring

Common coding devices provide powerful tools for interindividual alignment. Since they use the same resources for representing own and foreign action, they offer themselves for both: own action resonating to foreign action and understanding foreign action as resonating to own action. To grasp what this alignment entails, let us look briefly at episodes of social mirroring (Prinz 2012, chaps. 4–6; Schütz-Bosbach and Prinz 2015).

Episodes

Social mirroring episodes have two sides: the target individual whose acting is being mirrored and the mirror individual who mirrors the target's action. The mirror individual functions for the target individual like a mirror in the target individual's environment. Here, two basic types of mirroring episodes can be discerned: reciprocal and complementary.

In episodes of reciprocal mirroring, the target individual sees her own action imitated or replicated by the mirror individual. In such a setting, the mirror individual acts as a mirror for the target individual in a more or less literal sense. Still, social mirrors are fundamentally different from physical mirrors. Even if the mirror individual tries to provide as-perfect-as-possible copies of the target's action, these copies are always delayed in time and their kinematics will never be as perfectly correlated with the target's acting as spectacular images are. We may speak of reciprocal mirroring as long as the target is in a position to understand that the mirror's acting is a delayed copy of the target's own preceding action. Hence, the constitutive feature of reciprocal mirroring is the target's *understanding* of the mirror's action as a copy of the target's foregoing own action.

In episodes of complementary mirroring, the target sees her own action continued and carried out by the mirror individual, rather than replicated. This, of course, is entirely different from what physical mirrors do. Nevertheless, what complementary mirroring has in common with reciprocal mirroring is that the mirror individual's action is strongly contingent upon the target's preceding action and that the target may perceive and understand this contingency. In this case, too, the reach of effective mirroring goes as far as the target is in a position to *understand* the mirror individual's acting as a continuation of her own acting.

Not surprisingly, episodes of action-based mirroring play an important role in interactions between young infants and their caretakers. Babies and their caretakers often find themselves involved in episodes of action-based, proto-conversational interaction and communication (Trevarthen 1998). They take turns in imitating or continuing each other's action. Most of this work concentrates on the baby in the role of the mirror (i.e., mirroring the caretaker's actions), not the target (i.e., perceiving herself being mirrored by the caretaker). To understand the power of these episodes for interindividual alignment, however, we need to take both roles into account: not only that of the mirror individual but also that of the target individual who perceives her own actions being mirrored by the other.

Action-based mirroring is not limited to interactions with young infants. Mirroring episodes are likewise widespread among adults. For instance, an individual may, in a conversation, shrug his arms in response to his conversation partner doing the same (reciprocation). Likewise, an individual may take up another individual's action when the other temporarily withdraws (continuation). Such mirror episodes may often reflect automated habits rather than controlled and deliberate actions. Still, they act to align individuals through production and perception of closely related actions. Mirror episodes help individuals match their own actions to others' actions and others' actions to their own.

Practices

Social mirroring depends on functional mechanisms that instantiate common coding. At the same time, it also depends on social practices in which individuals must engage to exploit the potential that is inherent in these mechanisms. These practices can be viewed at a local and a global level. At the local level, episodes of social mirroring require that two individuals interact in a particular way. They need to engage in *mirror games*. Mirror games are designed to align actions through mutual reciprocation and continuation, deliberately or automatically. One may speculate that engaging in such mirror games is a human universal, at least as far as interactions with young babies are concerned.

At a more global level, mirror games are embedded in *mirror policies*. These policies reflect strategies that govern individuals' participation in mirror games. Individuals may be quite selective in playing these games. They may mirror some behaviors but not others. They may engage in mirror games under some circumstances but not others. Most importantly, they may be selective with respect to the target individuals to whom they grant their mirroring. For instance, they may mirror their children, family, and peers but not strangers, disabled individuals, or the elderly. Mirror policies thus act to induce both social assimilation and dissimilation, and eventually even discrimination.

Conclusion

An action-oriented view of cognition is nice to have, but, at the same time, it is not enough. Viewing action as an extension of cognition captures the functional logic of bottom-up control and the relationships between cognition and action implied in that control mode. However, the cognitive approach to action fails to capture essential features that make up the signature of human action (e.g., top-down control and action alignment). Top-down control operates within individuals and requires a framework that includes the formation of motives, goals, and intentions as precursors of action selection and execution. Action alignment operates between individuals and necessitates a framework that includes shared representations for perception and production.

To accommodate these features, we need to embark on a path that will lead us to *action science*. The study of action must address all aspects and functionalities of action, including its cognitive as well as noncognitive foundations.

18

Learning Action-Perception Cycles in Robotics

A Question of Representations and Embodiment

Jeannette Bohg and Danica Kragic

Abstract

Since the 1950s, robotics research has sought to build a general-purpose agent capable of autonomous, open-ended interaction with realistic, unconstrained environments. Cognition is perceived to be at the core of this process, yet understan#ding has been challenged because cognition is referred to differently within and across research areas, and is not clearly defined. The classic robotics approach is decomposition into functional modules which perform planning, reasoning, and problem solving or provide input to these mechanisms. Although advancements have been made and numerous success stories reported in specific niches, this systems-engineering approach has not succeeded in building such a cognitive agent.

The emergence of an action-oriented paradigm offers a new approach: action and perception are no longer separable into functional modules but must be considered in a complete loop. This chapter reviews work on different mechanisms for action-perception learning and discusses the role of embodiment in the design of the underlying representations and learning. It discusses the evaluation of agents and suggests the development of a new *embodied* Turing Test. Appropriate scenarios need to be devised in addition to current competitions, so that abilities can be tested over long time periods.

Introduction

In June 2014, the University of Reading reported that a machine passed the famous Turing Test: a computer program impersonated a 13-year-old Ukrainian boy, called Eugene Goostman, and was able, through text interface, to make a sufficient number of interrogators believe that they were communicating with an actual human being. This news attracted quite a bit of attention but not as much as one might have thought, since passing the Turing Test had been perceived to be proof that machines could think.

Turing proposed a general test of intelligence to measure the competency of an artificial system "in all purely intellectual fields." He believed that by the year 2000, machines would be capable of this mental process, classically labeled cognition. He discussed the problems associated with deciding when a machine would convincingly reach this level and proposed that the ambiguous question of whether machines could think be replaced by an imitation game which the machine would have to win to prove cognitive competency (Turing 1950). This test, he imagined, would assess the intellectual capabilities of the agent *independent* of the actual mechanism or principle behind it. Turing's original proposal and subsequent versions of the test (e.g., as is used in the Loebner Prize) did not attract significant attention in the robotics community.

One possible explanation for this relative disinterest might be found in an interesting parallel (Russell and Norvig 2003) between the quest for artificial intelligence (AI) and artificial flight: Aeronautical engineering is not defined as making "machines that fly so exactly like pigeons that they can fool even other pigeons." Aeronautic researchers are interested in the principles of aerodynamics. Thus, by analogy, AI researchers seem to have been interested in uncovering the underlying principles of intelligence rather than in duplicating an exemplar.

Early Approaches in Artificial Intelligence

In addition to suggesting the test, Turing (1950) theorized about the underlying principles. He favored the idea of a learning machine whose brain would be similar to that of a child (i.e., a blank slate). Certain built-in rules of operation for logical inference were possible, but these would be subject to change during learning. Interestingly, however, he did not consider it necessary for the agent to have limbs or eyes.

Subsequent researchers have dedicated significant attention to the problems that a machine would face during the proposed imitation game: understanding and producing natural language text, representing general knowledge and information from the ongoing conversation, reasoning to answer questions or to draw novel conclusions, and learning from experience to adapt to new situations. Special symbolic rule-based planners were developed that rely on the existence of an internal world model. Given a certain world state and a goal, these planners could devise a strategy to attain this goal. The first planning system, STRIPS, was developed for the Shakey Robot at Stanford Research Institute (Fikes and Nilsson 1971) and functioned independently to how the robot built the world model, recognized certain objects, or executed planned actions. These problems were supposed to be solved independently by general-purpose, task-independent black-box modules. Significant progress was made early on in terms of these planning algorithms. Based on this work, supercomputers are now able to beat the best human players in chess or Jeopardy. Yet robots are still unable to demonstrate the autonomy and skill of a one-year-old child in terms of perception and motor control. We suggest that the greatest challenge to developing a general-purpose autonomous agent arises at the interface between the agent and the world, not at logical reasoning over readily given abstract symbols.

Emergence of the Action-Oriented Paradigm in Robotics

Moravec (1988) pointed out the following paradox: high-level symbolic reasoning, which requires relatively high effort by humans, seems to be relatively easy to automatize. However, tasks that humans can perform effortlessly (e.g., grasping of arbitrary objects or manipulation of tools) seem difficult for machines to achieve. While we are consciously aware of symbolic reasoning, these latter tasks are controlled by subconscious processes and thus they are much harder to reproduce. Moravec claims that these processes developed over thousands of years of evolution while abstract reasoning is a rather recent development.

Related to this, Brooks (1990, 1991b) proposed a new way to think about artificial intelligence. In contrast to Turing, he believed that a machine needs limbs and eyes to interact with a complex and dynamic environment. He rejected the focus on internal general-purpose representations of the world, symbolic reasoning, and a functional decomposition of intelligence. Instead, he defined intelligence in terms of a combination of simple behaviors, which were defined by directly connecting perception modules to controllers. These behaviors were combined in a structure that Brooks called the *subsumption architecture*. He could show that robots using this idea would expose intelligent behavior in dynamic and cluttered environments. They were even able to show simple grasps and navigate in mapped environments, without any need for complex internal representation and reasoning. These robots had no memory; instead they relied on sensors for continuous feedback from the world around them.

Parallels between Cognitive Science and Robotics

Almost simultaneously to Brooks' proposal, the early work of Varela et al. (1992) established the "enactive approach" to cognition. Similarly to Brooks, Varela and colleagues did not consider cognition to be the process of extracting general-purpose, task-independent representations of the world. Instead, they held that cognitive processes of internalizing the external and building structures are guided by action. This is further related to the internal simulation theory in which the brain simulates the environment and reasons on it before acting (Jeannerod 1988). Clark's action-oriented representation (Clark 1998) and O'Regan and Noë's (2001) sensorimotor contingency (SMC) theory both support this work. According to SMC theory, the agent's SMCs are constitutive for cognitive processes and are defined as law-like relations between movements and associated changes in sensory inputs that are produced by the agent's actions. Accordingly, "seeing" cannot be understood as the processing of an internal visual "representation"; seeing corresponds to being engaged in a visual exploratory activity, mediated by knowledge of SMCs. Additional evidence from psychology and neuroscience stipulates that action in biological systems participates as a generative model in perceptual processes and in structuring knowledge about the world (Gallese et al. 1996; Fadiga et al. 1999; Borroni et al. 2005).

Robotics Research Today

Although the proposal by Brooks is now widely considered to have marked a paradigm shift in robotics, it still remains to be shown whether these ideas can yield more high-level autonomous behavior than that of insects. However, the robotics community has placed more research effort on the interface between an agent and its environment. This does not mean that robotics has agreed on one approach. The classic approach to AI and new directions coexist and are potentially combined. As it happens, this situation is similar to the development that occurred in cognitive science (Engel et al. 2013).

Current research in robotics is largely shaped by the *systems engineering approach* (Brock 2011), which very often aims at solving problems related to a specific application. Within the current research funding landscape, progress has to be fast and verifiable. Today's robots are complex systems that require expertise in many different subjects. A roboticist may commonly be specialized in one of them and try to abstract away the others. For example, researchers who are experts in control may know little about visual perception. Therefore, these modules are abstracted away and treated as black boxes. Any potentially complex two-sided interaction between control and perception is replaced by a simplified interface. Representations may be treated as general purpose and task independent. Very often, symbolic planners sit at the center of these approaches and devise plans that are computed over symbols provided by the black-box perception modules. Resulting action sequences are often executed in an open-loop manner without checking to see whether the expected effect has actually been achieved.

The research area of computer vision is rooted in the demand of robotics research for general-purpose, task-independent representations of semantic entities (Horn 1986). Due to the difficulties inherent in this problem, research in computer vision has developed away from robotics and now has little to do with it. First and foremost, it rarely addresses challenges that arise in robotics, such as real-time requirements or the possibility to act in the environment for exploration. Similarly, in the area of control, the generation of movement is mainly studied in isolation. Feedback controllers usually close the loop around joint angles, velocities, or motor torques. Not as much focus has been placed on feedback about the environment structure. If this kind of feedback is required, it is often provided by precise motion capture systems in the hope that sometime in the future computer vision researchers will deliver the promised reliable general-purpose black boxes.

These approaches have brought tremendous progress in their associated research areas. However, when trying to unite them within a robotics system through the systems-engineering approach, they are only successful in restricted application scenarios that are not open-ended, largely static, and controlled. In general, current robots lag surprisingly far behind humans although they have faster and less noisy sensors and actuators and can perform rapid decision making and control (Wolpert et al. 2011). If we are still striving to discover the underlying fundamental principles between autonomous and purposeful behavior, we have not yet found the key.

Many people believe that the action-oriented paradigm offers the key to permit new insights. We have seen the emergence of the field of developmental robotics, which strives toward learning machines, already proposed by Turing. However, developmental roboticists also emphasize the importance of the learning agent being embodied (Lungarella et al. 2003). Below, we review a portion of the work that follows the action-oriented paradigm and focus mainly on mechanisms for action-perception learning. We focus on the role of embodiment in the design of the underlying representations as well as for the specific learning mechanism.

Representations

General-purpose autonomous robots cannot be preprogrammed for all the tasks they will be required to do; just like humans, they should be able to gather information from different sources and learn from the experiences of both humans and other robots. Thus, the ability to acquire new skills and adapt existing ones to novel tasks and contexts is a necessity. For some natural domains (e.g., cooking and meal preparation), models or plans for different tasks are already available. For example, web pages such as ehow.com and wikihow.com provide simple and detailed instructions on how to plant a tree or make lemon curd. These sites contain thousands of directives for everyday activities: about 45,000 on wikihow.com and more than 250,000 on ehow.com. Using written and structured instructions is common for humans. Many repetitive and dangerous tasks in factories and laboratories have natural language and graphic workflow specifications that are similar to task instructions in the World Wide Web.

A natural idea is to enable robots to do the same. However, this poses several challenges. To look at, listen to, and perceive an instruction, robots need to be able to understand text, video, spoken commands, or even all of them at the same time. They need to be able to understand concepts that are symbolic and relate them to sensory information. For example, an object such as "fork" in spoken or written instructions needs to relate to specific visual features that can be extracted from an image. In addition, the representation of a fork needs to be such that the robot can distinguish it from a knife.

A classic approach is to develop general-purpose, task-independent representations of semantic entities needed in these aforementioned tasks. Representations like this promise effectiveness through compression of a lot of information into a single symbol, which then is able to generalize to all possible situations and contexts.

Humans use visual and other sensory feedback extensively to plan and execute actions. However, this process is not a well-defined one-way stream: how we plan and execute actions depends on what we already know about (a) the environment in which we operate (context), (b) the action we are about to undertake (task), and (c) the result expected from our actions (effect). This insight has been picked up in robotics and resulted in many models that try to represent actions and percepts jointly instead of finding *the one* representation that matches all purposes. The concept of affordances, as proposed by Gibson (1977), has inspired representations, especially in grasping and interaction with objects.

To a certain degree, affordances can be observed in images. In several works (Bohg and Kragic 2009; Saxena et al. 2008; Stark et al. 2008), relations between visual cues and grasping affordances are learned from training data. In Stark et al. (2008), object grasping areas are extracted from short videos of humans interacting with the objects. In Bohg and Kragic (2009) as well as Saxena et al. (2008), a large set of two-dimensional object views are labeled with grasping points. Early work on functional object recognition (Rivlin et al. 1995; Stark and Bowyer 1996) can be seen as a first step toward recognizing affordances from images. Objects are modeled in terms of their functional parts (e.g., handle, hammerhead; Rivlin et al. 1995) or by reasoning about shape in association to function (Stark and Bowyer 1996). In these approaches, the relation between objects and action is usually predefined by humans.

As pointed out by Sloman (2001), we are not consciously aware of a significant amount of human visual processing: we do not experience using optical flow to control our posture nor are we aware of the saccades and fixational eye movements that allow us to negotiate with the complexity of everyday scenes (Koch and Ullman 1985). Therefore, these processes are not easy for us to reproduce in an artificial system. It has been argued that representations should only be constructed by the system itself through interaction with and exploration of the world rather than through *a priori* specification or programming (Granlund 1999). Thus, objects should be represented as invariant combinations of percepts and responses, where the invariances (which are not restricted to geometric properties) need to be learned through interaction rather than specified or programmed *a priori* (Granlund 1999). A system's ability to interpret the external world is dependent on its ability to interact with it. This interaction structures the relationship between perception and action.

In robotics, this can be a slow process, due to the challenges involved when extensive physical interaction is required. Over the last several years, however, advanced oculomotor and hand-eye systems have been demonstrated (e.g., Moren et al. 2008; Montesano et al. 2008; Kraft et al. 2008; Rasolzadeh et al. 2010). There are approaches that let the robot interact with its environment and learn through trial and error. One example is a cognitive model for grasp learning in infants (Oztop et al. 2005). A model for learning affordances using Bayesian networks embedded within a general developmental architecture has been proposed by Montesano et al. (2008). Kraft et al. (2008) proposed object-action complexes as semi-supervised procedures for encoding sensorimotor relations and showed how this can be used to improve the robot's inner model and behavior. The idea was further developed by Song et al. (2010), where the relationships between object, action, constraint features, and task were encoded using Bayesian networks.

These approaches often consider actions at discrete moments in time (e.g., a grasp when approaching but not yet making contact with the object) or as a discrete symbol (e.g., pushing, pulling, grasping, pouncing). The representation of movement over time and how to couple it to sensory input is also an active area of research. One popular representation of this kind has been dynamic movement primitives (Ijspeert et al. 2002; Schaal et al. 2007), proposed for both feedforward and feedback motor commands. Dynamic movement primitives relate to optimal control theory approaches such as minimum jerk trajectories (Flash and Hogan 1985), as well as machine learning approaches such as hidden Markov models (Billard et al. 2004; Inamura et al. 2004). Dynamic movement primitives have been coupled to sensory input through maintaining a visual representation of the goal point and adapting the goal's tracked position (Pastor et al. 2009). Other ways to shortcut perception have been to use motion capture systems or easy-to-detect fiducial markers (Calinon 2009). Only recently have we seen how low-level sensory feedback can define the goal directly (Pastor et al. 2011). During execution, a dynamic movement primitive is adapted such that the robot *feels* the same as when the movement was demonstrated.

Sensorimotor knowledge in humans is structured as of childhood, and this type of lifelong learning has been the focus of developmental approaches in robotics (Pfeifer and Scheier 1999; Kuniyoshi et al. 2003; Lungarella et al. 2003). Thus far, however, developmental approaches have been demonstrated on rather simplistic problems. If a large corpus of data is available, informed learning approaches, such as imitation learning (Schaal 1999), can be applied.

Learning and Priors

Much of the work on learning and priors was inspired by Piaget's ideas of assimilation and accommodation. These two complementary processes of adaptation enable the experiences of the external world to be internalized. The problem of assimilation has been addressed more widely, given that some predefined structure has been used for classification of new experiences. The problem of accommodation requires the representation of knowledge structure to be changed as the new data is gathered and requires more advanced learning techniques to be employed.

An organism cannot develop without some built-in ability. However, if all abilities are built in, the organism is unable to develop. There is an optimal level for how much phylogeny should provide versus how much needs to be acquired during the lifetime.

A human spends years interacting with its environment before it can master certain complex cognitive or motor tasks. At the same time, robots and the computational modules are often expected to learn from very little data. Imitation may be a very good way to bootstrap an artificial system. Even then, during its "lifetime," a robot will encounter so many more situations than what could possibly have been demonstrated to it by a human teacher. In fields such as computer vision or speech processing, "big data" (visual or auditory data annotated with strong or weak semantic labels) has become increasingly more available, impacting the very type of research that is being performed in these areas. Methods that can be trained on these massive amounts of data are currently outperforming previous state-of-the-art approaches (Halevy et al. 2009). In robotics, there are no labeled databases of this order of magnitude to help bootstrap the system. This is most likely due to the complexity of a robot system, which makes it hard to collect and label these massive amounts of data. In contrast to computer vision and speech-processing data, a data point in a robotics database also depends on an action. Therefore, the usual assumption of independent and identically distributed data cannot be as easily made in a robotics system.

Robotic systems receive a continuous stream of sensory (visual, haptic, auditory) data that is currently largely unused. Data is often extracted at arbitrary discrete points in time and then processed independently of the other data points in the time series. Exceptions to this are feedback controllers that enable robots to execute movements toward a goal or along a trajectory. Feedback on some state variable that is actively controlled is continuously gathered, and the appropriate action is computed to minimize the error between the actual and desired state. Most commonly, these state variables contain joint angles and velocities, forces, and torques which act on joints directly or on end-effectors. The state may also contain information from vision sensors, such as the pose of objects in the environment. For these quantities, good models (e.g., rigid body motion or dynamics) exist to help the controllers design and compute the next best action based on carefully selected sensory feedback. For more complex or multimodal data or more complex goals, modeling the mapping between the sensory state and next best action becomes much harder.

Although the "big data" paradigm holds great promise for learning some of these aspects, it is unclear how the data which a robotics system produces (time-series, multimodal and synchronized in time, structured) can be leveraged. Some examples exist for learning from time-series data, but the majority of work focuses on learning from discrete data samples.

Currently, the big data paradigm considers the problem of discovering *correlations* in data. However, robotics seems to be largely dominated by another structuring principle in data: causality. Discovering causality from data is difficult (Pearl 2009). Nonetheless, intuitively, understanding causality seems to be the key to predict the changes in sensory percepts after an agent executes an action.

Embodiment and Imitation

From the viewpoint of morphology, our bodies, actuators, and sensors exist to support effective action (Kuypers 1973) but there is nothing from the perspective of robotic systems that requires a cognitive system to take human shape. Ziemke's framework of embodied systems distinguishes five types of embodiment: structural coupling, historical embodiment, physical embodiment, organismoid embodiment, and organismic embodiment (Ziemke 2003). A single type of embodiment, however, cannot guarantee that the resultant cognitive behavior will be in any way consistent with human models or concepts.

Transfer of information between a teacher (human/robot) and a student (robot) requires a common knowledge representation. When the human and student have identical motor and sensory capabilities, the task may be simply to transform the action of one to the other by changing the frame of reference. Such transfers are not commonly possible, given that embodiments and associated capabilities often differ. To ensure compatibility with human concepts, there may be a need for higher similarity to humans regarding physical movement, interaction, exploration, and perhaps even human form (Brooks 2002).

In terms of object grasping and manipulation, the naïve approach to facilitate grasp transfer between different embodiments is to model the observed action of the teacher and map all the action parameters to the robot hand, which is commonly referred to as the "action-level" imitation (Alissandrakis et al. 2002). However, since different embodiments have different capabilities, the action required to achieve the goal may be different.

Imitation learning is an effective approach for teaching robots simple tasks (Billard et al. 2008). The learning paradigm based on an internal model (Wolpert and Kawato 1998) has received considerable attention. The work by Rao et al. (2007) implements an internal model through Bayesian networks.

Demiris and Johnson (2003) show that the internal models that represent the brain circuitry subserving sensorimotor control also participate in action recognition. They are used to predict the goal of observed behavior and activate the correct actions to maintain or achieve the "goal" state. Later work (Oztop et al. 2005) extends the use of an internal model to the domain of visual-manual tasks. We believe that future research in this area will address the interplay between the embodiment, knowledge representation, and learning in more detail. Abstraction from the embodiment may be a key, but one wonders to what extent this is reasonable to do.

Evaluation and Verification: An Embodied Turing Test

While the aforementioned proposals have been verified in specific scenarios and applications, we lack an understanding of how big their potential is toward the development of general-purpose cognitive agents capable of autonomous and open-ended interaction with realistic, unconstrained environments. How can an action-oriented approach actually be verified and compared to other approaches?

Several tests have been proposed to evaluate general cognitive capabilities of an artificial agent, the most famous being the Turing Test. Turing (1950) was interested in the potential mechanisms and principles behind rational human reasoning, which we nowadays summarize with the somewhat fuzzy term of cognition. Instead of evaluating these mechanisms themselves, he proposed to measure the resemblance of an agent to a real person in a dialogue scenario. In this way, he proposed a way to circumvent the difficult problem of defining precisely the mental process of *thinking*. Turing believed that the exact computational structure of the mechanism does not matter as long as the artificial agent is perceived to perform rational human reasoning. Furthermore, he believed that equipping the machine with a body was entirely beside the point. The actual Turing Test has played a significant role in the field of human-machine interaction. Although new versions of this test have been proposed (Harnad 1991; Marcus 2014), only the total Turing Test begins to test sensorimotor capabilities.

To verify the methods proposed by the action-oriented paradigm, do we need a new *embodied* Turing Test?

One option would be to take the Turing Test and use it to evaluate not the actual mechanisms but rather the resemblance of how an artificial agent acts in the world compared to how a person would act. An external observer would decide whether a robot that is performing certain tasks in an environment is acting autonomously or is teleoperated. The tasks for the robot could involve manipulation and locomotion tasks of different degrees of difficulty, in different environments (e.g., household or disaster relief scenarios). Tasks could also involve physical interaction or collaboration with other agents or humans (e.g., preparing a meal, clearing a dinner table, assembling furniture, rescuing a person from a disaster site, or collaborating with a person to perform assembly tasks). The advantage of this type of test is that a specific task would need to be autonomously performed in a fluid manner that resembles how a human would perform the task. We imagine this human *manner* to involve a certain level of dexterity, flexibility in the presence of a dynamically changing environment, as well as robustness to noise and failures.

It is debatable whether the criterion of resemblance to human fluidity when performing a task is desirable. It may be important in tasks where the robot is collaborating or interacting with humans such that its actions are predictable, but may have limited relevance when it comes to other (e.g., household) tasks.

A Robotics Challenge

Several robotics competitions have been set up to evaluate the performance of artificial agents, not only for purely intellectual tasks but also for tasks involving physical interaction: RoboCup, RoboCup@Home, DARPA Learning Locomotion, DARPA Autonomous Robotic Manipulation, and the DARPA Robotics Challenge. These challenges usually have well-defined goals that revolve around a specific scenario, such as soccer, and can easily be verified. The scenarios are usually formulated broadly so that the goals can be adapted and made more or less difficult from phase to phase. Furthermore, such competitions have the ability to bundle forces and focus them onto one goal (Marcus 2014). The spirit of competition seems to be a powerful source of motivation among researchers.

Do We Need a New Embodied Turing Test?

Without a doubt, it would be advantageous to have a test that could easily evaluate a set of well-defined goals. However, defining these goals poses the initial challenge. Although competitions can serve as a powerful motivator, experience shows that no matter how carefully such goals are defined or out of which original question they came, what counts in the end is winning. The hope or intention of discovering principles behind, for example, intelligence or autonomy, may be rejected in favor of *getting the task done*. Certainly this can be observed in earlier attempts to pass the Turing Test through the use of parlor tricks and purposeful deceit (Marcus 2014). All of the above-mentioned robot competitions encourage what is commonly referred to as *hacking*; that is, engineering solutions which exploit a fixed structure in a scenario, thus sacrificing the generality of solutions. Nevertheless, competitions do offer clear demonstrations of which type of task can be achieved, and some fundamental insights are inevitably gained, although less than what one would hope for or expect.

Once a problem has been solved and its inner workings computationally and algorithmically revealed, we often no longer believe that the associated artificial system is intelligent. It is *just* computation. We wonder whether it is appropriate to let a person judge the resemblance of an agent executing a task to a real human doing the same. This may shift the focus from fulfilling a specific task to that of doing it robustly and fluidly.

Conclusions

Much can be said about perception and action as well as the work that has been done over the last sixty years. Here we reviewed cases which show some of the relations across different fields of research. Importantly, cognition is a process that needs to be studied and approached as such. Proper representations and learning mechanisms are necessary to meet the goal of developing autonomous agents capable of open-ended interaction. Equally important is the issue of how to assess and verify that an agent has made the proper choices. To evaluate and verify the capabilities of a robot, we suggest that the well-known Turing Test be reworked into an *embodied* Turing Test. Many scenarios can be envisioned for such a test; however, we believe short, competition-like scenarios are insufficient. Appropriate scenarios need to be devised that will test a robot's ability over long periods of time.

19

Action-Oriented Cognition and Its Implications

Contextualizing the New Science of Mind

Tony J. Prescott and Paul F. M. J. Verschure

Abstract

The action-oriented paradigm in cognitive science is emerging alongside a broader movement toward a more contextualized, pragmatic, and socially distributed science. This synergy between the view of the mind as practice and the practice of the science of mind bodes well for the development of a new, robust, and socially useful understanding of human experience through which scientific insights can connect with broader intellectual traditions and refocus on societal impact as opposed to impact factors. Through its emphasis on action, the paradigm is well-suited to address real-world problems while advancing fundamental understanding. This chapter explores promising domains in which applications of the action-oriented view are being pursued, including biomimetics, enactive approaches to design, and immersive technologies. Research that has real-world impacts entails social risk; therefore, to be ethical, research in action-oriented cognition should be performed openly and in dialogue with the wider public.

Introduction

This Forum was convened to consider the possibility of a "paradigm shift" in brain and behavioral sciences: one aimed at a more action-oriented view of cognition. In this chapter we consider the notion of paradigm shift within the wider context of how current science is practiced and supported. We argue that, within a societal and international context, science is increasingly judged by its ability to advance solutions to real-world problems. Moreover, this is not necessarily a bad thing: a focus on societal challenges can promote understanding and resolve the false dichotomy between pure and applied science. A strong test of the action-oriented view, then, is whether it is able to produce artifacts that can act effectively in the real world. This is the notion of the "pragmatic turn" applied, not just to how we understand brains and minds, but also to how we *do* our science. While it is beyond the scope of this contribution to review the practical impacts of the action-oriented approach fully, we will try to point to some areas where we think such impacts are happening, or where they might happen with some additional effort. Finally, we will argue that scientists can no longer afford to ignore, if they ever could, consideration of the potential negative impacts of their research. To pursue science in an ethical way obliges us, therefore, to think carefully about where our science and technology might lead, and engage in an open and public debate about our goals.

Paradigm Shifts and the Social and Political Context of Science

Our discussion begins with a consideration of the broader scientific context within which the action-oriented paradigm is being pursued and with which we see interesting parallels.

When Thomas Kuhn (1962/1970) developed the notion of a "paradigm shift," he did so in the context of a post-positivist philosophy of science that had hitherto focused more on methodology (e.g., Popper 1935/2005) and on the role of the individual scientist (e.g., Polanyi 1958). Kuhn's thesis, by contrast, recognized that the "scientific view," the dominant paradigm at any given point in time, is determined by a consensus within the community of research scientists and he sought to analyze specifically the dynamics of how this consensus can change. His approach brought a sociological perspective to the understanding of science, emphasizing how science *is* practiced rather than how, in some ideal decontextualized way, science *should* progress.

Kuhn remained convinced that science advances to a better understanding of nature over time, but it is easy to find support in his thesis for a more relativistic view. For instance, Kuhn's notion of *incommensurability*[1] asserted that competing paradigms rely on incompatible conceptual frameworks, such that measurement (how we obtain data), epistemic standards (the rules we apply when reasoning about data), and meaning (the semantics of the concepts we use to interpret and explain data) can each differ between paradigms. As a consequence, the truth of statements conceived within one paradigm cannot easily be assessed within another. Figure 19.1, reproduced from Froese et al. (2012), illustrates that there may be a degree of incommensurability between the action-oriented view and other approaches in cognitive and brain science. However, if truth depends on your scientific framework then the preference for one view over another can become a matter of taste. Following in the wake of Kuhn, Feyerabend (1975/2010) developed epistemological anarchism, the view that there is no such thing as the scientific method, which led him to later

---

[1] Kuhn's version of incommensurability is probably the best known, but the idea was also central to Feyerabend's philosophy of science, and its roots can be traced to the earlier writings of physicists, such as Einstein and Bohr, and Gestaltists such as Köhler.

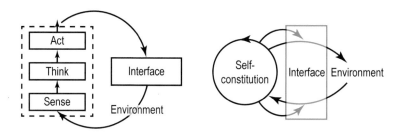

Figure 19.1 Contrast of a cognitivist (left) and an enactivist (right) view of a sensory augmentation technology (Froese et al. 2012). The cognitive view is characterized by an agent implementing a classical sense–think–act loop mediated by the interface device which comes between the agent and the environment. The enactive view sees the agent as an autonomous system engaged in an activity called "sense-making," which is an emergent property of the agent-environment interaction. The interface device alters the dynamics of this interaction resulting in "augmented sense-making" such that the device itself may be "experientially transparent." The diagram illustrates that because the cognitivist and enactivist paradigms employ quite different semantics and conceptual frameworks, they may be incommensurable in the Kuhnian sense. (Figure from Froese et al. 2012; reprinted with permission from IEEE Transactions on Haptics.)

endorse a relativist understanding of the history of science. For the wider postmodern movement, it became possible to characterize scientific knowledge as the belief system of an intellectual elite, a view that still has resonance in popular culture and which questions the authority that has been assigned to science traditionally. We believe that the pragmatic turn in the study of mind and brain must also address the question of how this paradigm shift and its practical consequences can be assimilated by society and induce positive change.

Research into the societal, political, and cultural context of science has expanded in recent decades, giving rise to the field of *science and technology studies*, and has come to have a significant impact on science policy, especially in Europe. In particular, the view of science as a process of discovery, pursued by dispassionate researchers who are neutral with respect to the broader context, already criticized by Polanyi and Kuhn, has been further challenged. In a series of influential books and articles, Nowotny and colleagues have contrasted this classical idea of science, characterized as mode-1, with a view of science, as practiced at the end of the twentieth and beginning of the twenty-first centuries, as "socially distributed," denoted as mode-2 (Gibbons et al. 1994; Nowotny 1999; Nowotny et al. 2001, 2005).

Mode-2 science is said to have several key characteristics. First, the traditional distinction between pure and applied research is modified by the contextualizing of even basic science relative to identified application domains: societal, economic, or technological. Second, rather than operating within traditional disciplines, research happens within transdisciplinary groupings which emerge to address identified priorities, then dissolve as trends or opportunities move elsewhere. This mixing of transient research cultures can, dependent on the incentive structures, provoke new forms of scientific creativity or opportunistic exploitation of resources. However, the temporary nature of such consortia rarely leads to lasting structures, such as the formation of new disciplines. Finally, science is pursued by increasingly diverse actors involving a broad set of stakeholders. Universities and government-funded research centers are still key players, but companies, large and small, get in on the act, as do nongovernmental organizations such as user groups, charities, and think tanks. There is little space for the individual researcher independent of their brilliance.

This changing climate for research is thought to be reflected in a number of identifiable trends including the top-down steering of research priorities, the commercialization of research, and the increase in scientific accountability (the "audit explosion"; Power 1997). Furthermore, the rise of open access publishing as well as the increasing role of social media in promoting and evaluating scientific ideas has served to democratize science as well as to dilute scientific authority—a commentator with many Twitter followers, or a popular blog, can be more influential than a scientist with a high H-factor.

The weakening of the traditional scientific hegemony, the legacy of relativist views in the philosophy of science, the growing coupling of research to innovation, and the emergence of alternative actors and of new forms of knowledge exchange have undermined the authority of the scientific view. To reestablish its validity, proponents of mode-2 science suggest that it is no longer sufficient for knowledge to be "reliable" in the sense of reflecting a consensus among "competent, well-informed scientists" (Ziman 1978). As Nowotny (1999:253) states:

> A 21st century view of science must not only embrace the wider societal context, but be prepared for the context to begin to talk back. Reliable knowledge will no longer suffice, at least in those cases, where the consensuality reached within the scientific community will fail to impress those outside. In a 21st century view of science, more will be demanded from science: a decisive shift toward a more extended notion of scientific knowledge, namely a shift toward socially robust or context-sensitive knowledge.

Social robustness thus implies the inclusion of nontraditional players in efforts to build consensus around a scientific view. This implies that science communication and dissemination becomes as important as the process of discovery and its results. As a paradigm that looks to develop a scientific understanding of human experience, the action-oriented view has many external audiences with whom common ground can be sought: engineering, the humanities, the creative arts, health, and spiritual traditions concerned with mindfulness. This possibility was also very much in the minds of some of the originators of the field (Varela et al. 1992). With the notion of socially distributed science coming more to the fore in science policy, the action-oriented view thus seems well-placed to build a science of human experience that is contextualized, robust, and connects to these wider intellectual traditions through its intrinsic commitment to the real-world relevance of knowledge.

Building Artifacts That Solve Real-World Problems As a Strategy for Research in Action-Oriented Cognition

Mode-2 science has been presented as a modern phenomenon; however, science has always been exposed to societal pressures and, in the past, has looked to push back. In 1850, in a presidential address to the American Association for the Advancement of Science (AAAS), Joseph Henry, the first Secretary of the Smithsonian Institute, sought to defend pure science from the intrusion of applied concerns:

> The incessant call in this country for practical results and the confounding of mechanical inventions with scientific discoveries has a very prejudicial influence on science....A single scientific principle may include a thousand applications and is therefore, though if not of immediate use, of vastly more importance even in a practical view (Rothenberg 1998:101–102).

A generation later, Henry Rowland (1883), first president of the American Physical Society, asserted in his speech to the AAAS that "to have the application of science, the science itself must exist"(Rowland 1883:242). Both of these leading figures of nineteenth century science considered it important to dissociate pure from applied research and to assert that the flow of ideas goes from discovery to innovation, not vice versa. Indeed, for Rowland, there was something morally admirable about the self-sacrificing nature of the scientist who avoided applied topics (similar in spirit to Henry V, when Shakespeare wrote of "the few, the very few, who, in spite of all difficulties, have kept their eyes fixed on the goal"). The notion of a pure science base as "scientific capital" that would allow technological innovation to flourish, and economic and societal benefits to flow, was later placed at the center of an influential report from the U.S. Office for Scientific Research and Development (Bush 1945); this led to the establishment of the National Science Foundation. Indeed, a key aim of that report was to protect basic science from potential erosion by excess focus on societal need, echoing the assumption of Henry and Rowland that applied research would pay for itself by generating revenue whereas pure science was too long-term and high-risk to be left to market forces. (For an assessment of the impact and legacy of the Bush report fifty years later, see Cole et al. 1994.)

Today, the dichotomy between pure and applied as well as the notion of a one-way flow of causality appears oversimplistic and may be holding back science (Brooks 1967; Stokes 1992; Nowotny et al. 2001). Further, the characterization of basic science as disconnected from application may have contributed to the perception of science as detached from, and thus insufficiently concerned with, the human condition. It may also have led to an unhelpful devaluing of mission-led scientific research that addresses societal needs that are not commercial (e.g., Sarewitz 2012). The history of science contradicts the notion that good research cannot serve both basic and applied goals. Pasteur, for instance, spent most of his life working on practical problems (e.g., sugar fermentation and the diseases of domestic animals), yet his research led to fundamental discoveries concerning the germ theory of disease as well as to the birth of a new research field (microbiology). Consideration of such examples led Brooks (1967) to propose a spectrum of research from pure to applied and Stokes (1992) to describe "Pasteur's quadrant," a two-dimensional conceptual plane where research efforts can emphasize fundamental understanding, societal use, or some productive blend of the two.

To locate the sweet spot in Pasteur's quadrant, for the domain of action-oriented cognition, we can also look for inspiration to the eighteenth century Neapolitan philosopher Giambattista Vico, who famously proposed that we can only understand that which we create: *Verum et factum reciprocantur seu convertuntur* (Verschure 2016). Vico's dictum can be viewed as an instruction to apply a synthetic approach, that is to build models, as well as to create technologies that validate and make practical use of scientific ideas. Following Vico, we can adapt the notion of the "pragmatic turn" to a methodological use—applied research as embodied proof-of-principle, or to turn it around the other way, "the theory as a machine" (Verschure 2013). Below we consider specific application domains where research in action-oriented cognition is using (or could take) this approach and where there is potential for bidirectional (fundamental understanding *and* societal use) impact.

Application Domains for Action-Oriented Cognition

As put forth by Engel et al. in the introduction to this volume, key principles of the action-oriented paradigm that can be exploited to develop applications include:

- Understanding cognition as the capacity to generate structure by action
- The immersion of the cognitive agent in its task domain
- The significance of the body and the possibility to export some aspects of the problem of generating appropriate actions to the body or to the environment
- The dynamic, context-sensitive, and adaptive nature of behavior-generating systems
- The ability to extend cognitive tasks into the environment

Here we discuss a number of key domains where these principles can or are being usefully applied. We do not attempt to be exhaustive but rather focus on domains where we have some experience and interest. Further discussion of possible application domains is provided by Dominey et al. (this volume).

Biomimetics

Biomimetics is the development of novel technologies through the distillation of principles from the study of biological systems (Bar-Cohen 2005). Biomimetic research operates in three directions (Prescott et al. 2014):

1. It promotes a flow of ideas from the biological sciences into engineering.
2. It provides physical models of biological counterparts that can serve as experimental platforms to understand them (Rosenblueth and Wiener 1945).
3. It creates a new class of technologies that can then be advanced toward innovation and direct application.

Since the action-oriented approach stems from the study of biological cognition, it seems natural that biomimetic artifacts are developed to embody and evaluate these principles.

Within biomimetics, one example domain where the action-oriented paradigm has been particularly influential is in the control of locomotion in biomimetic robots. The importance of embodiment in generating coordinated movement is beautifully illustrated in passive walking machines that exploit the natural dynamics of their parts for periodic motion. For example, bipedal walking machines, with no control system whatsoever, can generate a stable walking gait on a suitably sloped surface by relying on the passive dynamics of suitably configured mechanical parts (Collins et al. 2005). Animals provide control through their nervous systems in a manner that complements their natural body dynamics (Chiel and Beer 1997; Ijspeert 2014). In walking or running, for example, many-legged animals exploit the pendulum-like natural motion of jointed limbs to help generate a suitable cyclic pattern. By relinquishing some aspects of control to the body, they also benefit from the energy-recycling capability of elastic tissues. Muscles and tendons, for instance, convert kinetic energy to potential energy as the foot hits the ground, providing a store of energy to be released in the next step cycle. Designing controllers modeled on animal locomotion pattern generators that exploit these principles provides a very promising path for building efficient legged robots (Ijspeert 2014). Such controllers can be simpler (e.g., have fewer control parameters) than more traditional forms of continuous robot control, and will entrain themselves to the dynamics of the body, making them highly adaptable. Prosthetic limbs are being developed which similarly reduce the need for control through well-designed natural dynamics (Carrozza et al. 2005). Similar principles—extending cognition into the body and exploiting natural dynamics to simplify control—have been applied to a range of other motor tasks in robotics including reach and grasp, as well as to the control of hyper-redundant soft robots (Pfeifer and Bongard 2006; Trivedi et al. 2008).

Enactive Approaches to Design

One of the most productive domains for the application of action-oriented principles appears to be in the design of artifacts that are used in close conjunction with the body, such as sensory substitution or augmentation devices. Examples include the Enactive Torch (Froese et al. 2012), the Feelspace Belt (Nagel et al. 2005), and the vOICe (Auvray et al. 2007). A key design aim is to make the device "experientially transparent" (see Figure 19.1) such that the goal-directed behavior of the user naturally incorporates properties of the artifact, including its capacity to transform from one sensory modality to another. Design can benefit from an understanding of the sensorimotor contingencies (O'Regan and Noë 2001) to which sensing in a given modality is attuned. Applications include assistive technologies for people with sensory impairments and sensory augmentation systems for use in safety services, construction, and defense. Commercial devices such as the *Nintendo Wii* controller also take advantage of some of the principles identified by the enactive approach.

Immersive Technologies

With the development of next generation virtual reality and telepresence technologies, experiencing the world from a point-of-view other than that from behind our own eyes is becoming a possibility for all of us (Sanchez-Vives and Slater 2005). Psychologists have long-known that our conceptions of our physical selves are very flexible; however, recent advances in immersive technology now allow us directly to examine and test our expectations about the limits of the self concept along different dimensions such as the physical, temporal, and social (Blanke 2012; Prescott 2015). Theoretical constructs emanating from the action-oriented perspective are helping to understand the experience of immersion and the capacity of the brain to adapt to a virtual or remote body (Stoffregen et al. 2006). Enactive principles can also improve the design of immersive technologies to increase the feeling of presence. For example, studies suggest that the immersive experience is more compelling when actions in a virtual environment result in expected sensorimotor contingencies of objects (e.g., bending down to see the underside of a horizontal surface) and agents, and so modulate presence (Inderbitzin et al. 2013).

The technology for immersive virtual reality is advancing very rapidly, largely due to its importance to the entertainment and games industries. In addition to leisure, however, there are also important applications of immersive technologies in civil and commercial domains: teleoperation of remote equipment in hazardous environments; telepresence for delivering health care or conducting business; and augmented reality for applications in areas such as design, tourism, and retail. In health care, studies are demonstrating the potential for the use of virtual reality in reducing pain (Hoffman et al. 2001), treating eating disorders (Riva 2011), and in assisting recovery from stroke (Jack et al. 2001; Cameirão et al. 2012) and posttraumatic stress disorder (Difede and Hoffman 2002). With an aging population in developed countries, ever greater burdens are being placed on health care systems, and new tools for deployment in both the hospital and home are required to maintain healthy aging. Telehealth is emerging as an important element of health service plans to streamline and improve services, and a key element of this is the use of telepresence technologies for health workers and carers to interact remotely with patients or those in care (Riva 2000). Immersive technologies are also growing in importance as technologies through which people interact with family and friends, and could offer an important means for reducing psychological stress.

The above examples show that an action-oriented paradigm not only holds promise in advancing our understanding of mind and brain but also provides a prime example of twenty-first century science. These examples also illustrate that the action-oriented approach must promise to establish a closer synergy between basic and applied science beyond Pasteur's quadrant. We can view the application in health care, for instance, as a direct validation of basic science hypotheses. Hence, by following a more deductive approach toward applications, they in turn become experiments that can advance theory. This has also been called Vico's loop (Verschure 2016). Pursuing this model, however, implies that scientists themselves must also be aware of the implications of applied research, including ethics.

Ethical Issues

Mode-2 science has heralded a more critical shift in evaluating the motivations for scientific research and the potential societal impacts of its consequences. Research in the action-oriented domain should recognize that although scientific knowledge is neither good nor bad, the use of knowledge is not ethically neutral. Thus, scientists have a responsibility to think about how the results of their research might be deployed.

Expectations about the positive societal benefits of advanced technologies, derived from research in cognitive systems, vary from the unconditionally positive (Roco and Bainbridge 2003) to more guarded and cautious (Nordmann 2004; Kjølberg et al. 2008). The guarded perspective worries that benefits will be for the few, not the many—and contribute to a more unequal society— or that we risk dehumanizing ourselves by advancing too far down a path of self-modification and enhancement. This broad debate is set to continue and become even more pressing as applications move from science fiction toward technological fact.

It is known that the public is generally well-disposed toward science but that unease is often greatest at the intersection between science and technology, where advances are often disruptive and therefore perceived to entail risk (Wynne 2007).

Our view is that assessment of societal risk is an undertaking far too important for those engaged in the research not to be involved. This may mean that researchers need to take a different approach, away from the more traditional approach in science, which ensures that research plans conform with established codes (which govern, e.g., research involving animals and humans) but leaves consideration of the broader and longer-term risks to others, such as professional ethicists. The issue for science is this: if researchers do not get involved, they may find that lines of enquiry are shut down on the basis of "future scoping" activities undertaken by people who are less than fully informed. Because research in action-oriented cognition is often at the intersection of science and technology, it is both relevant and important, but also potentially hazardous. We advocate that the research plans of the community involved in action-oriented research should therefore integrate the analysis, understanding, and management of risk from an early stage.

A general approach to a risk-based science ethics is illustrated in Figure 19.2. Here, ethics in research is positioned along two complementary dimensions. The societal impact dimension (vertical) is concerned with the effects of technology on individuals and on society, projected along the dimension of time from the short-term to very long-term. Equally important, however,

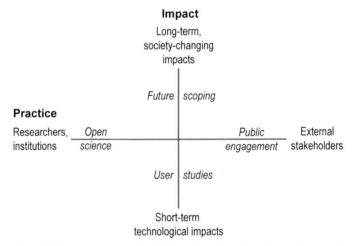

Figure 19.2   Ethics in research. The "Ethics Cross" divides the challenge of addressing ethical issues in research along two dimensions: investigating potential societal impacts (vertical) and pursuing research practices that foster a meaningful exchange, an analysis of research risks, through dialogue with the wider society (horizontal). The cross is based on Buchanan's (1985) analysis of product design as reinterpreted by Illah Nourbakhsh.

is the dimension of research practice (the horizontal axis in Figure 19.2) projected between researchers, and their institutions, and external stakeholders, including the general public. The proposal here is that the research community engage in a broad dialogue with potential stakeholders continually presenting its goals, methods, and achievements. Key mechanisms include adopting an open science strategy—including free dissemination of results, a willingness to conduct research in public view, and public engagement—proactively looking to disseminate ideas and outcomes. Engagement should be bidirectional and include a willingness to adapt research aims to address concerns that are well founded, and potentially to abandon lines of research that are identified as being too high risk.

Conclusion

These are exciting times for action-oriented cognitive science, as this volume demonstrates. However, because science is a human activity, its pursuit is subject to societal and political constraints, and in recent times, the value of science for its own sake is increasingly being questioned. The number of potential questions that can be asked of science is infinite. For publicly funded research, it is reasonable to require research to address important questions whose answers have the potential for substantial societal benefit. Our research field is fortunate because its core questions concern the human condition, and advances should lead to insights as to how to improve it. Taking Pasteur and Vico as our role models, we can better understand ourselves through action-oriented cognition, and through this "pragmatic turn" also do science that makes a difference.

First column: (top to bottom): Peter Dominey, Wolfgang Prinz, Andreas Engel, Tony Prescott, Peter Dominey, Jeannette Bohg, Günther Knoblich
Second column: Tony Prescott, Andrew Schwartz and Miriam Kyselo, Matej Hoffmann and Marti Sánchez-Fibla, Wolfgang Prinz, Shaun Gallagher, Matej Hoffmann, Andrew Schwartz and Tony Prescott
Third column: Andreas Engel, Günther Knoblich, Jeannette Bohg, Shaun Gallagher, Tobias Heed, Tobias Heed and Günther Knoblich, Peter Dominey

20

Implications of Action-Oriented Paradigm Shifts in Cognitive Science

Peter F. Dominey, Tony J. Prescott, Jeannette Bohg,
Andreas K. Engel, Shaun Gallagher,
Tobias Heed, Matej Hoffmann, Günther Knoblich,
Wolfgang Prinz, and Andrew Schwartz

Abstract

An action-oriented perspective changes the role of an individual from a passive observer to an actively engaged agent interacting in a closed loop with the world as well as with others. Cognition exists to serve action within a landscape that contains both. This chapter surveys this landscape and addresses the status of the pragmatic turn. Its potential influence on science and the study of cognition are considered (including perception, social cognition, social interaction, sensorimotor entrainment, and language acquisition) and its impact on how neuroscience is studied is also investigated (with the notion that brains do not passively build models, but instead support the guidance of action).

A review of its implications in robotics and engineering includes a discussion of the application of enactive control principles to couple action and perception in robotics as well as the conceptualization of system design in a more holistic, less modular manner. Practical applications that can impact the human condition are reviewed (e.g., educational applications, treatment possibilities for developmental and psychopathological disorders, the development of neural prostheses). All of this foreshadows the potential societal implications of the pragmatic turn. The chapter concludes that an action-oriented approach emphasizes a continuum of interaction between technical aspects of cognitive systems and robotics, biology, psychology, the social sciences, and the humanities, where the individual is part of a grounded cultural system.

Embodied Cognition and Enactivism

The concept of embodied cognition generally considers extra-neural bodily structures and processes important for cognition. There are a number of different theories of embodied cognition. Wilson (2002) distinguishes between theories that emphasize:

- Situatedness
- Online or real-time processes
- Off-loading cognitive processing onto the environment (also referred to as embedded or scaffolded cognition)
- The idea that the environment itself is part of the cognitive system (sometimes called extended mind or distributed cognition)
- The idea that cognition is for action (sometimes called action-oriented or enactive cognition)
- The idea that offline cognition is body-based

We focus on enactive or action-oriented cognition, but we do include concepts that involve embedded/scaffolded and extended/distributed cognition. Extended mind approaches include the idea of action-oriented representations, whereas more radical versions of enactivism eschew representationalism (Hutto and Myin 2013).

Action-oriented theories that emphasize sensorimotor contingencies (O'Regan and Noë 2001; Noë 2004) are usually put under the heading of enactivism. More radical forms of enactivism include the idea of sensorimotor contingencies, but also emphasize other aspects of embodiment such as affectivity (interoceptive, autonomic, emotional aspects), reward, interest (motivation), and embodied social interaction (Varela et al. 1992). Both enactive and extended approaches endorse the idea that cognition is not just "in the head." Most embodied approaches accept the Gibsonian idea of affordances. Despite certain common elements, however, it is safe to say that not all embodied cognitive theories agree on all issues.

Action-oriented theories of embodied cognition (including enactive and extended) are prefigured in the work of American pragmatists such as Peirce (1887) and Dewey (1896, 1916, 1938). Dewey, for example, spoke of the organism environment as a single unit of explanation, suggesting that the environment is not only physical but also social, and emphasizing movement and our use of tools and instruments as part of cognition (see Menary 2007, 2010). Enactive theories find important sources in the European philosophical tradition of phenomenology (Husserl, Heidegger, and Merleau-Ponty) where perception is characterized as pragmatic, guided by what the agent can do (i.e., by embodied skills and motor possibilities). Phenomenological contributions to cognitive science champion embodied-enactive aspects. Such approaches include neurophenomenology (Varela 1996), where experimental subjects are trained in phenomenological methods that focus on first-person experience, and "front-loaded" phenomenology, where phenomenological concepts or distinctions (e.g., sense of agency vs. sense of ownership for actions) are incorporated into experimental design (Gallagher and Zahavi 2012).

The enactive approach can be characterized by the following background assumptions (Varela et al. 1992; McGann 2007; Di Paolo et al. 2010; Engel 2010; Engel et al. 2013):

1. Cognition is considered as the exercise of skillful know-how in situated and embodied action.
2. Cognition structures the world of the agent, which is not pregiven or predefined.
3. Cognition is not viewed as happening only in the brain; it emerges from processes in the agent-environment loop. The intertwinement between agent and world is constitutive for cognition.
4. System states are thought to acquire meaning by their functional role in the context of action, rather than through a representational mapping from a stimulus domain.
5. The approach aims at grounding more complex cognitive functions in sensorimotor coordination.
6. In contrast to classical cognitive science, which has an individualistic and disembodied view of cognition, the enactive approach strongly refers to the embodied, extended, and socially situated nature of cognitive systems.
7. The approach implies strong links to dynamical systems theory and emphasizes the key relevance of dynamic coupling and dynamic coordination.

As illustrated in Figure 20.1, the pragmatic turn can be characterized in terms of various brands of action-oriented cognition. Parallel to the advent of enactivism, such a turn has independently occurred in classical representational approaches as well. Importing action into the study of cognition cannot be considered a unique signature of enactivist agendas. Three major moves that one may discern in representational approaches to cognition and action are:

1. Theoretical: Cognition-for-action takes a new look at cognitive functions (e.g., perception and attention) and emphasizes their role in action. The aim is to find out how these functions are constrained and shaped by the requirements of action control.
2. Experimental: Action is viewed as an object of study itself (with regard to production and perception as well as their mutual relationships). The aim is to investigate the representational underpinnings of action representation, with particular emphasis on scenarios of social interaction and communication (where action production and perception are combined).

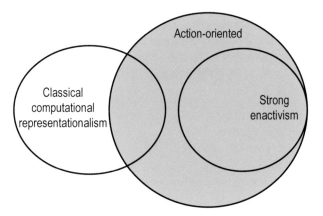

Figure 20.1   The primacy of action is constitutive of enactive and action-oriented approaches. Representational approaches address aspects of cognition for action. This common ground constitutes the pragmatic turn.

3.   Theoretical and experimental: Action is seen as a constitutive ingredient of cognition (particularly high-level cognition). The aim is to trace the latent and implicit involvement of action in cognition (coming fairly close to what nonrepresentational enactivism claims as well).

Action-related approaches to cognition have important characteristics. First, they allow for actions to generate and modulate perception, which allows for much more dynamic and active perception. Second, they allow for an ontogenetic perspective that can account for dynamic changes of the cognitive system over time and development. This approach is thus much more flexible than the more or less invariant information-processing machinery assumed by the cognitive approach. Third, the assumption that cognition is for action allows for a reinterpretation of various cognitive phenomena, especially limitations of processing. For instance, considering that attention may ultimately serve action allows one to understand that the purpose of selection is not to shield a capacity-limited system against stimulus overflow but to allow actions to be optimally fed with actual relevant information. If there is a limitation, it is on the action side—not on the input side.

Does This Pragmatic Turn Constitute a New Paradigm?

Kuhn (1962/1970) did not invent the term paradigm but he is largely responsible for how it is used today. A *paradigm* describes a collection of ideas—theories, concepts, laws, and experimental methods—that are shared by a community of scientists as a basis for describing the current status of knowledge about their domain and plans to advance knowledge within that domain. Although Kuhn initially talked about "pre-paradigmatic science," he later changed his view and proposed that at all stages in the development of a field there are paradigms around which scientific communities cohere. For Kuhn, the history of science is described by the changing status of different paradigms, including periods of more incremental research ("normal science"), where scientists work to develop the ideas within a paradigm, as well as more radical events (paradigm shifts), where a previously dominant paradigm crumbles and is replaced by another. Paradigm shifts happen when results that do not fit within the existing approach (so-called anomalies) accumulate and the new paradigm better addresses the anomalies and is consistent with much of the previous data accounted for under the earlier view. Experiments which identify key anomalies can be viewed as critical ones, although attribution of such significance is always easier in hindsight. It is the nature of scientific paradigms that ideas in one do not necessarily translate into another. For example, Einstein's theory of relativity reformulates fundamental concepts in physics in ways that cannot be fully expressed in terms of classical Newtonian physics. As a result of the changed view that results from a paradigm shift, scientific understanding of the world is qualitatively different in some important ways from how it was previously.

Looking at the action-oriented view in cognitive science, it is difficult to argue that it constitutes one paradigm. Instead, we might say that there are a number of related paradigms which share some common emphases on the role of action. Do we have a paradigm shift? Twenty years ago, not many people thought motor activity could influence perception. Now there is a consensus that the motor system contributes to understanding. Perhaps the important point is that to make progress from this proposed paradigm shift, we must focus not only on high-level definitions but also on how these concepts will lead to new *experimental paradigms,*[1] both in the natural sciences and medicine as well as in the engineering sciences. To understand cognition, it has to be viewed in the context of meaningful behavior. Independently of whether the resulting dominant paradigm might be strong enactivism or a hybrid paradigm, the crucial point is to take action into account as a major constituent of cognition.

How Does the Pragmatic Turn Change How We Do Science?

A New Perspective on Perception

The implications of the pragmatic turn can be illustrated by referring to one of the most discussed versions of an action-oriented approach: the sensorimotor contingency (SMC) theory developed by O'Regan and Noë (2001). In this framework, the agent's acquired knowledge of SMCs (i.e., the rules governing sensory changes produced by motor actions) are critical for both development

---

[1] The term "experimental paradigm" refers to specific procedures and protocols and should not to be confused with the term "paradigm" defined above.

and maintenance of cognitive capacities. Vision, as an example, involves much more than the processing of retinal information. According to this view, seeing corresponds to a way of acting, to exploratory activity mediated by deployment of specific skills and knowledge of SMCs. Neural activity patterns that emerge in visual cortex do not in themselves constitute seeing. Rather, the brain supports vision by enabling exercise and mastery of SMCs. This concept is in strong opposition to classical representation-centered approaches of vision, such as those of Marr (1982) or Biederman (1987), which assume that perception consists essentially in building context-neutral descriptions of objects that are stored in the brain, irrespective of any actions that might be performed by the agent. In contrast, according to the sensorimotor account, the concept of an object would consist in a family of related SMCs, and thus clustering across these contingencies provides the basis of learning object structures. One of the interesting implications of this account is that the distinction between declarative and procedural knowledge, which figures prominently in cognitivist approaches, dissolves in favor of the notion that different types of knowledge might be seen as repertoires of contingencies with different degrees of complexity. The SMC approach also stipulates interesting predictions for changes in patients with brain disorders. A key prediction is that patients with motor deficits (e.g., Parkinson disease or amyotrophic lateral sclerosis) should exhibit perceptual and cognitive deficits that result from an impoverished repertoire of SMCs that can still be utilized by the patient.

Concept of Attention as Action-Constrained Sensory Refocusing

Research on neural mechanisms of attention serve as an example of how an action-oriented account may inform the analysis of presumed "high-level" cognitive capacities. The "premotor theory of attention" (Rizzolatti et al. 1987) introduced the idea that the selection of sensory information should be modulated and focused by constraints arising from current action planning and execution. In agreement with this prediction, several studies have shown that movement preparation can lead to attentional shifts (Collins et al. 2008, 2010) and to changes in the acquisition of object-related information (Craighero et al. 1999; Eimer and van Velzen 2006; Fagioli et al. 2007). Functional imaging studies and neurophysiological recordings have provided evidence that the modulatory bias imposed by attention may indeed arise from premotor regions, in particular the frontal eye fields (Donner et al. 2000; Moore and Fallah 2001). MEG experiments on visual attention have shown that premotor regions like the frontal eye field are involved in top-down modulation of sensory processing through selective enhancement of dynamic coupling, expressed by phase coherence of fast neuronal oscillations between premotor, parietal, and sensory regions (Siegel et al. 2008). Similar evidence has also been obtained in a recent EEG study using an ambiguous audiovisual stimulus. Analysis of neural coherence revealed that large-scale interactions in a network of premotor, parietal, and temporal regions modulate the perception of the ambiguous stimulus (Hipp et al. 2011). Taken together, these findings provide clear support for an action-oriented account of attention. Actually, these studies suggest a shift in the concept of attention, which may indeed most appropriately be described as a bias in sensory processing that is procured by the current action context.

A New Perspective on Social Cognition

Different versions of the action-oriented view have had a major impact on social cognition research. Previously, social cognition research focused on explicit symbolic communication, on mind reading as theoretical inference, and on social categorization. One focus was on person perception, for example, in the mechanisms of face identification (Kanwisher et al. 1997), assigning traits to a person or categorization of others as in-group/out-group (Macrae and Bodenhausen 2000). Action-oriented views have brought social interaction into the foreground (Prinz 2012; Rizzolatti and Craighero 2004): simulationist models emphasize common coding and mirroring, two related representational and brain mechanisms for matching one's own actions and others' actions. This has sparked new experiments on imitation as well as the planning and execution of joint actions, such as carrying a table or playing a piano duet (Knoblich et al. 2011). Dynamic systems approaches highlight the importance of temporal entrainment in interpersonal action coordination (Riley et al. 2011). Enactivist approaches have brought to the foreground peoples' experiences during joint action (De Jaegher et al. 2010). Together, the different action-centered mechanisms have defined a new perspective on social cognition, and they have started to influence views of how individual cognition works (e.g., Pickering and Garrod 2013b).

Studying Social Interaction through Sensorimotor Entrainment

In classical representation-oriented approaches of social cognition, agents are thought to interact with conspecifics based on their capacity to develop a "theory of mind"; that is, to derive complex models of the intentions, beliefs, and personalities of other agents (Carruthers and Smith 1996). In this framework, which has also been termed the "spectator theory" of social cognition (Schilbach et al. 2013), the primary mode of interaction with the social environment is that of a detached observer who theorizes and produces inferences about other participants. The pragmatic turn inspires an alternative view which assumes that even complex modes of social interaction may be grounded in basic sensorimotor patterns enabling the dynamic coupling of agents (Di Paolo and De Jaegher 2012). Such sensorimotor patterns, or contingencies, are known to be highly relevant in cognition (O'Regan and Noë 2001; Engel et al. 2013; Maye and Engel, this volume). A key hypothesis that follows from this approach is that learning and mastery of action-effect contingencies may also be critical for predicting consequences of the action of others and, thus, to enable effective coupling of agents in social contexts. According to this notion, social interaction depends on dynamic coupling of agents; the interaction dynamics provides a clue to social understanding and shares aspects with the interactionist concept of social cognition (Di Paolo and De Jaegher 2012; Gallagher 2004). This concept also agrees well with the joint action model by Knoblich and colleagues, who predict that shared intentionality can arise from joint action (Sebanz et al. 2006).

Action-Oriented Word Learning in Language Acquisition

The advantage of considering an enactive approach in cognitive development has been clearly demonstrated in the domain of child language acquisition.[2] Word learning has often been characterized in terms of the potential ambiguity when a word is uttered and the referent could be any one of a number of possible objects in the visual scene. In this characterization, the infant passively attends to the words and the cluttered scene. How would this scene change in an enactive setting? There is evidence that gestures (especially pointing gestures) by caregivers correlate with gestures by infants and, critically, that gestures produced by infants at 14 months predict vocabulary at later age (Özçalışkan et al. 2009; Rowe and Goldin-Meadow 2009). This indicates the importance of the action-oriented communicative context for language acquisition. In this context, Yu and Smith (2013) examined child-parent dyads during word learning, while the children wore head-mounted cameras that indicated where they were looking. The data revealed that children were not passive, but grasped and held the objects in question, creating conditions where referential ambiguity was eliminated, and word learning efficacy was maximized. This action-oriented shift in the study of language acquisition will likely find useful application as well in the context of the social dynamics that will similarly reduce referential ambiguity (Dominey and Dodane 2004).

Action-Centered Approach to Neuroscience

Peter König said: "If you only pose simple questions, you only get simple answers." In this spirit, we need to reformat questions about the nervous system as a complex system in an action-centered context that also involves extra-neural components. A reductionist approach misses elements of the system that were designed or which evolved to work in the real world. The problem is that mainstream neuroscience is stuck in the old model where subjects are often

---

[2] Embodied language processing is another hallmark area where the action centeredness of cognition (in this case language) is primary.

passive or anesthetized. In classic vision and sensory experiments, at best, subjects report what they sensed passively; at worst, they are stimulated while anesthetized and unable to generate behavior that determines what they sense. Results from these open-loop experiments have been interpreted with static maps consisting of neurons that perform single, discrete functions. This points to another shortcoming of conventional neuroscience. It is as if each cell has one function and "causes" a change in a single output region. This is far from accurate. We now know that each cell encodes many parameters simultaneously and sends output to many places. These neurons "operate" continuously, not just when they receive input from a single source. The "causal" output (defined as the ability to change another neuron's probability of discharge) is typically very weak. It takes many neurons acting together to generate an action potential in another neuron. The "causal" chain is very noisy, can only be defined statistically, and is likely context dependent. It is difficult to decode information from a single neuron because its firing rate-parameter correlation is very weak; the change in discharge rate associated with any given parameter is small. However, across a population, if these small changes are consistent, the representation of the parameter value will emerge clearly. Describing causality in a complex system is difficult and can best be described as statistical structure. Common input to the population generates correlation between members of the population, and this structure can be recognized with proper analyses.

Changes in Conceptual and Methodological Orientation of Neuroscience Suggested by the Pragmatic Turn

A pragmatic turn in cognitive science will profoundly change our view of the brain and its function. Brains will no longer be considered only as devices for building models of the world (the "representationalist" view), but will instead be hypothesized to support the guidance of action. A key assumption is that action shapes brain structure and is constitutive for brain function. Importantly, this would hold not only during development, but also for the functioning of the adult brain.

These considerations suggest a refocusing of the conceptual premises of neuroscientific research. The following premises contribute to defining a framework for action-oriented, pragmatic neuroscience:

1. The primary concern of the experimenter is not the relation of neural activity patterns to stimuli, but to the action at hand and the situation in which the subject under study is currently engaged.
2. The functional roles of neural states might be viewed as supporting the capacity of structuring situations through action, rather than as "encoding" information about pregiven objects or events in the world.
3. Investigation of neural function encompasses the view that cognition is a highly active, selective, and constructive process.

4. Sensory processing should be considered in a holistic perspective, as being subject to strong top-down influences that continuously create predictions about forthcoming sensory events and eventually reflect constraints from current action.
5. The function of neurons and neural modules might not be considered in isolation, but with proper reference to other subsystems and the actions of the whole cognitive system.
6. Investigating the intrinsic dynamics of the brain becomes increasingly important, because interactions within and across neural assemblies are constitutive for the operations of the cognitive system.

Reorienting the focus of scientific investigation of cognitive processing will likely require the repertoire of neuroscientific methodology to be geared toward action-oriented experimental strategies. Some requirements of such methodological reorientation include:

1. Experiments must avoid studying passive subjects and, instead, allow for active exploration (e.g., free viewing, manual exploration).
2. Improved technologies are needed to track the actions of one or several interacting subjects.
3. Current neuroscientific methods (e.g., EEG, MEG, and fMRI) are limited with respect to subjects' ability to execute movements, either due to the design of the measurement apparatus or to artifacts being introduced into the signal by movement. Consequently, novel and improved technologies for acquiring neural activity and biosignals during movement are needed, as well as for analysis strategies that separate movement artifact from true signal.
4. Exploration of the relationship of neural activity with action parameters calls for the recording of nonneural, action-related signals. These may include, but are not restricted to, EMG, displacements, forces, heart and breathing rate, and possibly changes of the surroundings.
5. New data analysis techniques will be required to (a) examine correlations between behavior and high-dimensional neurophysiological signals and (b) develop methodologies that consider massively distributed coding both within and across brain structures.
6. Technologies allowing controlled manipulation of SMCs will likely become increasingly important. Virtual reality setups may become crucial experimental tools.

In summary, the pragmatic turn has had an important influence in changing the way that science is done. But what can we draw from these examples? From a strong enactivist view, it appears that as soon as time is critical for immediate interactions in a joint task, representational theories may have nothing to say. This suggests that within the framework of action there is a particularly highlighted status for interpersonal action. Interpersonal action is different, in the relational aspect: the object of my interaction is also interacting, generating reciprocity in prediction when interacting with others.

Implications for Robotics and Engineering

In defining a shift in perspective on the status of agents acting in the world, the pragmatic turn has immense potential to change the fields of robotics and engineering. One could say that robots have always been embodied and pragmatic in the sense that they are physical devices designed to perform useful actions in the world. However, with respect to high-level tasks, roboticists adopted the representationalist stance which came from good old-fashioned artificial intelligence (GOFAI; Haugeland 1985). This yielded the classical sense-think-act control architectures which emphasized on the "think" part that involved building and updating world models and planning the next action using AI techniques operating on these models. Sensing and acting were initially regarded as straightforward interfaces with the real world and were thus considered less interesting or challenging research problems. A good example is the Stanford Cart (Moravec 1983). As a result, real-time responsiveness was lost.

Embodiment, Compliance, and Soft Robotics

The pragmatic turn goes hand in hand with an embodied turn. The properties of physical bodies were completely neglected in the strand of robotics that came out of the GOFAI tradition. Even in more recent and much more impressive examples that successfully interact with the world in real time, like the DARPA Grand Challenge winner autonomous car Stanley (Thrun et al. 2006), a clear separation between body and brain is apparent. Indeed, the design philosophy in the DARPA challenge was: "*treat autonomous navigation as a software problem*" (Thrun et al. 2006, their emphasis). This philosophy is in stark contrast with the embodied perspective that posits tight coupling between brain/controller, body, and environment (e.g., Pfeifer and Scheier 1999).

A classic and extreme example of the fact that behavior may be generated by a completely brain-less mechanical system are the passive dynamic walkers (McGeer 1990). Although such contraptions are highly dependent on their ecological niche (e.g., slope of particular inclination), to some extent "pure physics walking" can already display simple "adaptive" properties, like robustness to perturbations. This is achieved through mechanical feedback and has been called self-stabilization (e.g., Blickhan et al. 2007). Another ingenious demonstration from robotics where morphology decisively contributes to the generation of behavior is the "universal gripper" (Brown et al. 2010): A bag of ground coffee is pressed onto an object and let to conform to its shape. Afterward, a vacuum pump evacuates air from the gripper, making the granular material jam and stabilizing the grasp. Another concerns the amazing climbing capabilities of geckos, which rely on van der Waals forces between their feet and the surface on which they are climbing; these are strong enough thanks to a hierarchical structure of compliance from centimeter to 500 nanometer scales (Autumn et al. 2002). These findings inspired the design of Stickybot, a robot that can climb smooth vertical surfaces (Kim et al. 2007).

Unlike "high-level" robotics capitalizing on the GOFAI (thus overlooking the importance of the body and closed-loop real-time interaction with the environment), control engineers would consider exactly these aspects as their "bread and butter." They would understand the robot as a dynamic system governed by differential equations. It responds to control input in a way that is defined by the body's structure and its interaction with the environment. Furthermore, adding a feedback loop allows shaping this response to achieve a desired behavior or goal state. The challenge includes identifying and modeling the dynamic system (i.e., finding the linear or nonlinear differential equations, deciding around which sensors to place the feedback loops, and designing the controller itself). For linear systems, a large body of powerful and well-understood mathematical tools exists to analyze the system's response and design appropriate controllers. However, the majority of real systems are governed by nonlinear dynamics. While a number of techniques exist to cope with these kinds of systems, they are usually far more complex to compute, apply only to a subset of systems, or approximate the nonlinear with linear dynamics.

As a result, the plant in a control system is typically treated as fixed, and the overall tendency has been to suppress its complex nonlinear dynamics in favor of stiff and linear behavior. Often this approach works surprisingly well. However, "linear systems are not rich enough to describe many commonly observed phenomena" (Sastry 1999:2). Complex and soft bodies offer new possibilities which can be exploited; however, they also pose difficulties for classic control approaches (see Hoffmann and Müller 2014).

Complementary to *passive compliance*,[3] there is an interest in *active compliance* where a controller mimics elasticity using otherwise stiff actuators. Although these kinds of systems are not as energy efficient as passive systems, they offer the possibility to study variable stiffness as well as to study where it is beneficial to introduce compliance in a system. The importance of compliance has been demonstrated for robotic locomotion (e.g., Kalakrishnan et al. 2010; Semini et al. 2013) and manipulation (e.g., Righetti et al. 2014; Deimel et al. 2013) where it increases robustness against perturbations that may be due to inaccuracies and noise in perception and actuation. Instead of precisely planning a movement in joint space, a feedback loop is closed around the

---

[3] Compliance is the opposite of stiffness. A stiff system can move along a desired trajectory or toward a goal position and will remain in this state no matter what external forces are applied to it. A compliant system allows deviations from these desired positions that may be introduced by external forces.

interaction forces between robot and environment such that it can gracefully give in when experiencing unexpected contact with the environment. These challenges are taken up by the growing field of "soft robotics" (e.g., Albu-Schäffer et al. 2008; Pfeifer et al. 2012; Trimmer 2013).

Action in Robotic Perception

One of the interesting claims of the enactive approach is that there may be no principle difference between "high-level" cognitive processes and "low-level" sensorimotor functions. Instead, the former is seen as grounded in and emerging from the latter. Thus perception cannot be treated as a passive and disembodied process: it is critically shaped by the morphological properties of the whole agent (including, of course, its sensory apparatus) as well as by the actions executed by the agent. This is in contrast with the prevailing approaches to robotic perception, vision in particular (e.g., Horn 1986).

Object concepts and object recognition may be considered to illustrate this. From an enactive perspective, object concepts do not consist of feature-based descriptions stored in memory agnostic to any action contexts. Objects would be defined by sets of possible actions that can be performed on them, and knowledge of an object concept would consist in the mastery of the relevant object-related SMCs (Engel et al. 2013; Maye and Engel, this volume). This view is supported by studies on object recognition in humans and robotic systems. Evidence in humans clearly demonstrates a dependence of exploratory eye movements on the specific task given prior to viewing an image (Yarbus 1967). The influence of semantic information, also in terms of object identity, is visible in the results. Furthermore, there is evidence that visual object recognition depends on exploratory eye movements during free viewing of images. Using ambiguous images, a recent study showed that eye movements performed prior to conscious object recognition predict the object identity recognized later (Kietzmann et al. 2011). For artificial vision systems, these ideas were first explored in the area of active vision and perception (Ballard 1991; Bajcsy 1988). They focused on implementing systems both in hardware and software that have the ability to choose actively what to sense. More recently, the area of interactive perception goes beyond the mere selection of percepts toward actively changing the state of the environment to increase information gain. In studies on robot vision, it has been shown that a visual scene can be disambiguated by actively manipulating objects in the scene through a robotic arm (Fitzpatrick and Metta 2002; Björkman et al. 2013; Högman et al. 2013). Similar implications apply to other sensory modalities. For example, a quadruped robot running on different ground is critically able to improve terrain discrimination if a history of the actions taken (gaits used) is considered together with sensory stimulations induced in tactile, proprioceptive, and inertial sensors (Hoffmann et al. 2012).

A Holistic Distributed Approach to Control: An Action-Oriented Systems Engineering Perspective

The representationalist view of cognition has some commonalities with traditional systems engineering[4] in the sense that it heavily relies on top-down design and modularization. Processing is often central and sequential, and different modules are recruited to perform their part in a pipeline-like manner (like sense-think-act). This approach is dominant in the engineering disciplines, because it allows for efficient separation of subtasks and distribution of work among different units.

However, as has already been argued extensively, humans and animals do not seem to rely on the same type of architecture. Instead, an effective and smooth interaction with their environment emerges from the interplay of a plethora of physical and informational processes that operate in parallel and have different couplings between them. This was articulated by Rodney Brooks when he openly attacked the GOFAI position in seminal papers: "Intelligence without representation" (Brooks 1991b) and "Intelligence without reason" (Brooks 1991a). Through building robots that interact with the real world, such as insect robots (Brooks 1989), Brooks realized that "when we examine very simple level intelligence we find that explicit representations and models of the world simply get in the way. It turns out to be better to use the world as its own model" (Brooks 1991b:396). Inspired by biological evolution, Brooks created a decentralized control architecture consisting of different layers: every layer a more or less simple coupling of sensors to motors. The levels operated in parallel, but were built in a hierarchy, hence the term subsumption architecture (Brooks 1986).

This approach proved to be effective in "low-level" tasks, such as walking or obstacle avoidance. However, to our knowledge, it has not demonstrated, to date, how it could scale to more "cognitive" tasks, beyond the immediate "here-and-now" timescale of the agent. Growing evidence from biology suggests that this approach has been adopted by humans and animals and thus can be scaled up. Therefore, if we take the enactive approach seriously, we need to revise the way robots are designed. Here, we elaborate on a few points toward this goal:

1. *From centralized modules to parallel, loosely coupled processes*: "Intelligence is emergent from an agent-environment interaction based on a large number of parallel, loosely coupled processes that run asynchronously and are connected to the agent's sensorimotor apparatus."

---

[4] Systems engineering was initially developed in the context of complex spacecraft system design (including the spacecraft and the associated ground data systems). Our perspective here is to in no way call into question the validity of this methodology, but rather to recognize that it was developed for a class of systems which do not at all have the same interaction requirements that one finds in real-time social interaction.

(Pfeifer and Scheier 1999:303). This contrasts with the classical approach consisting of centralized, functional modules operating sequentially. Furthermore, unlike the classical scheme, all processes operate essentially in a closed loop through the environment and on different timescales. In addition, as we are learning from the brain, process specialization is much weaker than representationalist views of the mind envisioned: different circuits are dynamically recruited for different tasks when need arises. This insight should flow into action-oriented and embodied design methodologies. Rather than starting out with a modularization of the system into functional blocks, it may be more beneficial to begin thinking about the interfaces. Modules (in a weaker sense) can subsequently be plugged in. This may have a profound influence on how the modules are designed by the end of the process. An even stronger version of this methodology would first be to design entire feedback loops. This would shift the focus away from modules toward combinations of different feedback loops.

2. *Timing is important*: Real-time interaction with the world and with other agents is critical. For example, in human-robot interaction, we are still very far from the natural "fluid" interaction that humans have among themselves. A similar observation can be made for physical contact interactions between the robot and its environment. When it comes to manipulation or locomotion in unstructured environments, the abilities of robots lag far behind those of humans in terms of dexterity, robustness, and speed. The low-level bodily and sensorimotor levels with fast feedback loops as well as their combination, integration, and coupling should play an important role in solving this lack of fluidity. We realize that for temporal coordination (either during verbal interaction with a person or physical interaction with the environment), underlying processes cannot be sequential and feedforward but have to be parallel, asynchronous, and predictive.

3. *Self-organized task solving*: Imagine the task of cleaning a table. If a household robot were to solve it in the classical approach, it would perform image segmentation first and then try to identify all the objects in the scene. Based on a model acquired this way, it could set itself a goal state (e.g., clear the table) and then apply search techniques from AI to obtain a complete plan on how to proceed. Intuitively, it is evident that this is not how a human would solve such a task: a human would most likely remove objects within reach without fully planning out the remaining steps. In the end, paraphrasing Brooks, the table is there and one does not need a model to clean it. Thus, seemingly complex problems can often be greatly simplified through sensorimotor coordination and by off-loading complex planning to the brain-body-environment interaction.

In summary, we are convinced that the embodied, action-oriented, or enactive approach has important consequences for the way robots and their control systems are designed. However, for some of the implications—in particular, to move away from specialized, clearly separated modules—the path to their adoption by industry may not be easy.

Practical Applications of the Action-Oriented Approach

We have come to be convinced of the potential impact of this change in perspective. Clearly, such a change in perspective must generate practical applications. These practical applications can be seen in domains that include education, clinical therapy, and the development of enactive prosthetic devices.

Education

Historically, the pragmatic approach has had a significant impact on education. While we will not treat this subject in detail, we note that John Dewey, in particular, had a profound commitment to and influence on public education. Dewey (1897, 1900, 1902, 1916, 1938) advocated the importance of education proceeding in a context of interaction, where curiosity and discovery motivated the student's active search for knowledge. Similarly, action-oriented education would emphasize the importance of participatory aspects of learning situations. Instead of "sitting in place" in a classroom, listening or working with books or even desktop technologies, action-oriented education would emphasize the use of one's whole body for learning. In contrast to conceptions of learning and cognition modeled on information processing and amodal problem solving, for example, the use of computer-generated interactive simulations (mixed reality immersive technologies)—where students can enter into a whole-body engagement with a subject matter—has been shown to increase learning speed and accuracy compared to desktop computer use (Lindgren and Moshell 2011; Lindgren and Johnson-Glenberg 2013).

Clinical Applications in Aphasia

Research in action-centered cognitive neuroscience has led to the development of a new translational method for the treatment of language deficits, or aphasias. This method, called intensive language action therapy, has yielded significant improvements in language and communication abilities, even at chronic stages of poststroke aphasia, in contrast to conventional approaches, which have not demonstrated comparable effects (Berthier and Pulvermüller 2011). Further translational progress may be achieved by developing similar methods for other neurological deficits affecting language and action.

Enactive Approaches to Developmental Disorders

Sensorimotor problems can be found in a number of developmental disorders. In autism spectrum disorder (ASD), for example, infants who are later diagnosed with autism display sensorimotor problems (e.g., postural stability, gait, timing and coordination of motor sequences, anticipatory adjustments and face expression) before they reach the developmental age associated with theory of mind, when typically developing children engage in joint attention and joint action with others and are learning to communicate (Trevarthen and Delafield-Butt 2013; Gallese et al. 2013; Cattaneo et al. 2007; Cook et al. 2013; Gallagher 2004; Hilton et al. 2012; Fabbri-Destro et al. 2009; Whyatt and Craig 2013; David et al. 2014). In individuals with ASD, Torres and colleagues show the occurrence of disrupted patterns in re-entrant (afferent, proprioceptive) sensory feedback that usually contributes to the autonomous regulation and coordination of motor output, and supports volitional control and fluid, flexible transitions between intentional and spontaneous behaviors (Torres 2013; Torres et al. 2013). In ASD, as well as in other developmental disorders (e.g., Down Syndrome), disruptions in motor processes may partly explain why individuals show difficulties in distinguishing goal-directed from goal-less movement (Torres 2013; Brincker and Torres 2013), anticipating the consequences of their own impending movements and applying fine-tuned discriminations to the actions and emotional facial expressions of others during real-time social interactions. These studies hold important implications for future research and therapeutic interventions, although further research is required to understand differences among sensorimotor problems in the different disorders and what precisely they contribute to each one (Gallagher and Varga 2015).

Action-Oriented Approaches to Schizophrenia

The early work by Elaine Walker and colleagues (e.g., Walker and Lewine 1990) has motivated investigations of motor activity in people with schizophrenia. Ford and coworkers showed that 2-year-old children, who became schizophrenic in adulthood, differed from their healthy siblings by exhibiting motor awkwardness and social withdrawal. Ford argues that this could be due to an inaccurate, slow, or faulty forward model of motor control starting in infancy. Using EEG-based methods that allow excellent temporal assessment of neural processes, they found abnormal premotor activity in patients with schizophrenia in the ~100 ms preceding talking (Ford et al. 2007), with abnormalities being greater in patients with more severe auditory hallucinations and worse amotivation and avolition, respectively.

Several studies have investigated action-perception loops in schizophrenia with particular reference to the understanding of delusions of control, and the belief that one's actions are controlled by another agent (C. D. Frith 2012; Blakemore, Smith et al. 2000; Frith et al. 2000a). The idea stems from

Helmholtz who pointed out that when we move our eyes, the world remains stationary, but if we poke our eye with our finger, the world appears to move. The normal attenuation of movement depends on relating the intention to move (or speak) with the anticipated sensory changes (the forward model). This is why we are unable to tickle ourselves. Schizophrenic patients with delusion of control can tickle themselves, presumably because something has gone wrong with this perception-action loop. The same framework has been applied to other disorders in the experience and production of action.

Exploiting the Link between Movement Disorders and Cognitive Disorders

Based on the foregoing examples and related data, a view has emerged that movement disorders and cognitive disorders are linked. Thus, a framework that considers the crossover of action and cognition would be useful. For example, in schizophrenia and depression, disturbed notions of agency are evident in individuals with these syndromes, but movement components are considered less frequently. In Parkinson disease, one finds the opposite: a bias toward movement.

The application of principles that are consistent with the enactive agenda can be found in the domain of neurorehabilitation. One example is the rehabilitation gaming system (RGS), which exploits virtual reality-based interaction to incorporate embodiment, first-person perspective, and goal-oriented action in rehabilitation protocols. RGS has been used by over 400 stroke patients in controlled clinical impact studies, and significant impact on functional recovery has been demonstrated (Cameirão et al. 2011, 2012). In a virtual reality therapy setup, Cameirão et al. have demonstrated the increased effectiveness of embodied versus disembodied therapy, noting the particular effectiveness of embodiment with a first-person perspective, and that goal-directed action is more effective than repetitive action. Likewise, effective results are also seen when patients are placed in a social context (e.g., working with others), thus improving self-image.

Neural Prosthetics

Neural prosthetics offers a clear example of action-based learning and control (Schwartz 2004). In the scheme illustrated in Figure 20.2, we consider brain operation to be the creation of behavior and movement to be behavioral output. The cost function is generation of the desired movement.

One aspect of this approach is the need to map each neuron's firing rate to movement. With intact subjects, this is relatively straightforward; movement parameters (e.g., arm direction) can be regressed against firing rate to capture a tuning function for each unit. This is more difficult with a paralyzed subject. Schwartz and colleagues have developed a "calibration" procedure

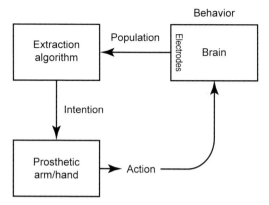

Figure 20.2 Neural prosthetic scheme: Electrodes in the brain record single-unit activity. Many units are recorded simultaneously and this population is processed with an extraction algorithm. This algorithm is a model of the transformation between neural activity and movement intention. The model generates an estimate of the intended movement signal, and this serves as the control signal for an external device (here, a prosthetic replica of the arm and hand). The prosthetic device moves according to the estimated intention signal, and its action is visualized by the subject who registers any errors in the desired movement. The error drives a learning process in the brain which modifies the neural output to reduce the error signal. Essentially, the brain is learning the extraction algorithm. This closed loop induces powerful learning, leading to high prosthetic performance.

based on observation-driven cortical activity (Velliste et al. 2008; Tkach et al. 2008; Collinger et al. 2012). By demonstrating different movements to the subject who is actively monitoring the motion of the prosthetic device, neural activity in the motor cortex is elicited that can be used to estimate an initial tuning function. After an initial estimate, the process is iterated; the subject's activity is gradually used to control the device by mixing their volitional signal with a set of parameters that attenuate errors which would move the device away from the displayed target. The amount of assistance (error attenuation) is gradually reduced while the proportion of volitional drive is increased until the control signal is composed only of the subject's extracted neural output.

The method of mixing volitional and autonomous control is called "shared mode" control (Clanton et al. 2013). For instance, suppose a paralyzed subject is operating the arm/hand to reach, grasp, and manipulate an object. Advanced robotic algorithms have been developed to generate autonomous manipulation (Katz et al. 2013), and these autonomous signals can also be mixed with those derived from brain activity. This could be used to test the idea of agency (e.g., by querying the subjects as to whether they were controlling the device, or if it was being controlled automatically as different ratios of automatic/volitional signals were mixed together within a neural prosthetic experiment). An open question is how the conscious experience of an artificial limb as one's own adds to the (motor) abilities of the agent over and above tool use, in which the tool is usually experienced as clearly separate from the body (e.g., a hammer, a fork). Ehrsson et al. (2004) suggest that parietal regions mediate multisensory matching of the rubber hand, but that premotor cortex is related to the conscious experience of embodiment. In contrast, current work (Seth, pers. comm.) is trying to identify how action may determine whether an object is perceived as embodied.

These experiments also show that single neurons encode multiple parameters in their firing rate patterns. For instance, many neurons in the motor cortices of human and nonhuman primates have been found to contain neurons with tuning functions describing 10 degrees of freedom for hand and arm movement (Wodlinger et al. 2015). These tuning functions are not clustered in parameter space but seem to be distributed uniformly. Furthermore, neurons with different tuning functions are located together anatomically in the same region of cortex. Finally, neurons with the same type of tuning function (cosine shaped) for movement can be found throughout the neural axis (Van Hemmen and Schwartz 2008). This suggests that the principle of directional tuning is widely distributed and not localized to specific structures.

Societal Implications: How Can We Connect to Everyday Human Life?

A central goal of enactivist theory as proposed by Varela et al. (1992) was to address what it means to be human in everyday life, the ultimate aim being to address existential and ethical questions about the human condition: Who am I? What am I? How can I live a good life? Following Merleau-Ponty, the human body can be viewed as both a physical structure and as a "lived-in" phenomenological one. Thus, to move from a science of embodied cognition to an enactive understanding of human nature, it is necessary to bridge this gap. Varela et al. thought that progress in this direction could be made by connecting ideas which emanate from cognitive science—embodiment, distributed cognition (e.g., connectionism and Minsky's "society of mind"), and self-organization and emergence—with insights external to science (e.g., from traditions such as Buddhist views on the groundless nature of the self). How does this enterprise to apply action-oriented cognitive science to the challenge of the human condition stand today?

Within contemporary action-oriented views, we discern a number of positions. For some, significant progress toward addressing existential issues can be identified through recent advances in our understanding of consciousness and the self (see Seth et al., this volume). By translating these theoretical insights into more everyday conceptual language, action-oriented cognitive science could play a more direct role in, for instance, the ongoing debate about the relationship between science and religion. For others, the action-oriented approach, despite its focus on subjectivity, is seen as providing only third-person insights into consciousness and personhood, which are then of no immediate help to understanding the first-person human predicament. From this perspective, action-oriented cognitive science can only impact indirectly on existential considerations through the filter of nonscientific frameworks. Although the range of views on these issues is broad, there is general agreement that through its emphasis on embodiedness, extended cognition, and intersubjectivity, the action-oriented approach opens up new pathways for dialogue between the biological and psychological sciences, on one hand, and the social sciences and humanities, on the other.

Availability of Phenomenological Expertise

That phenomenology can contribute to cognitive science is a relatively new notion (Gallagher and Varela 2003). If the task of cognitive science is to explain how we experience, engage, and interact with the world, it is claimed that phenomenology provides a controlled descriptive method that is able to characterize the explanandum, without necessarily specifying the cognitive mechanism that explains it. This is not so much a division of labor—since the two tasks (description and explanation) depend on each other—but rather more a case of mutual enlightenment. This raises the following questions: Even if all people have access to experience, are they all equally good at describing it? Do we need experts in phenomenology to provide such descriptions? Do such descriptions differ between cultures?

While it is likely that there are clearly some aspects of variability related to culture and experience, we may ask whether there are some universal invariants. If "experts" perceive things that novices cannot perceive, as numerous studies in athletics show (see, e.g., Mann et al. 2007), this suggests that the depth and precision of access to experience can be increased through practice, and that a new vocabulary can be developed to allow precise communication. Thus, for example, Buddhists who practice mindfulness mediation have a precise language for discussing detailed notions of phenomenological experience of different aspects of self (Trungpa 2002). Phenomenology, as well as Buddhist mindfulness, may offer important tools to allow us to develop insight into our everyday experiences.

The claim here is not that phenomenology is primary. Rather, according to what Varela (1996) calls the methodological principle of "mutual constraints," phenomenology and cognitive science should maintain some consistency, so that if there is some disagreement between them about some particular phenomenon or process, further research will be necessary to resolve the difference.

Phenomenology has informed enactivist approaches to cognition. Phenomenological philosophers have argued that world or environment is not independent of mind or organism. We need to think of the organism-environment (or the mind-world) as an epistemic system. The dynamic systems perspective of enactive thinking provides descriptions of agent-environment systems in which basic processes, including problems and pathologies, are described in relational terms: where problems involve some tension or conflict in the relation between the agent and environment. Brain-centered approaches tend to identify these problems entirely as occurring within the organism. An enactive approach would spread this out to examine the relations between brain, body, and environment and consider treatments in terms of those relations. Remedies could then be achieved by manipulating or intervening with the brain, the body, the environment, or some mix of factors. We note that many domains of practical application already use such a systems view without it being inspired by enactive thinking per se. Enactive researchers would like to push such approaches into much wider use than is currently the case.

Enactive Autonomy and Embodied Psychosocial Existence

As an example of how the enactive view could be applied to everyday life, we conclude this section by summarizing Kyselo's theory of the human self as embodied psychosocial existence (Kyselo 2014), as it incorporates some key principles of the broader action-oriented approach. According to Kyselo, the enactive self can be operationally defined as a socially enacted autonomous system, whose systemic network identity emerges as a result of an ongoing engagement in processes of *distinction* (which promotes the existence of the individual in his/her own right) and *participation* (which promotes connectedness with others). Some implications for understanding and improving everyday human life are as follows.

This view emphasizes that human identity depends on others. The enactive view presupposes a deep dynamic interrelation between agent and world—a relation of mutual co-constitution. For the self, this means that the social world is not merely a context or developmentally relevant, but that without continuously engaging with the social world of other people, without their contribution, no individual self can be generated or maintained. Human nature is not egocentric but genuinely relational. We are not embodied islands but existentially dependent on each other as long as we live. Further, because we care for the maintenance of our identities, we continuously and adaptively evaluate ourselves and our interactions with others accordingly.

Self-maintenance involves tension and vulnerability. Cognitive identity is generated and maintained under precarious circumstances. For the human self, this means that both kinds of network processes—those that enable distinction as well as those that enable participation—are required together to bring about the individual as a network of autonomous self-other organization. Without distinction, the individual risks dissolving in social interaction dynamics. Without participation—acts of openness toward others—the individual eschews structural renewal, risking isolation. Both goals are in opposition, bringing about a deep tension for the self that has to be negotiated. A useful insight for everyday life is that this tension is a necessity that cannot be avoided. Vulnerability and openness to others are appreciated as enabling the second dimension of self: the sense of being connected.

Human identity and understanding others requires continuous negotiation. Tension and conflict with others are to some extent unavoidable, since each individual is engaged in an interaction that strives to maintain this twofold sense of self. When two agents interact, their goal is not merely to reach consensus and harmony, but to reach it under particular conditions; namely, by acknowledging that their respective individual goals are met. Combining these ideas with dynamic systems theory, Kyselo and Tschacher (2014) propose an approach for understanding relationship dynamics and the negotiation of closeness and distance within couples, and what this means for personal well-being and the likelihood of relationship maintenance.

Finally, the twofold structure of the self, as distinct and participating, entails a further useful insight: individuals are limited in their personal control over their self-constitution. Other people have a say in the construction of our identities and, given that they have perspectives and interests of their own, others do not always comply with what we would like or need. The enactive view contrasts with the increasing emphasis on individualism in Western society—that we should seek to be the omnipotent creators of our own lives—and suggests that a better understanding of the role of others in constructing the self could help people to be more at ease with themselves.

Concluding Remarks

An action-oriented perspective changes the role of an individual from a passive observer to an actively engaged agent interacting in a closed loop with the world as well as with others. Cognition exists to serve action within a landscape that contains both. Here we have offered an overview of this landscape, where action is not just the output of the system but also where the system is there for action.

This perspective emphasizes a continuum of interaction between biology, psychology, the social sciences, and the humanities and has already had an impact on science by changing the way we consider perception, social cognition, and interaction, and the bases of their neurophysiological implementation. Other impacts can also be expected. For example, approaches in engineering need to change to what we refer to as action-centered systems engineering. Social implications are equally visible. From an action-oriented perspective, a human being is not an isolated individual responsible alone for his/her destiny, but rather a member of a grounded cultural system. The true test of these proposed implications cannot be fully evaluated today but will require the test of time to determine if we have seen clearly into the future.

Acknowledgments

We would like to thank our distinguished colleagues from the Forum for providing additional input to the chapter, including Judith M. Ford, Jürgen Jost, Miriam Kyselo, Friedemann Pulvermüller, Gottfried Vosgerau, Peter König, Paul Verschure, Marek McGann, Chris Frith, and Gabriella Vigliocco.

Bibliography

Note: Numbers in square brackets denote the chapter in which an entry is cited.

Abeles, M. 1991. Corticonics. Neural Circuitry of the Cerebral Cortex. Cambridge: Cambridge Univ. Press. [8]

Adams, R. A., S. Shipp, and K. J. Friston. 2013. Predictions Not Commands: Active Inference in the Motor System. *Brain Struct. Funct.* **218**:611–643. [2, 4, 6, 7]

Adams, R. A., K. E. Stephan, H. R. Brown, C. D. Frith, and K. J. Friston. 2013. The Computational Anatomy of Psychosis. *Front. Psychiatry* **4**:47. [6]

Aglioti, S. M., P. Cesari, M. Romani, and C. Urgesi. 2008. Action Anticipation and Motor Resonance in Elite Basketball Players. *Nat. Neurosci.* **11**:1109–1116. [2]

Alais, D., and D. Burr. 2004. The Ventriloquist Effect Results from Near-Optimal Bimodal Integration. *Curr. Biol.* **14**:257. [7]

Albu-Schäffer, A., O. Eiberger, M. Grebenstein, et al. 2008. Soft Robotics. *IEEE Robot. Autom. Mag.* **15**:20–30. [20]

Alissandrakis, A., C. L. Nehaniv, and K. Dautenhahn. 2002. Imitation with ALICE: Learning to Imitate Corresponding Actions across Dissimilar Embodiments. *IEEE Trans. Syst. Man Cybern. A Syst. Hum.* **32**:482–496. [18]

Allport, A. 1987. Selection for Action: Some Behavioral and Neurophysiological Considerations of Attention and Action, ed. H. Heuer and A. F. Sanders, pp. 395–419. Hillsdale, NJ: Lawrence Erlbaum. [2]

Alsmith, A. J. T., and F. de Vignemont. 2012. Embodying the Mind and Representing the Body. *Rev. Phil. Psych.* **3**:1–13. [7, 16]

Alston, W. P. 1964. Philosophy of Language. Englewood Cliffs, NJ: Prentice-Hall. [9]

Andersen, R. A., L. H. Snyder, D. C. Bradley, and J. Xing. 1997. Multimodal Representation of Space in the Posterior Parietal Cortex and Its Use in Planning Movements. *Annu. Rev. Neurosci.* **20**:303–330. [3]

Anderson, J. R. 1983. The Architecture of Cognition. Cambridge, MA: Harvard Univ. Press. [2]

Anderson, M. L. 2010. Neural Reuse: A Fundamental Organizational Principle of the Brain. *Behav. Brain Sci.* **33**:245–266. [2]

Apperly, I. A., and S. A. Butterfill. 2009. Do Humans Have Two Systems to Track Beliefs and Belief-Like States? *Psychol. Rev.* **116**:953. [16]

Araújo, D., K. Davids, and R. Hristovski. 2006. The Ecological Dynamics of Decision Making in Sport. *Psychol. Sport Exerc.* **7**:653–676. [2, 4]

Arbib, M. A., A. Billard, M. Iacoboni, and E. Oztop. 2000. Synthetic Brain Imaging: Grasping, Mirror Neurons and Imitation. *Neural Netw.* **13**:975–997. [10]

Arevalo, A. L., J. V. Baldo, and N. F. Dronkers. 2012. What Do Brain Lesions Tell Us About Theories of Embodied Semantics and the Human Mirror Neuron System? *Cortex* **48**:242–254. [9]

Artola, A., and W. Singer. 1993. Long-Term Depression of Excitatory Synaptic Transmission and Its Relationship to Long-Term Potentiation. *Trends Neurosci.* **16**:480–487. [9]

Aspell, J. E., L. Heydrich, G. Marillier, et al. 2013. Turning Body and Self Inside Out: Visualized Heartbeats Alter Bodily Self-Consciousness and Tactile Perception. *Psychol. Sci.* **24**:2445–2453. [7, 15]

Austin, J. L. 1962. How to Do Things with Words. Oxford: Clarendon Press. [9]

Austin, J. L. 1975. How to Do Things with Words (2nd edition). Oxford: Oxford Univ. Press. [10]

Autumn, K., M. Sitti, Y. A. Liang, et al. 2002. Evidence for Van Der Waals Adhesion in Gecko Setae. *PNAS* **99**:12252–12256. [20]

Auvray, M., S. Hanneton, and K. O'Regan. 2007. Learning to Perceive with a Visuo-Auditory Substitution System: Localisation and Object Recognition with "the Voice." *Perception* **36**: [19]

Avdiyenko, L., N. Bertschinger, and J. Jost. 2015. Adaptive Information-Theoretical Feature Selection for Pattern Classification. *Comp. Intell.* **577**:279–294. [8]

Ay, N., H. Bernigau, R. Der, and M. Prokopenko. 2012. Information-Driven Self-Organization: The Dynamical System Approach to Autonomous Robot Behavior. *Theory Biosci.* **131**:161–179. [8]

Aydede, M., and P. Robbins. 2009. The Cambridge Handbook of Situated Cognition. Cambridge: Cambridge Univ. Press. [5]

Aziz-Zadeh, L., T. Sheng, S.-L. Liew, and H. Damasio. 2012. Understanding Otherness: The Neural Bases of Action Comprehension and Pain Empathy in a Congenital Amputee. *Cereb. Cortex* **22**:811–189. [4]

Baars, B. J. 1988. A Cognitive Theory of Consciousness: Cambridge Univ. Press. [12, 14]

Bachmann, S., C. Bottmer, and J. Schroder. 2005. Neurological Soft Signs in First-Episode Schizophrenia: A Follow-up Study. *Am. J. Psychiatry* **162**:2337–2343. [15]

Bahrami, B., K. Olsen, P. E. Latham, et al. 2010. Optimally Interacting Minds. *Science* **329**:1081–1085. [12]

Bahrick, L. E., and J. S. Watson. 1985. Detection of Intermodal Contingency as a Potential Basis Proprioceptive-Visual of Self-Perception in Infancy. *Dev. Psychol.* **21**:963–973. [3]

Bajcsy, R. 1988. Active Perception. *IEEE Proc.* **76**:966–1005 [20]

Bak, T. H. 2013. The Neuroscience of Action Semantics in Neurodegenerative Brain Diseases. *Curr. Opin. Neurol.* **26**:671–677. [9]

Baker, G. P., and P. M. S. Hacker. 2009. Wittgenstein: Understanding and Meaning, Part 1, Essays. An Analytical Commentary on the Philosophical Investigations. Oxford: Wiley-Blackwell. [9]

Baldassarre, G., and M. Mirolli, eds. 2013. Intrinsically Motivated Learning in Natural and Artificial Systems. Heidelberg: Springer. [2]

Ballard, D. H. 1991. Animate Vision. *Artif. Intell.* **48**:57–86. [2, 20]

Balleine, B. W., and A. Dickinson. 1998. Goal-Directed Instrumental Action: Contingency and Incentive Learning and Their Cortical Substrates. *Neuropharmacology* **37**:407–419. [4]

Ballester, B. R., J. Nirme, E. Duarte, A. Cuxart, S. Rodriguez, P. Verschure, and A. Duff. 2015. The visual amplification of goal-oriented movements counteracts acquired non-use in hemiparetic stroke patients. *J. Neuroeng. Rehabil.* **12**:50. [15]

Banerjee, K., and P. Bloom. 2015. Everything Happens for a Reason: Children's Beliefs About Purpose in Life Events. *Child Dev.* **86**:503–518. [14]

Bang, D., R. Fusaroli, K. Tylen, et al. 2014. Does Interaction Matter? Testing Whether a Confidence Heuristic Can Replace Interaction in Collective Decision-Making. *Conscious. Cogn.* **26**:13–23. [12]

Bar, M. 2007. The Proactive Brain: Using Analogies and Associations to Generate Predictions. *Trends Cogn. Sci.* **11**:280–289. [14]

Baranes, A. F., P.-Y. Oudeyer, and J. Gottlieb. 2014. The Effects of Task Difficulty, Novelty and the Size of the Search Space on Intrinsically Motivated Exploration. *Front. Neurosci.* **8**:317. [4]

Bar-Cohen, Y., ed. 2005. Biomimetics: Biologically-Inspired Technologies. Boca Raton: CRC Press. [19]

Barkow, J. H., L. E. Cosmides, and J. E. Tooby. 1992. The Adapted Mind: Evolutionary Psychology and the Generation of Culture. New York: Oxford Univ. Press. [13]

Barlaam, F., C. Fortin, M. Vaugoyeau, C. Schmitz, and C. Assaiante. 2012. Development of Action Representation During Adolescence as Assessed from Anticipatory Control in a Bimanual Load-Lifting Task. *Neuroscience* **221**:56–68. [3]

Barlow, H. B. 1961. Possible Principles Underlying the Transformation of Sensory Messages. In: Sensory Communication, ed. W. Rosenblith, vol. 13, pp. 217–234. Cambridge, MA: MIT Press. [4]

Barrett, L. F. 2006. Solving the Emotion Paradox: Categorization and the Experience of Emotion. *Pers. Soc. Psychol. Rev.* **10**:20–46. [5]

———. 2012. Emotions Are Real. *Emotion* **12**:413–429. [15]

Barrett, L. F., and M. Bar. 2009. See It with Feeling: Affective Predictions During Object Perception. *Phil. Trans. R. Soc. B* **364**:1325–1334. [16]

Barsalou, L. W. 1991. Deriving Categories to Achieve Goals. In: Psychology of Learning and Motivation, ed. G. H. Bower, vol. 27, pp. 1–64. New York: Academic Press. [5]

———. 1999. Perceptual Symbol Systems. *Behav. Brain Sci.* **22**:577–600. [2, 4, 5]

———. 2003. Situated Simulation in the Human Conceptual System. *Lang. Cogn. Process.* **18**:513–562. [5]

———. 2008. Grounded Cognition. *Annu. Rev. Psychol.* **59**:617–645. [2, 4, 9, 14, 17]

———. 2009. Simulation, Situated Conceptualization, and Prediction. *Phil. Trans. R. Soc. B* **364**:1281–1289. [4, 5]

———. 2011. Integrating Bayesian Analysis and Mechanistic Theories in Grounded Cognition. *Behav. Brain Sci.* **34**:191–192. [5]

———. 2012. The Human Conceptual System. In: The Cambridge Handbook of Psycholinguistics, ed. K. M. M. Spivey, and M. F. Joanisse, pp. 239–258. New York: Cambridge Univ. Press. [5]

Barsalou, L. W., P. M. Niedenthal, A. K. Barbey, and J. A. Ruppert. 2003. Social Embodiment. In: Psychology of Learning and Motivation: Advances in Research and Theory, ed. B. H. Ross, vol. 43, pp. 43–92. San Diego: Academic Press. [5]

Bart, O., D. Hajami, and Y. Bar-Haim. 2007. Predicting School Adjustment from Motor Abilities in Kindergarten. *Infant Child Dev.* **16**:597–615. [3]

Bašnáková, J., K. Weber, K. M. Petersson, J. van Berkum, and P. Hagoort. 2014. Beyond the Language Given: The Neural Correlates of Inferring Speaker Meaning. *Cereb. Cortex* **24**:2572–2578. [9]

Bastos, A. M., W. M. Usrey, R. A. Adams, et al. 2012. Canonical Microcircuits for Predictive Coding. *Neuron* **76**:695–711. [6]

Baumard, N., J.-B. André, and D. Sperber. 2013. A Mutualistic Approach to Morality: The Evolution of Fairness by Partner Choice. *Behav. Brain Sci.* **36**:59–78. [4]

Baumeister, R. F., and E. Masicampo. 2010. Conscious Thought Is for Facilitating Social and Cultural Interactions: How Mental Simulations Serve the Animal–Culture Interface. *Psychol. Rev.* **117**:945. [14]

Baumeister, R. F., E. J. Masicampo, and C. N. Dewall. 2009. Prosocial Benefits of Feeling Free: Disbelief in Free Will Increases Aggression and Reduces Helpfulness. *Pers. Soc. Psychol. Bull.* **35**:260–268. [12]

Baumeister, R. F., E. J. Masicampo, and K. D. Vohs. 2011. Do Conscious Thoughts Cause Behavior? *Annu. Rev. Psychol.* **62**:331–361. [14]

Bayne, T. 2010. The Unity of Consciousness. Oxford: Oxford Univ. Press. [14]

Beauchamp, M. S., and A. Martin. 2007. Grounding Object Concepts in Perception and Action: Evidence from fMRI Studies of Tools. *Cortex* **43**:461–468. [1]

Becchio, C., V. Manera, L. Sartori, A. Cavallo, and U. Castiello. 2012. Grasping Intentions: From Thought Experiments to Empirical Evidence. *Front. Hum. Neurosci.* **6**:117–117. [3]

Becchio, C., D. Zanatto, E. Straulino, et al. 2014. The Kinematic Signature of Voluntary Actions. *Neuropsychologia* **64C**:169–175. [15]

Bechtel, W. 2008. Mental Mechanisms: Philosophical Perspectives on Cognitive Neuroscience. New York: Routledge. [5]

Beck, D. M., G. Rees, C. D. Frith, and N. Lavie. 2001. Neural Correlates of Change Detection and Change Blindness. *Nat. Neurosci.* **4**:645–650. [12]

Bedny, M., and A. Caramazza. 2011. Perception, Action, and Word Meanings in the Human Brain: The Case from Action Verbs. *Ann. NY Acad. Sci.* **1224**:81–95. [9]

Beilock, S. L., T. H. Carr, C. MacMahon, and J. L. Starkes. 2002. When Paying Attention Becomes Counterproductive: Impact of Divided versus Skill-Focused Attention on Novice and Experienced Performance of Sensorimotor Skills. *J. Exp. Psychol. Appl.* **8**:6–16. [12]

Bergström, N., C. H. Ek, M. Björkman, and D. Kragic. 2011. Scene Understanding through Autonomous Interactive Perception. In: Proc. 8th Intl. Conf. on Computer Vision Systems, ed. J. L. Crowley et al., pp. 153–162. Heidelberg: Springer. [1, 11]

Berkes, P., and L. Wiskott. 2005. Slow Feature Analysis Yields a Rich Repertoire of Complex Cell Properties. *J. Vis.* **5**:9. [4]

———. 2007. Analysis and Interpretation of Quadratic Models of Receptive Fields. *Nat. Protoc.* **2**:400–407. [4]

Bertenthal, B. I., and M. R. Longo. 2008. Motor Knowledge and Action Understanding: A Developmental Perspective. In: Embodiment, Ego-Space, and Action, ed. R. L. Klatzky et al., pp. 323–368. New York: Psychology Press. [17]

Berthier, M. L., and F. Pulvermüller. 2011. Neuroscience Insights Improve Neurorehabilitation of Post-Stroke Aphasia. *Nat. Rev. Neurol.* **7**:86–97. [20]

Bertschinger, N., E. Olbrich, N. Ay, and J. Jost. 2008. Autonomy: An Information Theoretic Perspective. *BioSyst.* **91**:331–345. [8]

Best, J. R., and P. H. Miller. 2010. A Developmental Perspective on Executive Function. *Child Dev.* **81**:1641–1660. [3]

Bhat, A. N., J. C. Galloway, and R. J. Landa. 2012. Relation between Early Motor Delay and Later Communication Delay in Infants at Risk for Autism. *Infant. Behav. Dev.* **35**:838–846. [3]

Biederman, I. 1987. Recognition-by-Components: A Theory of Human Image Understanding. *Psychol. Rev.* **94**:115–147. [20]

Billard, A., S. Calinon, R. Dillmann, and S. Schaal. 2008. Robot Programming by Demonstration. In: Springer Handbook of Robotics, ed. B. Siciliano and O. Khatib, pp. 1371–1394. Heidelberg: Springer. [18]

Billard, A., Y. Epars, S. Calinon, S. Schaal, and G. Cheng. 2004. Discovering Optimal Imitation Strategies. *Rob. Auton. Syst.* **47**:69–77. [18]

Binder, J. R., R. H. Desai, W. W. Graves, and L. L. Conant. 2009. Where Is the Semantic System? A Critical Review and Meta-Analysis of 120 Functional Neuroimaging Studies. *Cereb. Cortex* **19**:2767–2796. [4]

Binder, J. R., C. F. Westbury, K. A. McKiernan, E. T. Possing, and D. A. Medler. 2005. Distinct Brain Systems for Processing Concrete and Abstract Concepts. *J. Cogn. Neurosci.* **17**:905–917. [9]

Björkman, M., Y. Bekiroglu, V. Högman, and D. Kragic. 2013. Enhancing Visual Perception of Shape through Tactile Glances. In: IEEE/RSJ Intl. Conf. on Intelligent Robots and Systems (IROS), pp. 3180–3186. Tokyo: IEEE. [11, 20]

Blake, D. T., N. N. Byl, and M. M. Merzenich. 2002. Representation of the Hand in the Cerebral Cortex. *Behav. Brain Res.* **135**:179–184. [11, 13]

Blakemore, S.-J., C. D. Frith, and D. Wolpert. 1999. Spatiotemporal Prediction Modulates the Perception of Self-Produced Stimuli. *J. Cogn. Neurosci.* **11**:551–559. [7]

———. 2001. The Cerebellum Is Involved in Predicting the Sensory Consequences of Action. *Neuroreport* **12**:1879–1884. [4]

Blakemore, S. J., J. Smith, S. Steel, E. C. Johnstone, and C. D. Frith. 2000. The Perception of Self-Produced Sensory Stimuli in Patients with Auditory Hallucinations and Passivity Experiences: Evidence for a Breakdown in Self-Monitoring. *Psychol. Med.* **30**:1131–1139. [15, 20]

Blakemore, S. J., D. Wolpert, and C. Frith. 2000. Why Can't You Tickle Yourself? *Neuroreport* **11**:R11–16. [15]

Blanke, O. 2012. Multisensory Brain Mechanisms of Bodily Self-Consciousness. *Nat. Rev. Neurosci.* **13**:556–571. [19]

Blanke, O., and T. Metzinger. 2009. Full-Body Illusions and Minimal Phenomenal Selfhood. *Trends Cogn. Sci.* **13**:7–13. [15]

Blanke, O., P. Pozeg, M. Hara, et al. 2014. Neurological and Robot-Controlled Induction of an Apparition. *Curr. Biol.* **24**:2681–2686. [15]

Bleyenheuft, Y., and J.-L. Thonnard. 2010. Grip Control in Children before, During, and after Impulsive Loading. *J. Mot. Behav.* **42**:169–177. [3]

Blickhan, R., A. Seyfarth, H. Geyer, et al. 2007. Intelligence by Mechanics. *Phil. Trans. R. Soc. A* **365**:199–220. [20]

Block, N. 2005. Two Neural Correlates of Consciousness. *Trends Cogn. Sci.* **9**:46–52. [15]

———. 2007. Consciousness, Accessibility, and the Mesh between Psychology and Neuroscience. *Behav. Brain Sci.* **30**:481–499; discussion 499–548. [14]

Bloomfield, L. 1933. Language. New York: Holt, Rinehart & Winston. [9]

Bobzien, S. 2006. Moral Responsibility and Moral Development in Epicurus' Philosophy. In: The Virtuous Life in Greek Ethics, ed. B. Reis, pp. 206–299. New York: Cambridge Univ. Press. [7, 12]

Bohg, J., and D. Kragic. 2009. Grasping Familiar Objects Using Shape Context. ICAR 2009, pp. 1–6. [18]

Bohl, V., and N. Gangopadhyay. 2013. Theory of Mind and the Unobservability of Other Minds. *Philos. Explor.* **2**:1–20. [6]

Boly, M., M. I. Garrido, O. Gosseries, et al. 2011. Preserved Feedforward but Impaired Top-Down Processes in the Vegetative State. *Science* **332**:858–862. [14]

Bor, D., and A. K. Seth. 2012. Consciousness and the Prefrontal Parietal Network: Insights from Attention, Working Memory, and Chunking. *Front. Psychol.* **3**:63. [15]

Borg, E. 2013. More Questions for Mirror Neurons. *Conscious. Cogn.* **22**:1122–1131. [9]

Borghi, A. M., and F. Binkofski. 2014. Word Learning and Word Acquisition: An Embodied View on Abstract Concepts. New York: Springer. [2, 4]

Boria, S., M. Fabbri-Destro, L. Cattaneo, et al. 2009. Intention Understanding in Autism. *PLoS One* **4**:e5596–e5596. [3]

Borroni, P., M. Montagna, G. Cerri, F. Baldissera, and D. Kragic. 2005. Cyclic Time Course of Motor Excitability Modulation During the Observation of a Cyclic Hand Movement. *Brain Res.* **1065**:115–124. [18]

Botvinick, M., and J. Cohen. 1998. Rubber Hands "Feel" Touch That Eyes See. *Nature* **391**:756–756. [3, 7]

Boulenger, V., O. Hauk, and F. Pulvermüller. 2009. Grasping Ideas with the Motor System: Semantic Somatotopy in Idiom Comprehension. *Cereb. Cortex* **19**:1905–1914. [9]

Boulenger, V., A. C. Roy, Y. Paulignan, et al. 2006. Cross-Talk between Language Processes and Overt Motor Behavior in the First 200 msec of Processing. *J. Cogn. Neurosci.* **18**:1607–1615. [9]

Boulenger, V., Y. Shtyrov, and F. Pulvermüller. 2012. When Do You Grasp the Idea? MEG Evidence for Instantaneous Idiom Understanding. *NeuroImage* **59**:3502–3513. [9]

Boyd, R., and P. J. Richerson. 1985. Culture and the Evolutionary Process. Chicago: Univ. of Chicago Press. [13]

Braitenberg, V. 1986. Vehicles: Experiments in Synthetic Psychology. Cambridge, MA: MIT Press. [17]

Braitenberg, V., and A. Schüz. 1992. Basic Features of Cortical Connectivity and Some Considerations on Language. In: Language Origin: A Multidisciplinary Approach, ed. J. Wind et al., pp. 89–102. Dordrecht: Kluwer Academic Publ. [9]

———. 1998. Cortex: Statistics and Geometry of Neuronal Connectivity. Berlin: Springer. [9]

Braver, T. S., D. M. Barch, and J. D. Cohen. 1999. Cognition and Control in Schizophrenia: A Computational Model of Dopamine and Prefrontal Function. *Biol. Psychiatry* **46**:312–328. [6]

Breidbach, O., and J. Jost. 2006. On the Gestalt Concept. *Theory Biosci.* **125**:19–36. [8]

Brincker, M., and E. B. Torres. 2013. Noise from the Periphery in Autism. *Front. Integr. Neurosci.* **7**:Article 34. [20]

Brock, O. 2011. Is Robotics in Need of a Paradigm Shift? Berlin Summit on Robotics, pp. 1–10. [18]

Brooks, H. 1967. Applied Science and Technological Progress. *Science* **136**:1706–1712. [19]

Brooks, R. A. 1986. A Robust Layered Control System for a Mobile Robot. *IEEE J. Robot. Autom.* **2**:14–23. [20]

———. 1989. A Robot That Walks: Emergent Behaviors from a Carefully Evolved Network. *Neural Comput.* **1**:153–162. [20]

———. 1990. Elephants Don't Play Chess. *Robot. Auton. Syst.* **6**:3–15. [18]

———. 1991a. Intelligence without Reason. In: Proc. 12th Intl. Joint Conf. on Artificial Intelligence, vol. 1, pp. 569–595 Sydney: Morgan Kaufmann. [20]

———. 1991b. Intelligence without Representation. *Artif. Intell.* **47**:139–159. [1, 7, 18, 20]

———. 2002. Flesh and Machines: How Robots Will Change Us. New York: Pantheon Books. [18]

Brown, E., N. Rodenberg, J. Amend, et al. 2010. From the Cover: Universal Robotic Gripper Based on the Jamming of Granular Material. *PNAS* **107**:18809–18814. [20]

Brown, E. C., and M. Brüne. 2012. The Role of Prediction in Social Neuroscience. *Front. Hum. Neurosci.* **6**:147. [6]

Brown, H., R. A. Adams, I. Parees, M. Edwards, and K. J. Friston. 2013. Active Inference, Sensory Attenuation and Illusions. *Cogn. Process.* **14**:411–427. [6, 7]

Broz, F., C. L. Nehaniv, T. Belpaeme, et al. 2014. The Italk Project: A Developmental Robotics Approach to the Study of Individual, Social, and Linguistic Learning. *Top. Cogn. Sci.* **6**:534–544. [4]

Bruce, N., and J. Tsotsos. 2006. Saliency Based on Information Maximization. *Adv. Neural Inform. Processing Syst.* **18**:155–162. [8]

Bruineberg, J., and E. Rietveld. 2014. Self-Organization, Free Energy Minimization, and Optimal Grip on a Field of Affordances. *Front. Hum. Neurosci.* **8**:599. [7]

Buchanan, R. 1985. Declaration by Design: Rhetoric, Argument and Demonstration in Practice. *Design Issues* **2**:4–22. [19]

Buckner, R. L., J. R. Andrews-Hanna, and D. L. Schacter. 2008. The Brain's Default Network: Anatomy, Function, and Relevance to Disease. *Ann. NY Acad. Sci.* **1124**:1–38. [5]

Bullock, D., S. Grossberg, and F. H. Guenther. 1993. A Self-Organizing Neural Model of Motor Equivalent Reaching and Tool Use by a Multijoint Arm. *J. Cogn. Neurosci.* **5**:408–435. [15]

Buon, M., P. Jacob, E. Loissel, and E. Dupoux. 2013. A Non-Mentalistic Cause-Based Heuristic in Human Social Evaluations. *Cognition* **126**:149–155. [4]

Burkeman, O. 2015. Why Can't the World's Greatest Minds Solve the Mystery of Consciousness? *The Guardian*, Jan. 21. [14]

Bush, V. 1945. Science: The Endless Frontier, U.S. Office of Scientific Research and Development, Report to the President on a Program for Postwar Scientific Research. Washington, D.C.: GPO. [19]

Butz, M. V. 2008. How and Why the Brain Lays the Foundations for a Conscious Self. *Constructivist Foundations* **4**:1–42. [15]

Butz, M. V., R. Thomaschke, M. J. Linhardt, and O. Herbort. 2010. Remapping Motion across Modalities: Tactile Rotations Influence Visual Motion Judgments. *Exp. Brain Res.* **207**:1–11. [15]

Buzsáki, G. 2010. Neural Syntax: Cell Assemblies, Synapsembles, and Readers. *Neuron* **68**:362–285. [9]

Buzsáki, G., and E. I. Moser. 2013. Memory, Navigation and Theta Rhythm in the Hippocampal-Entorhinal System. *Nat. Neurosci.* **16**:130–138. [2]

Buzsáki, G., A. Peyrache, and J. Kubie. 2015. Emergence of Cognition from Action. In: Cold Spring Harbor Symposia on Quantitative Biology, p. 024679. Cold Spring Harbor: CSH Laboratory Press. [2]

Byrge, L., O. Sporns, and L. B. Smith. 2014. Developmental Process Emerges from Extended Brain-Body-Behavior Networks. *Trends Cogn. Sci.* **18**:395–403. [2, 4]

Caggiano, V., L. Fogassi, G. Rizzolatti, P. Thier, and A. Casile. 2009. Mirror Neurons Differentially Encode the Peripersonal and Extrapersonal Space of Monkeys. *Science* **324**:403–406. [16]

Calinon, S. 2009. Robot Programming by Demonstration: A Probabilistic Approach. Lausanne: EPFL Press. [18]

Cameirão, M. da Silva, S. B. Bermudez i Badia, E. D. Oller, and P. F. Verschure. 2011. Virtual Reality Based Rehabilitation Speeds up Functional Recovery of the Upper Extremities after Stroke: A Randomized Controlled Pilot Study in the Acute Phase of Stroke Using the Rehabilitation Gaming System. *Restor. Neurol. Neurosci.* **29**:1–12. [20]

Cameirão, M. S., S. B. Badia, E. Duarte, A. Frisoli, and P. F. Verschure. 2012. The Combined Impact of Virtual Reality Neurorehabilitation and Its Interfaces on Upper Extremity Functional Recovery in Patients with Chronic Stroke. *Stroke* **43**:2720–2728. [19, 20]

Campos, J. J., D. I. Anderson, M. A. Barbu-Roth, et al. 2000. Travel Broadens the Mind. *Infancy* **1**:149–219. [3]

Cangelosi, A., and S. Harnad. 2001. The Adaptive Advantage of Symbolic Theft over Sensorimotor Toil: Grounding Language in Perceptual Categories. *Evol. Comm.* **4**:117–142. [9]

Cangelosi, A., and M. Schlesinger. 2014. Developmental Robotics: From Babies to Robots. Cambridge, MA: MIT Press. [4]

Cannon, E. N., and A. L. Woodward. 2012. Infants Generate Goal-Based Action Predictions. *Dev. Sci.* **15**:292–298. [3]

Cannon, T. D., I. M. Rosso, C. E. Bearden, L. E. Sanchez, and T. Hadley. 1999. A Prospective Cohort Study of Neurodevelopmental Processes in the Genesis and Epigenesis of Schizophrenia. *Dev. Psychopathol.* **11**:467–485. [15]

Caporale, N., and Y. Dan. 2008. Spike Timing-Dependent Plasticity: A Hebbian Learning Rule. *Annu. Rev. Neurosci.* **31**:25–46. [9]

Caramazza, A., S. Anzellotti, L. Strnad, and A. Lingnau. 2014. Embodied Cognition and Mirror Neurons: A Critical Assessment. *Annu. Rev. Neurosci.* **37**:1–15. [2]

Carey, S., and R. Gelman. 2014. The Epigenesis of Mind: Essays on Biology and Cognition. Oxford: Psychology Press. [4]

Carota, F., R. Moseley, and F. Pulvermüller. 2012. Body-Part-Specific Representations of Semantic Noun Categories. *J. Cogn. Neurosci.* **24**:1492–1509. [9]

Carrozza, M. C., G. Cappiello, G. Stellin, et al. 2005. A Cosmetic Prosthetic Hand with Tendon Driven Under-Actuated Mechanism and Compliant Joints: Ongoing Research and Preliminary Results. Proc. 2005 IEEE Intl. Conf. on Robotics and Automation (ICRA), pp. 2661–2666. [19]

Carruthers, P., and P. K. Smith. 1996. Theories of Theories of Mind. Cambridge: Cambridge Univ. Press. [20]

Cattaneo, L., M. Fabbri-Destro, S. Boria, et al. 2007. Impairment of Actions Chains in Autism and Its Possible Role in Intention Understanding. *PNAS* **104**:17825–17830. [3, 20]

Chalmers, D. J. 1995. Facing up to the Problem of Consciousness. *J. Conscious. Stud.* **2**:200–219. [14]

———. 2000. What Is a Neural Correlate of Consciousness? In: Neural Correlates of Consciousness, ed. T. Metzinger, A Bradford Book, pp. 17–39. Cambridge, MA: MIT Press. [12]

———. 2010. The Character of Consciousness. Oxford: Oxford Univ. Press. [14]

Chambon, V., E. Filevich, and P. Haggard. 2014. What Is the Human Sense of Agency, and Is It Metacognitive? In: The Cognitive Neuroscience of Metacognition, ed. S. M. Fleming and C. D. Frith, pp. 321–342. Heidelberg: Springer. [7]

Chambon, V., and P. Haggard. 2012. Sense of Control Depends on Fluency of Action Selection, Not Motor Performance. *Cognition* **125**:441–451. [7]

Chang, A. Y., R. Kanai, and A. K. Seth. 2015. Cross-Modal Prediction Changes the Timing of Conscious Access During the Motion-Induced Blindness. *Conscious. Cogn.* **31**:139–147. [15]

Chao, L. L., and A. Martin. 2000. Representation of Manipulable Man-Made Objects in the Dorsal Stream. *NeuroImage* **12**:478–484. [4]

Chen, M., and J. A. Bargh. 1999. Consequences of Automatic Evaluation: Immediate Behavioral Predispositions to Approach or Avoid the Stimulus. *Pers. Soc. Psychol. Bull.* **25**:215–224. [4]

Chiel, H. J., and R. D. Beer. 1997. The Brain Has a Body: Adaptive Behavior Emerges from Interactions of Nervous System, Body and Environment. *Trends Neurosci.* **20**:553–557. [19]

Chierchia, G., and S. McConnell-Ginet. 2000. Meaning and Grammar: An Introduction to Semantics. Cambridge, MA: MIT Press. [9]

Churchland, P. S. 1986. Neurophilosophy: Toward a Unified Science of the Mind-Brain. Cambridge, MA: MIT Press. [14]

Churchland, P. S., V. S. Ramachandran, and T. J. Sejnowski. 1994. A Critique of Pure Vision. In: Large-Scale Neuronal Theories of the Brain, ed. C. Koch and J. Davis, pp. 23–60. Cambridge, MA: MIT Press. [1]

Cisek, P. 2007. Cortical Mechanisms of Action Selection: The Affordance Competition Hypothesis. *Phil. Trans. R. Soc. B* **362**:1585–1599. [6, 14]

Cisek, P., and J. F. Kalaska. 2010. Neural Mechanisms for Interacting with a World Full of Action Choices. *Annu. Rev. Neurosci.* **33**:269–298. [2–4]

Clanton, S. T., A. J. C. McMorland, Z. Zohny, et al. 2013. Seven Degree of Freedom Cortical Control of a Robotic Arm. In: Brain-Computer Interface Research: A State-of-the-Art Summary, ed. C. Guger et al., pp. 73–81. Heidelberg: Springer. [20]

Clark, A. 1995. Moving Minds: Situating Content in the Service of Real-Time Success. *Phil. Perspect* **9**:89–104. [1]

———. 1998. Being There. Putting Brain, Body, and World Together Again. Cambridge, MA: MIT Press. [1, 2, 5, 11, 13, 16, 18]

———. 1999. An Embodied Cognitive Science? *Trends Cogn. Sci.* **3**:345–351. [9, 16]

———. 2008. Supersizing the Mind: Embodiment, Action, and Cognitive Extension. Oxford: Oxford Univ. Press. [5, 7, 8, 13]

———. 2013a. The Many Faces of Precision. *Front. Psychol.* **4**:270. [6]

———. 2013b. Whatever Next? Predictive Brains, Situated Agents, and the Future of Cognitive Science. *Behav. Brain Sci.* **36**:181–204. [2, 5–7, 13–15]

Clark, A., and D. Chalmers. 1998. The Extended Mind. *Analysis* **58**:7–19. [11, 16]

Clark, A., and R. Grush. 1999. Towards a Cognitive Robotics. *Adapt. Behav.* **7**:5–16. [2, 16]

Claxton, L. J., R. Keen, and M. E. McCarty. 2003. Evidence of Motor Planning in Infant Reaching Behavior. *Psychol. Sci.* **14**:354–356. [3]

Cohen, E. 2012. The Evolution of Tag-Based Cooperation in Humans. *Curr. Anthropol.* **53**:588–616. [4]

Cole, J., H. Brooks, D. Stokes, E., et al., eds. 1994. Science the Endless Frontier 1945–1995: Learning from the Past, Designing for the Future, Part I. New York: Columbia Univ. Press. [19]

Collinger, J. L., B. Wodlinger, J. E. Downey, et al. 2012. High-Performance Neuroprosthetic Control by an Individual with Tetraplegia. *Lancet* **38**:557–564. [20]

Collins, S., A. Ruina, R. Tedrake, and M. Wisse. 2005. Efficient Bipedal Robots Based on Passive-Dynamic Walkers. *Science* **307**:1082–1085. [19]

Collins, T., T. Heed, and B. Röder. 2010. Visual Target Selection and Motor Planning Define Attentional Enhancement at Perceptual Processing Stages. *Front. Hum. Neurosci.* **4**:14. [20]

Collins, T., T. Schicke, and B. Röder. 2008. Action Goal Selection and Motor Planning Can Be Dissociated by Tool Use. *Cognition* **109**:363–371. [20]

Conant, R. C., and W. R. Ashby. 1970. Every Good Regulator of a System Must Be a Model of That System. *Int. J. Syst. Sci.* **1**:89–97. [12, 15]

Contreras-Vidal, J. L., J. Bo, J. P. Boudreau, and J. E. Clark. 2005. Development of Visuomotor Representations for Hand Movement in Young Children. *Exp. Brain Res.* **162**:155–164. [3]

Cook, J. L., S. J. Blakemore, J. Smith, et al. 2013. Atypical Basic Movement Kinematics in Autism Spectrum Conditions. *Brain* **136**:2816–2824. [20]

Cook, R., G. Bird, C. Catmur, C. Press, and C. Heyes. 2014. Mirror Neurons: From Origin to Function. *Behav. Brain Sci.* **37**:177–192. [17]

Coricelli, G., H. D. Critchley, M. Joffily, et al. 2005. Regret and Its Avoidance: A Neuroimaging Study of Choice Behavior. *Nat. Neurosci.* **8**:1255–1262. [12]

Cotterill, R. 1998. Enchanted Looms: Conscious Networks in Brains and Computers. Cambridge: Cambridge Univ. Press. [2]

Cowie, D., T. R. Makin, and A. J. Bremner. 2013. Children's Responses to the Rubber-Hand Illusion Reveal Dissociable Pathways in Body Representation. *Psychol. Sci.* **24**:762–769. [3]

Craig, A. D. 2003. Interoception: The Sense of the Physiological Condition of the Body. *Curr. Opin. Neurobiol.* **13**:500–505. [15]

———. 2009. How Do You Feel—Now? The Anterior Insula and Human Awareness. *Nat. Rev. Neurosci.* **10**:59–70. [14]

Craighero, L., L. Fadiga, G. Rizzolatti, and C. Umilta. 1999. Action for Perception: A Motor-Visual Attentional Effect. *J. Exp. Psychol. Hum. Percept. Perform.* **25**:1673–1692. [20]

Crapse, T. B., and M. A. Sommer. 2008. Corollary Discharge across the Animal Kingdom. *Nat. Rev. Neurosci.* **9**:587–600. [1, 4, 11]

Crick, F. 1994. The Astonishing Hypothesis: The Scientific Search for the Soul. New York. New York: Scribner. [12]

Crick, F., and C. Koch. 1990. Towards a Neurobiological Theory of Conciousness. *Semin. Neurosci.* **2**:263–275. [14]

———. 1995. Are We Aware of Neural Activity in Primary Visual Cortex? *Nature* **375**:121–123. [12]

Critchley, H. D., S. Wiens, P. Rotshtein, A. Ohman, and R. J. Dolan. 2004. Neural Systems Supporting Interoceptive Awareness. *Nat. Neurosci.* **7**:189–195. [14]

Csibra, G. 2007. Action Mirroring and Action Understanding: An Alternative Account. In: Sensorimotor Foundations of Higher Cognition: Attention and Performance, ed. P. Haggard et al., vol. 12, pp. 453–459. Oxford: Oxford Univ. Press. [3, 9]

Csibra, G., and G. Gergely. 2011. Natural Pedagogy as Evolutionary Adaptation. *Phil. Trans. R. Soc. B* **366**:1149–1157. [4]

Custers, R., and H. Aarts. 2010. The Unconscious Will: How the Pursuit of Goals Operates Outside of Conscious Awareness. *Science* **329**:47–50. [14]

Dähne, S., N. Wilbert, and L. Wiskott. 2014. Slow Feature Analysis on Retinal Waves Leads to V1 Complex Cells. *PLoS Comput. Biol.* **10**:e1003564. [4]

Damasio, A. R. 2000. The Feeling of What Happens: Body, Emotion and the Making of Consciousness. New York: Random House. [5]

———. 2012. Self Comes to Mind: Constructing the Conscious Brain. New York: Vintage. [14]

Danziger, E. 2006. The Thought That Counts: Interactional Consequences of Variation in Cultural Theories of Meaning. In: The Roots of Human Sociality: Culture, Cognition and Human Interaction., ed. S. Levinson and N. Enfield, Wenner-Gren Intl. Symposium, pp. 259–278, L. C. Aiello, series ed. New York: Berg Press. [12]

D'Argembeau, A., S. Raffard, and M. Van der Linden. 2008. Remembering the Past and Imagining the Future in Schizophrenia. *J. Abnorm. Psychol.* **117**:247–251. [7]

D'Ausilio, A., F. Pulvermüller, P. Salmas, et al. 2009. The Motor Somatotopy of Speech Perception. *Curr. Biol.* **19**:381–385. [9]

David, N., A. Newen, and K. Vogeley. 2008. The "Sense of Agency" and Its Underlying Cognitive and Neural Mechanisms. *Conscious. Cogn.* **17**:523–534. [1, 7, 11]

David, N., J. Schultz, E. Milne, et al. 2014. Right Temporoparietal Gray Matter Predicts Accuracy of Social Perception in the Autism Spectrum. *J. Autism Dev. Disord.* **44**:1433–1446. [20]

Davidson, D. 1967. Truth and Meaning. *Synthese* **17**:304–323. [9]

Davidson, P. R., and D. M. Wolpert. 2003. Motor Learning and Prediction in a Variable Environment. *Curr. Opin. Neurobiol.* **13**:232–237. [3]

Davis, E. E., N. J. Pitchford, T. Jaspan, D. McArthur, and D. Walker. 2010. Development of Cognitive and Motor Function Following Cerebellar Tumour Injury Sustained in Early Childhood. *Cortex* **46**:919–932. [3]

Davis, E. E., N. J. Pitchford, and E. Limback. 2011. The Interrelation between Cognitive and Motor Development in Typically Developing Children Aged 4–11 Years Is Underpinned by Visual Processing and Fine Manual Control. *Br. J. Psychol.* **102**:569–584. [3]

Dawkins, R. 1982. The Extended Phenotype. Oxford: Oxford Univ. Press. [13]

Day, B. L., and I. N. Lyon. 2000. Voluntary Modification of Automatic Arm Movements Evoked by Motion of a Visual Target. *Exp. Brain Res.* **130**:159–168. [3]

Dayan, P., G. E. Hinton, and R. M. Neal. 1995. The Helmholtz Machine. *Neural Comput.* **7**:889–904. [6]

Dayanidhi, S., A. Hedberg, F. J. Valero-Cuevas, and H. Forssberg. 2013. Developmental Improvements in Dynamic Control of Fingertip Forces Last Throughout Childhood and into Adolescence. *J. Neurophysiol.* **110**:1583–1592. [3]

Dean, J. 1990. The Neuroethology of Perception and Action. In: Relationships between Perception and Action: Current Approaches, ed. O. Neumann and W. Prinz, pp. 81–132. Berlin: Springer. [17]

De Graef, P., K. Verfaillie, F. Germeys, V. Gysen, and C. V. Eccelpoel. 2001. Trans-Saccadic Representation Makes Your Porsche Go Places. *Behav. Brain Sci.* **24**:981–982. [5]

Dehaene, S. 2014. Consciousness and the Brain: Deciphering How the Brain Codes Our Thoughts. New York: Viking. [14]

Dehaene, S., and J. P. Changeux. 2011. Experimental and Theoretical Approaches to Conscious Processing. *Neuron* **70**:200–227. [14, 15]

Dehaene, S., and L. Cohen. 2007. Cultural Recycling of Cortical Maps. *Neuron* **56**:384–398. [2]

———. 2011. The Unique Role of the Visual Word Form Area in Reading. *Trends Cogn. Sci.* **15**:254–262. [4]

Dehaene, S., M. Kerszberg, and J. P. Changeux. 1998. A Neuronal Model of a Global Workspace in Effortful Cognitive Tasks. *PNAS* **95**:14529–14534. [14]

Dehaene, S., and L. Naccache. 2001. Towards a Cognitive Neuroscience of Consciousness: Basic Evidence and a Workspace Framework. *Cognition* **79**:1–37. [12]

Deimel, R., C. Eppner, J. Álvarez-Ruiz, M. Maertens, and O. Brock. 2013. Exploitation of Environmental Constraints in Human and Robotic Grasping. *Intl. Symp. Robot. Res.* **2013**:116. [20]

De Jaegher, H., E. Di Paolo, and S. Gallagher. 2010. Can Social Interaction Constitute Social Cognition? *Trends Cogn. Sci.* **14**:441–447. [16, 20]

de Klerk, C. C. J. M., M. H. Johnson, C. M. Heyes, and V. Southgate. 2014. Baby Steps: Investigating the Development of Perceptual-Motor Couplings in Infancy. *Dev. Sci.* **18**:1–11. [3]

Delgado, M. R., R. H. Frank, and E. A. Phelps. 2005. Perceptions of Moral Character Modulate the Neural Systems of Reward During the Trust Game. *Nat. Neurosci.* **8**:1611–1618. [12]

Demiris, Y., and M. Johnson. 2003. Distributed, Predictive Perception of Actions: A Biologically Inspired Robotics Architecture for Imitation and Learning. *Connect. Sci.* **15**:231–243. [18]

Dennett, D. C. 1988. The Intentional Stance. Cambridge, MA: Bradford Books/MIT Press. [14]

———. 1992. Consciousness Explained. New York: Little Brown. [14]

———. 2004. Freedom Evolves. London: Penguin Books. [16]

Desai, R. H., L. L. Conant, J. R. Binder, H. Park, and M. S. Seidenberg. 2013. A Piece of the Action: Modulation of Sensory-Motor Regions by Action Idioms and Metaphors. *NeuroImage* **83**:862–869. [9]

Desmurget, M., and A. Sirigu. 2009. A Parietal-Premotor Network for Movement Intention and Motor Awareness. *Trends Cogn. Sci.* **13**:411–419. [14]

de Vignemont, F., and P. Fourneret. 2004. The Sense of Agency: A Philosophical and Empirical Review of the "Who" System. *Conscious. Cogn.* **13**:1. [7]

Dewey, J. 1896. The Reflex Arc Concept in Psychology. *Psychol. Rev.* **3**:357–370. [9, 11, 13, 20]

———. 1897. My Pedagogic Creed. *The School Journal* **54**:77–80. [20]

———. 1900. The School and Society. Chicago: Univ. of Chicago Press. [20]

———. 1902. The Child and the Curriculum. Chicago: Univ. of Chicago Press. [20]

———. 1916. Essays in Experimental Logic. Chicago: Univ. of Chicago Press. [20]

———. 1929/1958. Experience and Nature (revised edition). New York: Dover. [13]

———. 1938. Logic: The Theory of Inquiry. New York: Holt, Rinehart & Winston. [20]

———. 1981. The Middle Works, 1899–1924, vol. 10. Carbondale: Southern Illinois Univ. Press. [13]

———. 2008. The Early Works, 1882–1898. Carbondale: Southern Illinois Univ. Press. [13]

de Wit, S., and A. Dickinson. 2009. Associative Theories of Goal-Directed Behaviour: A Case for Animal–Human Translational Models. *Psychol. Res.* **73**:463–476. [17]

Diamond, A. 2000. Close Interrelation of Motor Development and Cognitive Development and of the Cerebellum and Prefrontal Cortex. *Child Dev.* **71**:44–56. [3]

———. 2013. Executive Functions. *Annu. Rev. Psychol.* **64**:135–168. [3]

Difede, J., and H. G. Hoffman. 2002. Virtual Reality Exposure Therapy for World Trade Center Post-Traumatic Stress Disorder: A Case Report. *Cyberpsychol. Behav.* **5**:529–535. [19]

Dijksterhuis, A., and H. Aarts. 2010. Goals, Attention, and (Un) Consciousness. *Annu. Rev. Psychol.* **61**:467–490. [14]

Dijksterhuis, A., and J. A. Bargh. 2001. The Perception-Behavior Expressway: Automatic Effects of Social Perception on Social Behavior. *Adv. Exp. Soc. Psychol.* **33**:1–40. [14]

Dijksterhuis, A., and L. F. Nordgren. 2006. A Theory of Unconscious Thought. *Perspect. Psychol. Sci.* **1**:95–109. [12]

Di Paolo, E. A. 2005. Autopoiesis, Adaptivity, Teleology, Agency. *Phenom. Cogn. Sci.* **4**:429–452. [16]

Di Paolo, E. A., and H. De Jaegher. 2012. The Interactive Brain Hypothesis. *Front. Hum. Neurosci.* **6**:163. [1, 11, 20]

Di Paolo, E. A., M. Rohde, and H. De Jaegher. 2010. Horizons for the Enactive Mind: Values, Social Interaction and Play. In: Enaction: Towards a New Paradigm for Cognitive Science, ed. J. Stewart et al., pp. 33–87. Cambridge, MA: MIT Press. [11, 15, 20]

Dobzhansky, T. 1973. Nothing in Biology Makes Sense except in the Light of Evolution. *Am. Biol. Teach.* **35**:125–129. [14]

Dogge, M., M. Schaap, R. Custers, D. M. Wegner, and H. Aarts. 2012. When Moving without Volition: Implied Self-Causation Enhances Binding Strength between Involuntary Actions and Effects. *Conscious. Cogn.* **21**:501–506. [12]

Dominey, P. F., and C. Dodane. 2004. Indeterminacy in Language Acquisition: The Role of Child Directed Speech and Joint Attention. *J. Neuroling.* **17**:121–145. [20]

Donald, M. 1993. Precis of Origins of the Modern Mind: Three Stages in the Evolution of Culture and Cognition. *Behav. Brain Sci.* **16**:737–748. [5]

Donner, T., A. Kettermann, E. Diesch, et al. 2000. Involvement of the Human Frontal Eye Field and Multiple Parietal Areas in Covert Visual Selection During Conjunction Search. *Eur. J. Neurosci.* **12**:3407–3414. [20]

Doursat, R., and E. Bienenstock. 2007. Neocortical Self-Structuration as a Basis for Learning. In: Proc. 5th Intl. Conf. on Development and Learning (ICDL 2006), pp. 1–6. Bloomington: Indiana Univ. Press. [9]

Doya, K. 2008. Modulators of Decision Making. *Nat. Neurosci.* **11**:410–416. [6]

Drescher, G. L. 1991. Made-up Minds: A Constructivist Approach to Artificial Intelligence. Cambridge, MA: MIT Press. [2]

Dretske, F. 1988. Explaining Behavior. Cambridge: MIT Press. [15]

Dretske, F. I. 1995. Naturalizing the Mind. Cambridge, MA: MIT Press. [5]

Dreyfus, H. L. 1992. What Computers Still Can't Do: A Critique of Artificial Reason. Cambridge, MA: MIT Press. [1, 16]

Duff, A., and P. F. M. J. Verschure. 2010. Unifying Perceptual and Behavioral Learning with a Correlative Subspace Learning Rule. *Neurocomputing* **73**:1818–1830. [14]

Dutilh, G., J. Vandekerckhove, B. U. Forstmann, et al. 2012. Testing Theories of Post-Error Slowing. *Atten. Percept. Psychophys.* **74**:454–465. [12]

Edelman, G. M. 1987. Neural Darwinism: The Theory of Neuronal Group Selection. New York: Basic Books. [4]

———. 1989. The Remembered Present: A Biological Theory of Consciousness. New York: Basic Books. [14]

Edwards, L. A. 2014. A Meta-Analysis of Imitation Abilities in Individuals with Autism Spectrum Disorders. *Autism Res.* **7**:363–380. [3]

Egorova, N., F. Pulvermüller, and Y. Shtyrov. 2014. Neural Dynamics of Speech Act Comprehension: An MEG Study of Naming and Requesting. *Brain Topogr.* **27**:375–392. [9]

Egorova, N., Y. Shtyrov, and F. Pulvermüller. 2013. Early and Parallel Processing of Pragmatic and Semantic Information in Speech Acts: Neurophysiological Evidence. *Front. Hum. Neurosci.* **7**:1–13. [9]

Ehlich, K. 2007. Sprache und Sprachliches Handeln. Berlin: De Gruyter. [9]

Ehrsson, H. H. 2007. The Experimental Induction of Out-of-Body Experiences. *Science* **317**:1048. [7]

Ehrsson, H. H., C. Spence, and R. E. Passingham. 2004. That's My Hand! Activity in Premotor Cortex Reflects Feeling of Ownership of a Limb. *Science* **305**:875–877. [20]

Eimer, M., and J. van Velzen. 2006. Covert Manual Response Preparation Triggers Attentional Modulations of Visual but Not Auditory Processing. *Clin. Neurophysiol.* **117**:1063–1074. [20]

Ekornås, B., A. J. Lundervold, T. Tjus, and M. Heimann. 2010. Anxiety Disorders in 8–11-Year-Old Children: Motor Skill Performance and Self-Perception of Competence. *Scand. J. Psychol.* **51**:271–277. [3]

Elze, T., C. Song, R. Stollhoff, and J. Jost. 2011. Chinese Characters Reveal Impacts of Prior Experience on Very Early Stages of Perception. *BMC Neurosci.* **12**:14. [8]

Emck, C., R. J. Bosscher, P. C. W. Van Wieringen, T. Doreleijers, and P. J. Beek. 2011. Gross Motor Performance and Physical Fitness in Children with Psychiatric Disorders. *Dev. Med. Child Neurol.* **53**:150–155. [3]

Eng, K., R. J. Douglas, and P. F. M. J. Verschure. 2005. An Interactive Space That Learns to Influence Human Behavior. *IEEE Trans. Syst. Man Cybern. A Syst. Hum.* **35**:66–77. [14]

Engel, A. K. 2010. Directive Minds: How Dynamics Shapes Cognition. In: Enaction: Towards a New Paradigm for Cognitive Science, ed. J. Stewart et al., pp. 219–243. Cambridge, MA: MIT Press. [1, 11, 20]

Engel, A. K., A. Maye, M. Kurthen, and P. König. 2013. Where's the Action? The Pragmatic Turn in Cognitive Science. *Trends Cogn. Sci.* **17**:202–209. [1, 2, 4, 5, 8, 9, 11, 13, 15, 18, 20]

Engel, C., and W. Singer, eds. 2008. Better Than Conscious? Decision Making, the Human Mind, and Implications for Institutions. Strüngmann Forum Report, vol. 1. J. Lupp, series ed. Cambridge, MA: MIT press. [12]

Ernst, M. O., and M. S. Banks. 2002. Humans Integrate Visual and Haptic Information in a Statistically Optimal Fashion. *Nature* **415**:429–433. [3]

Evans, J. S. B. T. 2008. Dual-Processing Accounts of Reasoning, Judgment, and Social Cognition. *Annu. Rev. Psychol.* **59**:255–278. [14]

Fabbri-Destro, M., L. Cattaneo, S. Boria, and G. Rizzolatti. 2009. Planning Actions in Autism. *Exp. Brain Res.* **192**:521–525. [3, 20]

Fadiga, L., G. Buccino, L. Craighero, et al. 1999. Corticospinal Excitability Is Specifically Modulated by Motor Imagery: A Magnetic Stimulation Study. *Neuropsychologia* **37**:147–158. [18]

Fadiga, L., L. Craighero, G. Buccino, and G. Rizzolatti. 2002. Speech Listening Specifically Modulates the Excitability of Tongue Muscles: A TMS Study. *Eur. J. Neurosci.* **15**:399–402. [9]

Fagioli, S., B. Hommel, and R. I. Schubotz. 2007. Intentional Control of Attention: Action Planning Primes Action-Related Stimulus Dimensions. *Psychol. Res.* **71**:22–29. [20]

Faivre, N., L. Mudrik, N. Schwartz, and C. Koch. 2014. Multisensory Integration in Complete Unawareness: Evidence from Audiovisual Congruency Priming. *Psychol. Sci.* [12]

Fehr, E., and S. Gächter. 2002. Altruistic Punishment in Humans. *Nature* **415**:137–140. [12]

Feldman, A. G., D. J. Ostry, M. F. Levin, P. L. Gribble, and A. B. Mitnitski. 1998. Recent Tests of the Equilibrium-Point Hypothesis (Lambda Model). *Motor Control* **2**:189–205. [3]

Feldman, H., and K. J. Friston. 2010. Attention, Uncertainty, and Free-Energy. *Front. Hum. Neurosci.* **4**:215. [6, 7, 15]

Felleman, D. J., and D. C. Van Essen. 1991. Distributed Hierarchical Processing in the Primate Cerebral Cortex. *Cereb. Cortex* **1**:1–47. [12]

Feyerabend, P. K. 1975/2010. Against Method. New York: Verso Books. [19]

Fikes, R. E., and N. J. Nilsson. 1971. Strips: A New Approach to the Application of Theorem Proving to Problem Solving. Proc. 2nd Intl. Joint Conf. on Artificial Intelligence, pp. 608–620. San Francisco: Morgan Kaufmann. [18]

Filippetti, M. L., M. H. Johnson, S. Lloyd-Fox, D. Dragovic, and T. Farroni. 2013. Body Perception in Newborns. *Curr. Biol.* **23**:2413–2416. [3]

Filiz-Ozbay, E., and E. Y. Ozbay. 2007. Auctions with Anticipated Regret: Theory and Experiment. *Am. Econ. Rev.* **97**:1407–1418. [12, 15]

Fillmore, C. J. 1975. An Alternative to Checklist Theories of Meaning. Proc. Annual Meeting Berkeley Linguistics Society, pp. 123–131. [9]

Fitzpatrick, P. M., and G. Metta. 2002. Towards Manipulation-Driven Vision. IEEE/ RSJ Intl. Conf. on Intelligent Robots and Systems, pp. 43–48. Lausanne: IEEE. [20]

Flash, T., and N. Hogan. 1985. The Coordination of Arm Movements: An Experimentally Confirmed Mathematical Model. *J. Neurosci.* **5**:1688–1703. [18]

Fleming, S. M., B. Maniscalco, Y. Ko, et al. 2015. Action-Specific Disruption of Perceptual Confidence. *Psychol. Sci.* **26**:89–98. [15]

Fletcher, P. C., and C. D. Frith. 2009. Perceiving Is Believing: A Bayesian Approach to Explaining the Positive Symptoms of Schizophrenia. *Nat. Rev. Neurosci.* **10**:48–58. [6]

Fodor, J. A. 1981. Representations: Essays on the Foundations of Cognitive Science. Cambridge, MA: MIT Press. [1]

———. 1983. The Modulality of Mind. Cambridge, MA: MIT Press. [9]

Fodor, J. A., and Z. W. Pylyshyn. 1988. Connectionism and Cognitive Architecture: A Critical Analysis. *Cognition* **28**:3–71. [5]

Fogassi, L., P. F. Ferrari, B. Gesierich, et al. 2005. Parietal Lobe: From Action Organization to Intention Understanding. *Science* **308**:662–667. [9]

Ford, J. M., B. J. Roach, W. O. Faustman, and D. H. Mathalon. 2007. Synch before You Speak: Auditory Hallucinations in Schizophrenia. *Am. J. Psychiatry* **164**:458–466. [15, 20]

———. 2008. Out-of-Synch and out of Sorts: Dysfunction of Motor-Sensory Communication in Schizophrenia. *Biol. Psychiatry* **63**:736–743. [1, 15]

Forssberg, H., H. Kinoshita, A. C. Eliasson, et al. 1992. Development of Human Precision Grip. *Exp. Brain Res.* **90**:393–398. [3]

Fotopoulou, A. 2015. The Virtual Bodily Self: Mentalisation of the Body as Revealed in Anosognosia for Hemiplegia. *Conscious. Cogn.* **33**:500–510. [7]

Fouragnan, E., G. Chierchia, S. Greiner, et al. 2013. Reputational Priors Magnify Striatal Responses to Violations of Trust. *J. Neurosci.* **33**:3602–3611. [12]

Fourneret, P., and M. Jeannerod. 1998. Limited Conscious Monitoring of Motor Performance in Normal Subjects. *Neuropsychologia* **36**:1133–1140. [12]

Fournier, K. A., C. J. Hass, S. K. Naik, N. Lodha, and J. H. Cauraugh. 2010. Motor Coordination in Autism Spectrum Disorders: A Synthesis and Meta-Analysis. *J. Autism Dev. Disord.* **40**:1227–1240. [3]

Frank, M. J., A. Scheres, and S. J. Sherman. 2007. Understanding Decision-Making Deficits in Neurological Conditions: Insights from Models of Natural Action Selection. *Phil. Trans. R. Soc. B* **362**:1641–1654. [6]

Franklin, David W. W., and D. M. Wolpert. 2011. Computational Mechanisms of Sensorimotor Control. *Neuron* **72**:425–442. [3]

Frassle, S., J. Sommer, A. Jansen, M. Naber, and W. Einhauser. 2014. Binocular Rivalry: Frontal Activity Relates to Introspection and Action but Not to Perception. *J. Neurosci.* **34**:1738–1747. [15]

Frijda, N. H. 1986. The Emotions. Cambridge: Cambridge Univ. Press. [5]

Friston, K. J. 2005. A Theory of Cortical Responses. *Phil. Trans. R. Soc. B* **360**:815–836. [12, 14]

———. 2008. Hierarchical Models in the Brain. *PLoS Comput. Biol.* **4**:e1000211. [4, 6]

———. 2009. The Free-Energy Principle: A Rough Guide to the Brain? *Trends Cogn. Sci.* **13**:293–301. [15]

———. 2010. The Free-Energy Principle: A Unified Brain Theory? *Nat. Rev. Neurosci.* **11**:127–138. [1, 4, 5, 7, 12, 15]

———. 2013. Life as We Know It. *J. R. Soc. Interface* **10**:20130475. [7, 12, 15]

Friston, K. J., R. Adams, L. Perrinet, and M. Breakspear. 2012. Perceptions as Hypotheses: Saccades as Experiments. *Front. Psychol.* **3**:151. [4, 7]

Friston, K. J., J. Daunizeau, J. Kilner, and S. J. Kiebel. 2010. Action and Behavior: A Free-Energy Formulation. *Biol. Cybern.* **102**:227–260. [1, 2, 4]

Friston, K. J., and S. J. Kiebel. 2009. Predictive Coding under the Free-Energy Principle. *Phil. Trans. R. Soc. B* **364**:1211–1221. [6]

Friston, K. J., J. Kilner, and L. Harrison. 2006. A Free Energy Principle for the Brain. *J. Physiol. Paris* **100**:70–87. [6]

Friston, K. J., J. Mattout, and J. M. Kilner. 2011. Action Understanding and Active Inference. *Biol. Cybern.* **104**:137–160. [3, 6]

Friston, K. J., F. Rigoli, D. Ognibene, et al. 2015. Active Inference and Epistemic Value. *Cogn. Neurosci.* **6**:187–214. [2]

Friston, K. J., S. Samothrakis, and R. Montague. 2012. Active Inference and Agency: Optimal Control without Cost Functions. *Biol. Cybern.* **106**:523–541. [15]

Friston, K. J., P. Schwartenbeck, T. Fitzgerald, et al. 2013. The Anatomy of Choice: Active Inference and Agency. *Front. Hum. Neurosci.* **7**:598. [7, 13]

———. 2014. The Anatomy of Choice: Dopamine and Decision-Making. *Phil. Trans. R. Soc. B* **369**: [15]

Friston, K. J., T. Shiner, T. Fitzgerald, et al. 2012. Dopamine, Affordance and Active Inference. *PLoS Comput. Biol.* **8**:e1002327. [6]

Friston, K. J., and K. E. Stephan. 2007. Free Energy and the Brain. *Synthese* **159**:417–458. [1, 7, 13]

Friston, K. J., K. E. Stephan, B. Li, and J. Daunizeau. 2010. Generalised Filtering. *Math. Probl. Engin.* **2010**:621670. [6]

Friston, K. J., C. Thornton, and A. Clark. 2012. Free-Energy Minimization and the Dark Room Problem. *Front. Psychol.* **3**:130. [7]

Frith, C. D. 2008. Social Cognition. *Phil. Trans. R. Soc. B* **363**:2033–2039. [14]

———. 2012. Explaining Delusions of Control: The Comparator Model 20 Years on. *Conscious. Cogn.* **21**:52–54. [15, 20]

———. 2014. Action, Agency and Responsibility. *Neuropsychologia* **55**:137–142. [7, 15]

Frith, C. D., S.-J. Blakemore, and D. M. Wolpert. 2000a. Abnormalities in the Awareness and Control of Action. *Phil. Trans. R. Soc. B* **355**:1771–1788. [14, 15, 20]

———. 2000b. Explaining the Symptoms of Schizophrenia: Abnormalities in the Awareness of Action. *Brain Res. Rev.* **31**:357–363. [1, 7, 11]

Frith, C. D., and U. Frith. 2006. How We Predict What Other People Are Going to Do. *Brain Res.* **1079**:36–46. [15]

———. 2008. Implicit and Explicit Processes in Social Cognition. *Neuron* **60**:503–510. [2]

———. 2012. Mechanisms of Social Cognition. *Annu. Rev. Psychol.* **63**:287–313. [4]

Frith, U. 2012. Why We Need Cognitive Explanations of Autism. *Q. J. Exp. Psychol.* **65**:2073–2092. [3, 4]

Frith, U., J. M. Morton, and A. M. Leslie. 1991. The Cognitive Basis of a Biological Disorder: Autism. *Trends Neurosci.* **14**:433–438. [3]

Fritz, G. 2013. Dynamische Texttheorie. Gießen: Gießener Elektronische Bibliothek. [9]

Froese, T., M. McGann, W. Bigge, A. Spiers, and A. K. Seth. 2012. The Enactive Torch: A New Tool for the Science of Perception. *IEEE Trans. Haptics* **5**:363–375. [19]

Fusaroli, R., B. Bahrami, K. Olsen, et al. 2012. Coming to Terms: Quantifying the Benefits of Linguistic Coordination. *Psychol. Sci.* **23**:931–939. [12]

Fuster, J. M. 1990. Prefrontal Cortex and the Bridging of Temporal Gaps in the Perception-Action Cycle. *Ann. NY Acad. Sci.* **608**:318–329. [10]

———. 1995. Memory in the Cerebral Cortex. An Empirical Approach to Neural Networks in the Human and Nonhuman Primate. Cambridge, MA: MIT Press. [9]

———. 2003. Cortex and Mind: Unifying Cognition. Oxford: Oxford Univ. Press. [9]

Fuster, J. M., and G. E. Alexander. 1971. Neuron Activity Related to Short-Term Memory. *Science* **173**:652–654. [9]

Gallagher, S. 2000. Philosophical Conceptions of the Self: Implications for Cognitive Science. *Trends Cogn. Sci.* **4**:14. [7]

———. 2004. Understanding Interpersonal Problems in Autism: Interaction Theory as an Alternative to Theory of Mind. *Philos. Psychiatr. Psychol.* **11**:199–217. [20]

———. 2005. How the Body Shapes the Mind. Oxford: Oxford Univ. Press. [4, 14, 16]

———. 2006. Where's the Action? Epiphenomenalism and the Problem of Free Will. In: Does Consciousness Cause Behavior? An Investigation of the Nature of Volition, ed. W. Banks et al., pp. 109–124. Cambridge, MA: MIT Press. [16]

———. 2009. Philosophical Antecedents of Situated Cognition. In: The Cambridge Handbook of Situated Cognition, ed. P. Robbins and M. Aydede. Cambridge: Cambridge Univ. Press. [13]

———. 2013. A Pattern Theory of Self. *Front. Hum. Neurosci.* **7**:443. [5]

———. 2014. Pragmatic Interventions into Enactive and Extended Conceptions of Cognition. *Phil. Issues* **24**:110–126. [13, 16]

Gallagher, S., and F. Varela. 2003. Redrawing the Map and Resetting the Time: Phenomenology and the Cognitive Sciences. *Can. J. Philos.* **29**:93–132. [20]

Gallagher, S., and S. Varga. 2015. Conceptual Issues in Autism Spectrum Disorders. *Curr. Opin. Psychiatry* **28**:127–132. [20]

Gallagher, S., and D. Zahavi. 2012. The Phenomenological Mind. London: Routledge. [20]

Gallese, V. 2014. Bodily Selves in Relation: Embodied Simulation as Second-Person Perspective on Intersubjectivity. *Phil. Trans. R. Soc. B* **369**:2013–2177. [16]

Gallese, V., L. Fadiga, L. Fogassi, and G. Rizzolatti. 1996. Action Recognition in the Premotor Cortex. *Brain* **119**:593–609. [18]

Gallese, V., and C. Keysers. 2001. Mirror Neurons: A Sensorimotor Representation System. Behavioral and Brain Sciences. *Behav. Brain Sci.* **24**:983–984. [5]

Gallese, V., C. Keysers, and G. Rizzolatti. 2004. A Unifying View of the Basis of Social Cognition. *Trends Cogn. Sci.* **8**:396–403. [3, 15]

Gallese, V., and G. Lakoff. 2005. The Brain's Concepts: The Role of the Sensory-Motor System in Conceptual Knowledge. *Cogn. Neuropsychol.* **22**:455–479. [11]

Gallese, V., M. J. Rochat, and C. Berchio. 2013. The Mirror Mechanism and Its Potential Role in Autism Spectrum Disorder. *Dev. Med. Child Neurol.* **55**:15–22. [20]

Gallese, V., M. J. Rochat, G. Cossu, and C. Sinigaglia. 2009. Motor Cognition and Its Role in the Phylogeny and Ontogeny of Action Understanding. *Dev. Psychol.* **45**:103–113. [3]

Gallie, W. B. 1952. Peirce and Pragmatism. Harmondsworth: Penguin Books. [13]

Gallistel, C. R. 1980. The Organization of Action: A New Synthesis. Hillsdale, NJ: Erlbaum. [17]

Gallotti, M., and C. D. Frith. 2013. Social Cognition in the We-Mode. *Trends Cogn. Sci.* **17**:160–165. [12]

Garagnani, M., and F. Pulvermüller. 2013. Neuronal Correlates of Decisions to Speak and Act: Spontaneous Emergence and Dynamic Topographies in a Computational Model of Frontal and Temporal Areas. *Brain Lang.* **127(1)**:75–85. [9]

Garagnani, M., T. Wennekers, and F. Pulvermüller. 2008. A Neuroanatomically-Grounded Hebbian Learning Model of Attention-Language Interactions in the Human Brain. *Eur. J. Neurosci.* **27**:492–513. [9]

Gardner, H. 1987. The Mind's New Science: A History of the Cognitive Revolution. New York: Basic Books. [14]

Gazzaniga, M. S. 2011. Who's in Charge? Free Will and the Science of the Brain. New York: Harper Collins. [14]

Gelman, S. A. 2003. The Essential Child: Origins of Essentialism in Everyday Thought. Oxford: Oxford Univ. Press. [5]

Gentsch, A., A. Weber, M. Synofzik, G. Vosgerau, and S. Schütz-Bosbach. 2016. Towards a Common Framework of Grounded Action Cognition: Relating Motor Control, Perception and Cognition. *Cognition* **146**:81–89. [4]

Gergely, G., and P. Jacob. 2013. Reasoning About Instrumental and Communicative Agency in Human Infancy. In: Rational Constructivism in Cognitive Development, ed. F. Xu and T. Kushnir, pp. 59–94, J. B. Benson, series ed. Amsterdam: Elsevier. [15]

Gerlach, K. D., R. N. Spreng, K. P. Madore, and D. L. Schacter. 2014. Future Planning: Default Network Activity Couples with Frontoparietal Control Network and Reward-Processing Regions During Process and Outcome Simulations. *Soc. Cogn. Affect. Neurosci.* **9**:1942–1951. [5]

Gerstner, W., R. Kempter, J. v. Hemmen, and H. Wagner. 1996. A Neuronal Learning Rule for Sub-Millisecond Temporal Coding. *Nature* **383**:76–78. [8]

Gibbons, M., C. Limoges, H. Nowotny, et al. 1994. The New Production of Knowledge. The Dynamics of Science and Research in Contemporary Socities. London: Sage. [19]

Gibson, J. J. 1977. The Theory of Affordances. In: Perceiving, Acting, and Knowing: Toward an Ecological Psychology, ed. R. Shaw and J. Bransford, pp. 67–82. Hillsdale, NJ: Lawrence Erlbaum. [16, 18]

———. 1979. The Ecological Approach to Visual Perception. Mahwah, NJ: Lawrence Erlbaum. [1, 2, 11, 15]

Gidley Larson, J. C., A. J. Bastian, O. Donchin, R. Shadmehr, and S. H. Mostofsky. 2008. Acquisition of Internal Models of Motor Tasks in Children with Autism. *Brain* **131**:2894–2903. [3]

Gigerenzer, G., and P. Todd. 1999. Simple Heuristics That Make Us Smart. Oxford: Oxford Univ. Press. [8]

Gilovich, T., and V. H. Medvec. 1995. The Experience of Regret: What, When, and Why. *Psychol. Rev.* **102**:379–395. [12]

Glasel, H., F. Leroy, J. Dubois, et al. 2011. A Robust Cerebral Asymmetry in the Infant Brain: The Rightward Superior Temporal Sulcus. *NeuroImage* **58**:716–723. [4]

Gleitman, L. R., K. Cassidy, R. Nappa, A. Papafragou, and J. C. Trueswell. 2005. Hard Words. *Lang. Learn. Dev.* **1**:23–64. [4]

Glenberg, A. M. 1997. What Memory Is For. *Behav. Brain Sci.* **20**:1–19; discussion, 19–55. [2]

———. 2010. Embodiment as a Unifying Perspective for Psychology. *Wiley Interdiscip. Rev. Cogn. Sci.* **1**:586–596. [16]

Glenberg, A. M., and V. Gallese. 2012. Action-Based Language: A Theory of Language Acquisition, Comprehension, and Production. *Cortex* **48**:905–922. [9]

Glenberg, A. M., and M. P. Kaschak. 2002. Grounding Language in Action. *Psychon. Bull. Rev.* **9**:558–565. [2, 4]

Godfrey-Smith, P. 1996. Complexity and the Function of Mind in Nature. Cambridge: Cambridge Univ. Press. [13]

———. 2001. On the Status and Explanatory Structure of Developmental Systems Theory. In: Cycles of Contingency. Developmental Systems and Evolution., ed. S. Oyama et al., pp. 283–298. Cambridge, MA: MIT Press. [16]

Goldberg, A. E. 2006. Constructions at Work: The Nature of Generalisation in Language. Oxford: Oxford Univ. Press. [9]

Goldman, A. I. 2012. A Moderate Approach to Embodied Cognitive Science. *Rev. Phil. Psych.* **3**:71–88. [16]

Goldman, A. I., and F. de Vignemont. 2009. Is Social Cognition Embodied? *Trends Cogn. Sci.* **13**:154–159. [16]

Goldman-Rakic, P. S., M. S. Lidow, J. F. Smiley, and M. S. Williams. 1992. The Anatomy of Dopamine in Monkey and Human Prefrontal Cortex. *J. Neural Transm. Suppl.* **36**:163–177. [6]

Gollwitzer, P. M., and G. B. Moskowitz. 1996. Goal Effects on Action and Cognition. In: Social Psychology: Handbook of Basic Principles, ed. E. T. Higgins and A. W. Kruglanski, pp. 361–399. New York: Guilford Press. [17]

Gopnik, A., A. N. Meltzoff, and P. K. Kuhl. 1999. The Scientist in the Crib: Minds, Brains, and How Children Learn. New York: William Morrow. [13]

Gopnik, A., and L. Schulz. 2004. Mechanisms of Theory Formation in Young Children. *Trends Cogn. Sci.* **8**:371–377. [4]

Gori, M., M. Del Viva, G. Sandini, and D. C. Burr. 2008. Young Children Do Not Integrate Visual and Haptic Form Information. *Curr. Biol.* **18**:694–698. [3, 7, 15]

Gorniak, P., and D. Roy. 2007. Situated Language Understanding as Filtering Perceived Affordances. *Cogn. Sci.* **31**:197–231. [2]

Gottlieb, J., P.-Y. Oudeyer, M. Lopes, and A. F. Baranes. 2013. Information-Seeking, Curiosity, and Attention: Computational and Neural Mechanisms. *Trends Cogn. Sci.* **17**:585–593. [2, 4]

Gowen, E., and A. F. Hamilton. 2012. Motor Abilities in Autism: A Review Using a Computational Context. *J. Autism Dev. Disord.* **43**:323–344. [3]

Granlund, G. 1999. Does Vision Inevitably Have to Be Active? Proc. Scandinavian Conf. on Image Analysis, vol. 1, pp. 11–19. Kangerlussuaq, Greenland. [18]

Grant, E. R., and M. J. Spivey. 2003. Eye Movements and Problem Solving: Guiding Attention Guides Thought. *Psychol. Sci.* **14**:462–466. [2]

Graziano, M. S. A. 2013. Consciousness and the Social Brain. Oxford: Oxford Univ. Press. [14]

Graziano, M. S. A., and C. G. Gross. 1994. The Representation of Extrapersonal Space: A Possible Role for Bimodal, Visual-Tactile Neurons. In: The Cognitive Neurosciences, ed. M. Gazzaniga, pp. 1021–1034. Cambridge, MA: MIT Press. [11]

Graziano, M. S. A., and S. Kastner. 2011. Human Consciousness and Its Relationship to Social Neuroscience: A Novel Hypothesis. *Cogn. Neurosci.* 2:98–113. [12]

Green, D., T. Charman, A. Pickles, et al. 2009. Impairment in Movement Skills of Children with Autistic Spectrum Disorders. *Dev. Med. Child Neurol.* 51:311–316. [3]

Gregory, R. L. 1968. Perceptual Illusions and Brain Models. *Proc. R. Soc. Lond. B* 171:179–196. [6]

Grün, S., and S. Rotter, eds. 2010. Analysis of Parallel Spike Trains. Heidelberg: Springer. [8]

Grush, R. 2004. The Emulation Theory of Representation: Motor Control, Imagery, and Perception. *Behav. Brain Sci.* 27:377–396. [2, 4]

Gulberti, A., W. Hamel, C. Buhmann, et al. 2015. Subthalamic Deep Brain Stimulation Improves Auditory Sensory Gating Deficit in Parkinson's Disease. *Clin. Neurophysiol.* 126:565–574. [11]

Gürerk, O., B. Irlenbusch, and B. Rockenbach. 2006. The Competitive Advantage of Sanctioning Institutions. *Science* 312:108–111. [12]

Haggard, P. 2008. Human Volition: Towards a Neuroscience of Will. *Nat. Rev. Neurosci.* 9:934–946. [15]

Haggard, P., S. Clark, and J. Kalogeras. 2002. Voluntary Action and Conscious Awareness. *Nat. Neurosci.* 5:382–385. [12, 14, 15]

Haggard, P., and M. Eimer. 1999. On the Relation between Brain Potentials and the Awareness of Voluntary Movements. *Exp. Brain Res.* 126:128–133. [14]

Haggard, P., and B. Libet. 2001. Conscious Intention and Brain Activity. *J. Conscious. Stud.* 8:47–64. [16]

Hagura, N., R. Kanai, G. Orgs, and P. Haggard. 2012. Ready Steady Slow: Action Preparation Slows the Subjective Passage of Time. *Proc. R. Soc. Lond. B* 279:4399–4406. [15]

Halevy, A., P. Norvig, and F. Pereira. 2009. The Unreasonable Effectiveness of Data. *IEEE Intell. Syst.* 24:8–12. [18]

Hall, L., P. Johansson, and T. Strandberg. 2012. Lifting the Veil of Morality: Choice Blindness and Attitude Reversals on a Self-Transforming Survey. *PLoS One* 7:e45457. [12]

Hamilton, A. F. 2008. Emulation and Mimicry for Social Interaction: A Theoretical Approach to Imitation in Autism. *Q. J. Exp. Psychol.* 61:101–115. [3]

———. 2013a. The Mirror Neuron System Contributes to Social Responding. *Cortex* 49:2957–2959. [3]

———. 2013b. Reflecting on the Mirror Neuron System in Autism: A Systematic Review of Current Theories. *Dev. Cogn. Neurosci.* 3:91–105. [3]

Hamilton, A. F., R. M. Brindley, and U. Frith. 2007. Imitation and Action Understanding in Autistic Spectrum Disorders: How Valid Is the Hypothesis of a Deficit in the Mirror Neuron System? *Neuropsychologia* 45:1859–1868. [3]

Hamlin, J. K., N. Mahajan, Z. Liberman, and K. Wynn. 2013. Not Like Me: Bad Infants Prefer Those Who Harm Dissimilar Others. *Psychol. Sci.* 24:589–594. [4]

Hamlin, J. K., K. Wynn, P. Bloom, and N. Mahajan. 2011. How Infants and Toddlers React to Antisocial Others. *PNAS* 108:19931–19936. [4]

Happe, F., and U. Frith. 2006. The Weak Coherence Account: Detail Focused Cognitive Style in Autism Spectrum Disorders. *J. Autism Dev. Disord.* 36:5–25. [6]

Hardy, T. 1909. Before Life and After. http://www.readbookonline.net/readOn-Line/10445/. (accessed Oct. 12, 2015). [12]

Harless, E. 1861. Der Apparat Des Willens. *Zt. für Philosophie und Philosophische Kritik* **38**:50–73. [10]

Harnad, S. 1990. The Symbol Grounding Problem. *Phys. Nonlinear Phenom.* **42**:335–346. [4, 9, 14]

———. 1991. Other Bodies, Other Minds: A Machine Incarnation of an Old Philosophical Problem. *Minds and Machines* **1**:43–54. [18]

Hassabis, D., R. N. Spreng, A. A. Rusu, et al. 2014. Imagine All the People: How the Brain Creates and Uses Personality Models to Predict Behavior. *Cereb. Cortex* **24**:1979–1987. [15]

Hassin, R. R., H. Aarts, B. Eitam, R. Custers, and T. Kleiman. 2009. Non-Conscious Goal Pursuit and the Effortful Control of Behavior. In: Oxford Handbook of Human Action, ed. E. Morsella et al., pp. 549–568. New York: Oxford Univ. Press. [17]

Hatsopoulos, N. G., and A. J. Suminski. 2011. Sensing with the Motor Cortex. *Neuron* **72**:477–487. [1]

Haugeland, J. 1985. Artificial Intelligence: The Very Idea. Cambridge, MA: MIT Press. [20]

Hauk, O., I. Johnsrude, and F. Pulvermüller. 2004. Somatotopic Representation of Action Words in the Motor and Premotor Cortex. *Neuron* **41**:301–307. [9]

Hauk, O., and F. Pulvermüller. 2004. Neurophysiological Distinction of Action Words in the Fronto-Central Cortex. *Hum. Brain Mapp.* **21**:191–201. [9]

Hayek, F. A. 1943. Scientism and the Study of Society. Part II. *Economica* 34–63. [14]

Hayhoe, M. M., T. McKinney, K. Chajka, and J. B. Pelz. 2011. Predictive Eye Movements in Natural Vision. *Exp. Brain. Res.* **217**:125–136. [11]

Hayhoe, M. M., A. Shrivastava, R. Mruczek, and J. B. Pelz. 2003. Visual Memory and Motor Planning in a Natural Task. *J. Vis.* **3**:49–63. [15]

Hazel, N. 2008. Cross-National Comparison of Youth Justice. London: Youth Justice Board. [12]

Hebb, D. O. 1949. The Organization of Behavior. A Neuropsychological Theory. New York: Wiley. [9]

Heckhausen, J., and H. Heckhausen, eds. 2008. Motivation and Action. New York: Cambridge Univ. Press. [17]

Heidegger, M. 1962. Being and Time, transl. J. Macquarrie and E. Robinson. New York: Harper. [16]

Heider, F. 1944. Social Perception and Phenomenal Causality. *Psychol. Rev.* **51**:358–374. [14]

Held, R., and A. Hein. 1963. Movement-Produced Stimulation in the Development of Visually Guided Behavior. *J. Comp. Physiol. Psychol.* **56**:872–876. [2, 10, 11, 13]

Helmholtz, H. 1866/1962. Concerning the Perceptions in General. In: Treatise on Physiological Optics, vol. 3. New York: Dover. [6]

Herbart, J. 1816. Lehrbuch zur Psychologie. Königsberg: Unzer. [7]

Hernik, M., and V. Southgate. 2012. Nine-Months-Old Infants Do Not Need to Know What the Agent Prefers in Order to Reason About Its Goals: on the Role of Preference and Persistence in Infants' Goal-Attribution. *Dev. Sci.* **15**:714–722. [3]

Herrera, E., and F. Cuetos. 2012. Action Naming in Parkinson's Disease Patients On/Off Dopamine. *Neurosci. Lett.* **513**:219–222. [11]

Herreros, I., and P. F. M. J. Verschure. 2013. Nucleo-Olivary Inhibition Balances the Interaction between the Reactive and Adaptive Layers in Motor Control. *Neural Netw.* **47**:64–71. [14]

Herschbach, M. 2008. Folk Psychological and Phenomenological Accounts of Social Perception. *Philos. Explor.* **11**:223–235. [6]

Hesslow, G. 2002. Conscious Thought as Simulation of Behaviour and Perception. *Trends Cogn. Sci.* **6**:242–247. [2, 14]

Heyes, C. 2001. Causes and Consequences of Imitation. *Trends Cogn. Sci.* **5**:253–261. [3]

Heyes, C. 2012. Grist and Mills: on the Cultural Origins of Cultural Learning. *Phil. Trans. R. Soc. B* **367**:2181–2191. [4]

Heyes, C., and U. Frith. 2014. The Cultural Evolution of Mind Reading. *Science* **344**:1243091. [4]

Hickok, G. 2009. Eight Problems for the Mirror Neuron Theory of Action Understanding in Monkeys and Humans. *J. Cogn. Neurosci.* **21**:1229–1243. [9]

———. 2010. The Role of Mirror Neurons in Speech Perception and Action Word Semantics. *Lang. Cogn. Process.* **25**:749–776. [9]

Hickok, G., and M. Hauser. 2010. (Mis)Understanding Mirror Neurons. *Curr. Biol.* **20**:R593–R594. [3, 9]

Hickok, G., and D. Poeppel. 2007. The Cortical Organization of Speech Processing. *Nat. Rev. Neurosci.* **8**:393–402. [9]

Hilgard, J. R. 1991. Learning and Maturation in Preschool Children. *J. Genet. Psychol.* **152**:528–548. [4]

Hilton, C., Y. Zhang, M. White, C. L. Klohr, and J. Constantino. 2012. Motor Impairment Concordant and Discordant for Autism Spectrum Disorders. *Autism* **16**:430–441. [20]

Hintikka, J. 1998. What Is Abduction? The Fundamental Problem of Contemporary Epistemology. *Trans. of the C. S. Peirce Soc.* **34**:503–533. [13]

Hipp, J. F., A. K. Engel, and M. Siegel. 2011. Oscillatory Synchronization in Large-Scale Cortical Networks Predicts Perception. *Neuron* **69**:387–396. [20]

Hödl, L. 1992. Reue. In: Historisches Wörterbuch Der Philosophie, vol. 8, ed. J. Ritter and K. Gründer, pp. 944–951. Basel: Schwabe. [12]

Hoffman, H. G., D. R. Patterson, G. J. Carrougher, and S. R. Sharar. 2001. Effectiveness of Virtual Reality-Based Pain Control with Multiple Treatments. *Clin. J. Pain* **17**:229–235. [19]

Hoffmann, H. 2007. Perception through Visuomotor Anticipation in a Mobile Robot. *Neural Netw.* **20**:22–33. [2]

Hoffmann, M., and V. C. Müller. 2014. Trade-Offs in Exploiting Body Morphology for Control: From Simple Bodies and Model-Based Control to Complex Bodies with Model-Free Distributed Control Schemes. In: E-book on Opinions and Outlooks on Morphological Computation, ed H. Hauser et al. http://www.merlin.uzh.ch/contributionDocument/download/7499 (accessed Oct. 15, 2015). [20]

Hoffmann, M., N. Schmidt, R. Pfeifer, A. K. Engel, and A. Maye. 2012. Using Sensorimotor Contingencies for Terrain Discrimination and Adaptive Walking Behavior in the Quadruped Robot Puppy. In: From Animals to Animats 12, ed. T. Ziemke et al., pp. 54–56. Heidelberg: Springer. [20]

Högman, V., M. Björkman, and D. Kragic. 2013. Interactive Object Classification Using Sensorimotor Contingencies. In: IEEE/RSJ Intl. Conf. on Intelligent Robots and Systems (IROS), pp. 2799–2805. Tokyo: IEEE. [11, 20]

Hohwy, J. 2011. Phenomenal Variability and Introspective Reliability. *Mind Lang.* **26**:261–286. [7]

———. 2012. Attention and Conscious Perception in the Hypothesis Testing Brain. *Front. Psychol.* **3**:96. [7]

———. 2013. The Predictive Mind. Oxford: Oxford Univ. Press. [6, 7, 13, 15, 16]

———. 2014. The Self-Evidencing Brain. *Nous* DOI: 10.1111/nous.12062. [7, 12]

Hohwy, J., and C. D. Frith. 2004. Can Neuroscience Explain Consciousness? *J. Conscious. Stud.* **11**:180–198. [7]

Hohwy, J., C. Palmer, and B. Paton. 2015. Distrusting the Present. *Phenom. Cogn. Sci.* DOI 10.1007/s11097-11015-19439-11096. [7]

Hollerman, J. R., and W. Schultz. 1998. Dopamine Neurons Report an Error in the Temporal Prediction of Reward During Learning. *Nat. Neurosci.* **1**:304–309. [12]

Hommel, B. 2013. Ideomotor Action Control: on the Perceptual Grounding of Voluntary Actions and Agents. In: Action Science: Foundations of an Emerging Discipline, ed. A. Herwig et al., pp. 113–136. Cambridge, MA: MIT Press. [7]

Hommel, B., J. Müsseler, G. Aschersleben, and W. Prinz. 2001. The Theory of Event Coding (TEC): A Framework for Perception and Action Planning. *Behav. Brain Sci.* **24**:849–878. [1, 5, 10, 17]

Horgan, J. 1997. The End of Science: Facing the Limits of Knowledge in the Twilight of the Scientific Age. New York: Random House. [14]

Horn, B. K. 1986. Robot Vision (1st edition). New York: McGraw-Hill. [18, 20]

Hrdy, S. B. 2009. Mothers and Others: The Evolutionary Origins of Mutual Understanding. Cambridge, MA: Harvard Univ. Press. [15]

Hubel, D. 1995. Eye, Brain, and Vision. New York: Scientific American Library. [9]

Hubel, D. H., and T. N. Wiesel. 1970. The Period of Susceptibility to the Physiological Effects of Unilateral Eye Closure in Kittens. *J. Physiol.* **206**:419–436. [4]

Hughes, C. 1996. Control of Action and Thought: Normal Development and Dysfunction in Autism: A Research Note. *J. Child Psychol. Psychiatry* **37**:229–236. [3]

Humphrey, N. 1999. A History of the Mind: Evolution and the Birth of Consciousness: Springer. [12]

———. 2002. Bugs and Beast before the Law. In: The Mind Made Flesh, pp. 235–254. Oxford: Oxford Univ. Press. [12]

———. 2006. Seeing Red: A Study in Consciousness. Cambridge, MA: Harvard Univ. Press. [14]

Humphries, M. D., R. Wood, and K. Gurney. 2009. Dopamine-Modulated Dynamic Cell Assemblies Generated by the Gabaergic Striatal Microcircuit. *Neural Netw.* **22**:1174–1188. [6]

Hurley, S. L. 1998. Consciousness in Action. Cambridge, MA: Harvard Univ Press. [7, 13]

Husserl, E. 1989. Ideas Pertaining to a Pure Phenomenology and to a Phenomenological Philosophy, Second Book. Studies in the Phenomenology of Constitution, trans. R. Rojcewicz and A. Schuwer. Dordrecht: Springer. [16]

Hutchins, E. 1995. Cognition in the Wild. Cambridge, MA: MIT Press. [13]

Hutto, D. 2012. Radicalizing Enactivism: Basic Minds without Content. Cambridge, MA: MIT Press. [15]

Hutto, D. D., and E. Myin. 2013. Radicalizing Enactivism: Basic Minds without Content. Cambridge, MA: MIT Press. [6, 7, 20]

Huxley, T. H. 1874. On the Hypothesis That Animals Are Automata, and Its History. *Nature* **10**:362–366. [12]

Iacoboni, M., I. Molnar-Szakacs, V. Gallese, et al. 2005. Grasping the Intentions of Others with One's Own Mirror Neuron System. *PLoS Biol.* **3**:e79. [9]

Ijspeert, A. J. 2014. Biorobotics: Using Robots to Emulate and Investigate Agile Locomotion. *Science* **346**:196–203. [19]

Ijspeert, A. J., J. Nakanishi, and S. Schaal. 2002. Movement Imitation with Nonlinear Dynamical Systems in Humanoid Robots. Proc. IEEE Intl. Conf. Robotics and Automation, pp. 1398–1403. Washington, D.C.: IEEE. [18]

Inamura, T., I. Toshima, H. Tanie, and Y. Nakamura. 2004. Embodied Symbol Emergence Based on Mimesis Theory. *Int. J. Rob. Res.* **23**:363–377. [18]

Inderbitzin, M. P., A. Betella, A. Lanatá, et al. 2013. The Social Perceptual Salience Effect. *J. Exp. Psychol. Hum. Percept. Perform.* **39**:62–74. [14, 19]

Indurkhya, B. 1992. Metaphor and Cognition: An Interactionist Approach. Dordrecht: Kluwer Academic Publ. [4]

Iriki, A., and M. Taoka. 2012. Triadic (Ecological, Neural, Cognitive) Niche Construction: A Scenario of Human Brain Evolution Extrapolating Tool Use and Language from the Control of Reaching Actions. *Phil. Trans. R. Soc. B* **367**:10–23. [2, 4]

Ito, M. 1993. Movement and Thought: Identical Control Mechanisms by the Cerebellum. *Trends Neurosci.* **16**:448–450. [2]

Ito, T., E. Z. Murano, and H. Gomi. 2004. Fast Force-Generation Dynamics of Human Articulatory Muscles. *J. Appl. Physiol.* **96**:2318–2324; discussion 2317. [3]

Iverson, J. M. 2010. Developing Language in a Developing Body: The Relationship between Motor Development and Language Development. *J. Child Lang.* **37**:229–261. [3]

Ivry, R. B., and S. W. Keele. 1989. Timing Functions of the Cerebellum. *J. Cogn. Neurosci.* **1**:136–152. [5]

Jack, D., R. Boian, A. S. Merians, et al. 2001. Virtual Reality-Enhanced Stroke Rehabilitation. *IEEE Trans. Neural Syst. Rehab. Eng.* **9**:308–318. [19]

Jackendoff, R. 2002. Foundations of Language: Brain, Meaning, Grammar, Evolution. Oxford: Oxford Univ. Press. [16]

———. 2011. What Is the Human Language Faculty? Two Views. *Language* **87**:586–624. [10]

Jackson, F. 1982. Epiphenomenal Qualia. *Philos. Q.* 127–136. [12]

Jacob, P., and M. Jeannerod. 2005. The Motor Theory of Social Cognition: A Critique. *Trends Cogn. Sci.* **9**:21–25. [9]

James, W. 1890. The Principles of Psychology. New York: Holt. [7, 10, 14, 17]

———. 1894. The Physical Basis of Emotion. *Psychol. Rev.* **1**:516–529. [15]

Jeannerod, M. 1988. The Neural and Behavioural Organization of Goal-Directed Movements, vol. 15. Oxford Psychology Series. Oxford: Clarendon Press. [18]

———. 1994. The Representing Brain: Neuronal Correlates of Motor Intention and Imagery. *Behav. Brain Sci.* **17**:187–202. [9]

———. 1997. The Cognitive Neuroscience of Action. Oxford: Blackwell. [17]

———. 2001. Neural Simulation of Action: A Unifying Mechanism for Motor Cognition. *NeuroImage* **14**:S103–S109. [1]

———. 2006. Motor Cognition: What Actions Tell to the Self. Oxford: Oxford Univ. Press. [2, 4, 9, 15, 17]

Jeannerod, M., M. A. Arbib, G. Rizzolatti, and H. Sakata. 1995. Grasping Objects: The Cortical Mechanisms of Visuomotor Transformation. *Trends Neurosci.* **18**:314–320. [10]

Jenni, O. G., A. Chaouch, J. Caflisch, and V. Rousson. 2013. Correlations between Motor and Intellectual Functions in Normally Developing Children between 7 and 18 Years. *Dev. Neuropsychol.* **38**:98–113. [3]

Jiang, J., C. Summerfield, and T. Egner. 2013. Attention Sharpens the Distinction between Expected and Unexpected Percepts in the Visual Brain. *J. Neurosci.* **33**:18438–18447. [6]

Job, V., C. S. Dweck, and G. M. Walton. 2010. Ego Depletion: Is It All in Your Head? Implicit Theories About Willpower Affect Self-Regulation. *Psychol. Sci.* **21**:1686–1693. [12]

Joffily, M., and G. Coricelli. 2013. Emotional Valence and the Free-Energy Principle. *PLoS Comput. Biol.* **9**:e1003094. [15]

Johansson, P., L. Hall, S. Sikstrom, and A. Olsson. 2005. Failure to Detect Mismatches between Intention and Outcome in a Simple Decision Task. *Science* **310**:116–119. [12]

Johansson, R. S., and K. J. Cole. 1992. Sensory-Motor Coordination During Grasping and Manipulative Actions. *Curr. Opin. Neurobiol.* **2**:815–823. [3]

Johnson, M. H. 2012. Executive Function and Developmental Disorders: The Flip Side of the Coin. *Trends Cogn. Sci.* **16**:454–457. [3]

Johnson, M. H., S. Dziurawiec, H. Ellis, and J. Morton. 1991. Newborns' Preferential Tracking of Face-Like Stimuli and Its Subsequent Decline. *Cognition* **40**:1–19. [4]

Johnson-Frey, S., M. E. McCarty, and R. Keen. 2004. Reaching Beyond Spatial Perception: Effects of Intended Future Actions on Visually Guided Prehension. *Vis. Cogn.* **11**:371–399. [3]

Jongbloed-Pereboom, M., M. W. Nijhuis-van der Sanden, N. Saraber-Schiphorst, C. Crajé, and B. Steenbergen. 2013. Anticipatory Action Planning Increases from 3 to 10 Years of Age in Typically Developing Children. *J. Exp. Child Psychol.* **114**:295–305. [3]

Jost, J. 2003. On the Notion of Fitness, Or: The Selfish Ancestor. *Theory Biosci.* **121**:331–350. [8]

———. 2004. External and Internal Complexity of Complex Adaptive Systems. *Theory Biosci.* **123**:69–88. [8, 10]

———. 2005. Dynamical Systems: Examples of Complex Behavior. Heidelberg: Springer. [8]

———. 2006. Temporal Correlation Based Learning in Neuron Models. *Theory Biosci.* **125**:37–53. [8]

Jost, J., K. Holthausen, and O. Breidbach. 1997. On the Mathematical Foundations of a Theory of Neural Representation. *Theory Biosci.* **116**:125–139. [8]

Jovanovic, B., and K. Drewing. 2014. The Influence of Intersensory Discrepancy on Visuo-Haptic Integration Is Similar in 6-Year-Old Children and Adults. *Front. Psychol.* **5**:57. [3]

Joyce, J. 1922. Ulysses. Paris: Skalespear and Company. [12]

Kahneman, D. 2011. Thinking, Fast and Slow. New York: Farrar, Straus and Giroux. [14]

Kalakrishnan, M., J. Buchli, P. Pastor, M. Mistry, and S. Schaal. 2010. Learning, Planning, and Control for Quadruped Locomotion over Challenging Terrain. *Int. J. Rob. Res.* **30**:236–258. [20]

Kanakogi, Y., and S. Itakura. 2011. Developmental Correspondence between Action Prediction and Motor Ability in Early Infancy. *Nat. Commun.* **2**:341–341. [3]

Kannape, O. A., L. Schwabe, T. Tadi, and O. Blanke. 2010. The Limits of Agency in Walking Humans. *Neuropsychologia* **48**:1628–1636. [12]

Kanwisher, N., J. McDermott, and M. M. Chun. 1997. The Fusiform Face Area: A Module in Human Extrastriate Cortex Specialized for Face Perception. *J. Neurosci.* **17**:4302–4311. [20]

Karasik, L. B., C. S. Tamis-LeMonda, and K. E. Adolph. 2011. Transition from Crawling to Walking and Infants' Actions with Objects and People. *Child Dev.* **82**:1199–1209. [3]

Kärcher, S. M., S. Fenzlaff, D. Hartmann, S. K. Nagel, and P. König. 2012. Sensory Augmentation for the Blind. *Front. Hum. Neurosci.* **6**:37. [4]

Karl, J. M., and I. Q. Whishaw. 2014. Haptic Grasping Configurations in Early Infancy Reveal Different Developmental Profiles for Visual Guidance of the Reach versus the Grasp. *Exp. Brain Res.* **232**:3301–3316. [3]

Karmiloff-Smith, A. 2012. Challenging the Use of Adult Neuropsychological Models for Explaining Neurodevelopmental Disorders: Developed versus Developing Brains. *Q. J. Exp. Psychol.* **66**:37–41. [3]

Kaspar, K., S. König, J. Schwandt, and P. König. 2014. The Experience of New Sensorimotor Contingencies by Sensory Augmentation. *Conscious. Cogn.* **28**:47–63. [4]

Katz, D., M. Kazemi, J. A. Bagnell, and A. Stentz. 2013. Clearing a Pile of Unknown Objects Using Interactive Perception. In: Proc. IEEE Intl. Conf. on Robotics and Automation, pp. 154–161. Taipei: IEEE. [20]

Kawato, M., and D. M. Wolpert. 1998. Internal Models for Motor Control. *Novartis Found. Symp.* **218**:291–304; discussion 304–307. [3]

Kelly, S., A. Ozyurek, and E. Maris. 2010. Two Sides of the Same Coin: Speech and Gesture Mutually Interact in Language Comprehension. *Psychol. Sci.* **21**:260–267. [10]

Kelso, J. A. S. 1995. Dynamic Patterns: The Self-Organization of Brain and Behavior. Cambridge, MA: MIT Press. [2, 4]

Kelso, J. A. S., B. Tuller, E. Vatikiotis-Bateson, and C. A. Fowler. 1984. Functionally Specific Articulatory Cooperation Following Jaw Perturbations During Speech: Evidence for Coordinative Structures. *J. Exp. Psychol. Hum. Percept. Perform.* **10**:812–832. [15]

Kemmerer, D., J. G. Castillo, T. Talavage, S. Patterson, and C. Wiley. 2008. Neuroanatomical Distribution of Five Semantic Components of Verbs: Evidence from fMRI. *Brain Lang.* **107**:16–43. [9]

Kemmerer, D., D. Rudrauf, K. Manzel, and D. Tranel. 2012. Behavioural Patterns and Lesion Sites Associated with Impaired Processing of Lexical and Conceptual Knowledge of Action. *Cortex* **48**:826–848. [9]

Kersten, D., P. Mamassian, and A. Yuille. 2004. Object Perception as Bayesian Inference. *Annu. Rev. Psychol.* **55**:271–304. [12]

Keysers, C., and V. Gazzola. 2009. Unifying Social Cognition. In: Mirror Neuron Systems: The Role of Mirroring Processes in Social Cognition, ed. J. A. Pineda, pp. 3–38. New York: Humana Press. [17]

Keysers, C., and D. Perrett. 2004. Demystifying Social Cognition: A Hebbian Perspective. *Trends Cogn. Sci.* **8**:501–507. [17]

Kiebel, S. J., K. von Kriegstein, J. Daunizeau, and K. J. Friston. 2009. Recognizing Sequences of Sequences. *PLoS Comput. Biol.* **5**:e1000464. [6]

Kiefer, M., and F. Pulvermüller. 2012. Conceptual Representations in Mind and Brain: Theoretical Developments, Current Evidence and Future Directions. *Cortex* **48**:805–825. [9]

Kierkegaard, S. 1843/1992. Either/Or: A Fragment of Life. London: Penguin Books. [12]

Kietzmann, T. C., S. Geuter, and P. König. 2011. Overt Visual Attention as a Causal Factor of Perceptual Awareness. *PLoS One* **6**:e22614. [11, 20]

Kilner, J. M., K. J. Friston, and C. D. Frith. 2007. Predictive Coding: An Account of the Mirror Neuron System. *Cogn. Process.* **8**:159–166. [2, 6]

Kim, S., M. Spenko, S. Trujillo, et al. 2007. Whole Body Adhesion: Hierarchical, Directional and Distributed Control of Adhesive Forces for a Climbing Robot. IEEE Intl. Conf. on Robotics and Automation (ICRA), pp. 1268–1273. Rome: IEEE. [20]

King, B. R., M. Oliveira, J. L. Contreras-Vidal, and J. E. Clark. 2012. Development of State Estimation Explains Improvements in Sensorimotor Performance across Childhood. *J. Neurophysiol.* **107**:3040–3049. [3]

King, B. R., M. M. Pangelinan, K. A. Kagerer, and J. E. Clark. 2010. Improvements in Proprioceptive Functioning Influence Multisensory-Motor Integration in 7- to 13-Year-Old Children. *Neurosci. Lett.* **483**:36–40. [3]

Kirsch, D., and P. Maglio. 1994. On Distinguishing Epistemic from Pragmatic Actions. *Cogn. Sci.* **18**:513–549. [13, 15]

Kiverstein, J., and A. Clark. 2009. Introduction: Mind Embodied, Embedded, Enacted: One Church or Many? *Topoi* **28**:1–7. [7]

Kjølberg, K., G. C. Delgado-Ramos, F. Wickson, and S. Strand. 2008. Models of Governance for Converging Technologies. *Technol. Anal. Strateg.* **20**:83–97. [19]

Klein, D. J., P. König, and K. P. Körding. 2003. Sparse Spectrotemporal Coding of Sounds. *EURASIP J. Adv. Signal Process.* **2003**:902061. [4]

Kleinfeld, D., E. Ahissar, and M. E. Diamond. 2006. Active Sensation: Insights from the Rodent Vibrissa Sensorimotor System. *Curr. Opin. Neurobiol.* **16**:435–444. [4]

Kleinfeld, D., R. W. Berg, and S. M. O'Connor. 1999. Anatomical Loops and Their Electrical Dynamics in Relation to Whisking by Rat. *Somatosens. Motor Res.* **16**:69–88. [4]

Klossek, U. M. H., J. Russell, and A. Dickinson. 2008. The Control of Instrumental Action Following Outcome Devaluation in Young Children Aged between 1 and 4 Years. *J. Exp. Psychol. Gen.* **137**:39. [4]

Klyubin, A., D. Polani, and C. Nehaniv. 2005. Empowerment: A Universal Agent-Centric Measure of Control. In: Proc. IEEE Congress on Evolutionary Computation, vol. 1, pp. 128–135. Piscataway: IEEE. [1, 8]

Knoblich, G., S. Butterfill, and N. Sebanz. 2011. Psychological Research on Joint Action: Theory and Data. *Psychol. Learn. Motiv.* **54**:59–101. [20]

Knoblich, G., and N. Sebanz. 2008. Evolving Intentions for Social Interaction: From Entrainment to Joint Action. *Phil. Trans. R. Soc. B* **363**:2021–2031. [11]

Kobayashi, H., and S. Kohshima. 1997. Unique Morphology of the Human Eye. *Nature* **387**:767–768. [15]

Koch, C. 2012. Consciousness: Confessions of a Romantic Reductionist. Cambridge, MA: MIT Press. [14]

Koch, C., and S. Ullman. 1985. Shifts in Selective Visual Attention: Towards the Underlying Neural Circuitry. *Hum. Neurobiol.* **4**:219–227. [18]

Koch, S., and D. E. Leary, eds. 1985. A Century of Psychology as Science. Washington, D.C.: American Psychological Assoc. [14]

Koechlin, E., and C. Summerfield. 2007. An Information Theoretical Approach to Prefrontal Executive Function. *Trends Cogn. Sci.* **11**:229–235. [12]

Kohler, E., C. Keysers, M. A. Umilta, et al. 2002. Hearing Sounds, Understanding Actions: Action Representation in Mirror Neurons. *Science* **297**:846–848. [9]

Kohler, I. 1951. Über Aufbau und Wandlungen Der Wahrnehmungswelt. Insbesondere Über Bedingte Empfindungen. Vienna: Rohrer. [4]

Konczak, J., and J. Dichgans. 1997. The Development toward Stereotypic Arm Kinematics During Reaching in the First 3 Years of Life. *Exp. Brain Res.* **117**:346–354. [3]

König, P., and N. Krüger. 2006. Symbols as Self-Emergent Entities in an Optimization Process of Feature Extraction and Predictions. *Biol. Cybern.* **94**:325–334. [2, 4]

König, P., K. Kuhnberger, and T. C. Kietzmann. 2013. A Unifying Approach to High- and Low-Level Cognition. In: Models, Simulations, and the Reduction of Complexity, ed. U. v. Gähde et al., vol. 4, pp. 117–139. Berlin: De Gruyter. [4]

Körding, K. P., C. Kayser, W. Einhäuser, and P. König. 2004. How Are Complex Cell Properties Adapted to the Statistics of Natural Stimuli? *J. Neurophysiol.* **91**:206–212. [4]

Koster-Hale, J., R. Saxe, J. Dungan, and L. L. Young. 2013. Decoding Moral Judgments from Neural Representations of Intentions. *PNAS* **110**:5648–5653. [4]

Kouider, S., C. Stahlhut, S. V. Gelskov, et al. 2013. A Neural Marker of Perceptual Consciousness in Infants. *Science* **340**:376–380. [14]

Kousta, S.-T., G. Vigliocco, D. P. Vinson, M. Andrews, and E. D. Campo. 2011. The Representation of Abstract Words: Why Emotion Matters. *J. Exp. Psychol.* **140**:14–34. [4, 9]

Kovács, Á. M., E. Téglás, and A. D. Endress. 2010. The Social Sense: Susceptibility to Others' Beliefs in Human Infants and Adults. *Science* **330**:1830–1834. [4, 14]

Koziol, L. F., D. Budding, N. Andreasen, et al. 2014. Consensus Paper: The Cerebellum's Role in Movement and Cognition. *Cerebellum* **13**:151–177. [2]

Kraft, D., N. Pugeault, E. Başeski, et al. 2008. Birth of the Object: Detection of Objectness and Extraction of Object Shape through Object-Action Complexes. *Int. J. Hum. Robotics* **5**:247–265. [18]

Krakauer, M. F. Ghilardi, C. Ghez, and J. W. Krakauer. 1999. Independent Learning of Internal Models for Kinematic and Dynamic Control of Reaching. *Nat. Neurosci.* **2**:1026–1031. [3]

Krüger, V., D. Kragic, A. Ude, and C. Geib. 2007. The Meaning of Action: A Review on Action Recognition and Mapping. *Adv. Robotics* **21**:1473–1501. [1]

Kuhn, T. S. 1962/1970. The Structure of Scientific Revolutions. Chicago: Univ. of Chicago Press. [14, 19]

Kuhtz-Buschbeck, J. P., H. Stolze, K. Jöhnk, A. Boczek-Funcke, and M. Illert. 1998. Development of Prehension Movements in Children: A Kinematic Study. *Exp. Brain Res.* **122**:424–432. [3]

Kuniyoshi, Q., Y. Yorozu, M. Inaba, and H. Inoue. 2003. From-Visuo Motor Self Learning to Early Imitation. In: Proc. IEEE Intl. Conf. on Robotics and Automation, pp. 3132–3139. Taipei: IEEE. [18]

Kuypers, H. 1973. The Anatomical Organization of the Descending Pathways and Their Contribution to Motor Control Especially in Primates. In: New Developments in Electromyography and Clinical Neurophysiology, ed. J. E. Desmedt, vol. 3, pp. 38–68. New York: S. Karger. [18]

Kyselo, M. 2014. The Body Social: An Enactive Approach to the Self. *Front. Psychol.* **5**:986. [12, 15, 20]

Kyselo, M., and W. Tschacher. 2014. An Enactive and Dynamical Systems Theory Account of Dyadic Relationships. *Front. Psychol.* **5**:452. [20]

Lakoff, G. 1987. Women, Fire, and Dangerous Things. What Categories Reveal About the Mind. Chicago: Univ. of Chicago Press. [9]

Laland, K., J. Odling-Smee, and M. Feldman. 2000. Niche Construction, Biological Evolution, and Cultural Change. *Behav. Brain Sci.* **23**:131–175. [13]

Lamme, V. A. 2010. How Neuroscience Will Change Our View on Consciousness. *Cogn. Neurosci.* **1**:204–240. [15]

———. 2015a. The Crack of Dawn: Perceptual Functions and Neural Mechanisms That Mark the Transition from Uncosnscious Processing to Conscious Vision. In: Open Mind, ed. T. K. Metzinger and J. M. Windt. Frankfurt: MIND Group. [12]

———. 2015b. Predictive Coding Is Unconscious, So That Consciousness Happens Now. In: Open Mind, ed. T. K. Metzinger and J. M. Windt. Frankfurt: MIND Group. [12]

Lamme, V. A., and P. R. Roelfsema. 2000. The Distinct Modes of Vision Offered by Feedforward and Recurrent Processing. *Trends Neurosci.* **23**:571–579. [15]

Landauer, T. K. 1999. Latent Semantic Analysis (LSA), a Disembodied Learning Machine, Acquires Human Word Meaning Vicariously from Language Alone. *Behav. Brain Sci.* **22**:624–625. [9]

Langacker, R. W. 2008. Cognitive Grammar: A Basic Introduction: Oxford Univ. Press. [9]

Lau, H., and D. Rosenthal. 2011. Empirical Support for Higher-Order Theories of Conscious Awareness. *Trends Cogn. Sci.* **15**:365–373. [14]

Lau, H. C., R. D. Rogers, and R. E. Passingham. 2007. Manipulating the Experienced Onset of Intention after Action Execution. *J. Cogn. Neurosci.* **19**:81–90. [12]

Lauro, L. J., M. Tettamanti, S. F. Cappa, and C. Papagno. 2008. Idiom Comprehension: A Prefrontal Task? *Cereb. Cortex* **18**:162–170. [9]

Lavelle, J. S. 2012. Theory-Theory and the Direct Perception of Mental States. *Rev. Phil. Psych.* **3**:213–230. [6]

Lawson, R. P., G. Rees, and K. J. Friston. 2014. An Aberrant Precision Account of Autism. *Front. Hum. Neurosci.* **8**:302. [6]

Lee, T. S., and D. Mumford. 2003. Hierarchical Bayesian Inference in the Visual Cortex. *J. Opt. Soc. Am. A Opt. Image Sci. Vis.* **20**:1434–1448. [6]

Lemus, L., A. Hernandez, R. Luna, A. Zainos, and R. Romo. 2010. Do Sensory Cortices Process More Than One Sensory Modality During Perceptual Judgments? *Neuron* **67**:335–348. [12]

Lenggenhager, B., T. Tadi, T. Metzinger, and O. Blanke. 2007. Video Ergo Sum: Manipulating Bodily Self-Consciousness. *Science* **317**:1096. [7]

Leonard, H. C., R. Bedford, T. Charman, et al. 2013. Motor Development in Children at Risk of Autism: A Follow-up Study of Infant Siblings. *Autism* **18**:281–2891. [3]

Leonard, H. C., and W. L. Hill. 2014. Review: The Impact of Motor Development on Typical and Atypical Social Cognition and Language: A Systematic Review. *Child Adolesc. Ment. Health* **19**:163–170. [3]

Lepora, N., and G. Pezzulo. 2015. Embodied Choice: How Action Influences Perceptual Decision Making. *PLoS Comput. Biol.* **11**:e1004110. [2]

Leroy, F., H. Glasel, J. Dubois, et al. 2011. Early Maturation of the Linguistic Dorsal Pathway in Human Infants. *J. Neurosci.* **31**:1500–1506. [4]

Levine, G. M., J. B. Halberstadt, and R. L. Goldstone. 1996. Reasoning and the Weighting of Attributes in Attitude Judgments. *J. Pers. Soc. Psychol.* **70**:230–240. [12]

Levine, J. 1983. Materialism and Qualia: The Explanatory Gap. *Pac. Philos. Q.* **64**:354–361. [12, 14]

Levinson, S. C. 2013. Recursion in Pragmatics. *Language* **89**:149–162. [9]

Levinthal, D. J., and P. L. Strick. 2012. The Motor Cortex Communicates with the Kidney. *J. Neurosci.* **32**:6726–6731. [15]

Levy, R., and P. S. Goldman-Rakic. 2000. Segregation of Working Memory Functions within the Dorsolateral Prefrontal Cortex. *Exp. Brain Res.* **133**:23–32. [5]

Lewontin, R. 1982. Organism and Environment. In: Learning Development and Culture: Essays in Evolutionary Epistemology, ed. H. Plotkin, pp. 151–170. New York: Wiley. [13]

———. 1983. The Organism as the Subject and the Object of Evolution. In: The Dialectical Biologist, ed. R. Levins and R. Lewontin, pp. 85–106. Cambridge, MA: Harvard Univ. Press. [13]

Libertus, K., and A. Needham. 2010. Teach to Reach: The Effects of Active vs. Passive Reaching Experiences on Action and Perception. *Vision Res.* **50**:2750–2757. [3]

Libet, B. 1985. Unconscious Cerebral Initiative and the Role of Conscious Will in Voluntary Action. *Behav. Brain Sci.* **8**:529–566. [8, 14, 16]

Libet, B., C. A. Gleason, E. W. Wright, and D. K. Perl. 1983. Time of Conscious Intention to Act in Relation to Cerebral Activities (Readiness Potential): The Unconscious Initiation of a Freely Voluntary Act. *Brain Mind* **106**:623–642. [12, 16]

Lidow, M. S., P. S. Goldman-Rakic, D. W. Gallager, and P. Rakic. 1991. Distribution of Dopaminergic Receptors in the Primate Cerebral Cortex: Quantitative Autoradiographic Analysis Using [3H]Raclopride, [3H]Spiperone and [3H] SCH23390. *Neuroscience* **40**:657–671. [6]

Lillard, A. 1998. Ethnopsychologies: Cultural Variations in Theories of Mind. *Psychol. Bull.* **123**:3–32. [4]

Lindgren, R., and M. Johnson-Glenberg. 2013. Emboldened by Embodiment Six Precepts for Research on Embodied Learning and Mixed Reality. *Educ. Res.* **42**:445–452. [20]

Lindgren, R., and J. M. Moshell. 2011. Supporting Children's Learning with Body-Based Metaphors in a Mixed Reality Environment. In: Proc. 10th Intl. Conf. on Interaction Design and Children, ed. T. Moher et al., pp. 177–180. New York: ACM. [20]

Liszka, J. 1996. A General Introduction to the Semeiotic Papers of Charles Sanders Pierce. Bloomington: Indiana Univ. Press. [13]

Little, D., and F. Sommer. 2013. Learning and Exploration in Action-Perception Loops. *Front. Neur. Circuits* **7**:Article 37. [8]

Logan, G. D., and M. J. Crump. 2010. Cognitive Illusions of Authorship Reveal Hierarchical Error Detection in Skilled Typists. *Science* **330**:683–686. [12]

Lotto, A. J., G. S. Hickok, and L. L. Holt. 2009. Reflections on Mirror Neurons and Speech Perception. *Trends Cogn. Sci.* **13**:110–114. [9]

Lotze, R. H. 1852. Medicinische Psychologie Oder Physiologie Der Seele. Leipzig: Weidmannsche Buchhandlung. [7, 10, 17]

Lungarella, M., G. Metta, R. Pfeifer, and G. Sandini. 2003. Developmental Robotics: A Survey. *Connect. Sci.* **15**:151–190. [2, 18]

Lutz, C. 1985. Ethnopsychology Compared to What? Explaining Behavior and Consciousness among the Ifaluk. In: Person, Self, and Experience, ed. G. White and J. Kirkpatrick, pp. 35–79. Berkeley: Univ. of California Press. [4]

MacNeil, L. K., and S. H. Mostofsky. 2012. Specificity of Dyspraxia in Children with Autism. *Neuropsychology* **26**:165–171. [3]

Macrae, C. N., and G. V. Bodenhausen. 2000. Social Cognition: Thinking Categorically About Others. *Annu. Rev. Psychol.* **51**:93–120. [20]

Mahon, B. Z., and A. Caramazza. 2008. A Critical Look at the Embodied Cognition Hypothesis and a New Proposal for Grounding Conceptual Content. *J. Physiol. Paris* **102**:59–70. [9]

Malt, B. C. 1995. Category Coherence in Cross-Cultural Perspective. *Cogn. Psychol.* **29**:85–148. [5]

Mann, D. T., A. M. Williams, P. Ward, and C. M. Janelle. 2007. Perceptual-Cognitive Expertise in Sport: A Meta-Analysis. *J. Sport Exerc. Psychol.* **29**:457. [20]

Maravita, A., and A. Iriki. 2004. Tools for the Body (Schema). *Trends Cogn. Sci.* **8**:79–86. [4, 11]

Marcel, A. J. 1993. Slippage in the Unity of Consciousness. *CIBA Found. Symp.* **174**:168–180; discussion 180–166. [15]

Marcus, G. 2014. What Comes after the Turing Test? *The New Yorker*, October 12, 2015. [18]

Markman, A. B., and E. Dietrich. 2000a. Extending the Classical View of Representation. *Trends Cogn. Sci.* **4**:470–475. [5]

———. 2000b. In Defense of Representation. *Cogn. Psychol.* **40**:138–171. [5]

Marr, D. 1982. Vision : A Computational Investigation into the Human Representation and Processing of Visual Information. New York: W. H. Freeman. [1, 3, 20]

Martin, A. 2007. The Representation of Object Concepts in the Brain. *Annu. Rev. Psychol.* **58**:25–45. [1, 2, 5, 11]

Martin, A., C. L. Wiggs, L. G. Ungerleider, and J. V. Haxby. 1996. Neural Correlates of Category-Specific Knowledge. *Nature* **379**:649–652. [9]

Martin, J.-R., and E. Pacherie. 2013. Out of Nowhere: Thought Insertion, Ownership and Context-Integration. *Conscious. Cogn.* **22**:111–122. [7]

Martin, M. G. F. 2002. The Transparency of Experience. *Mind Lang.* **17**:376–425. [4]

Martius, G., R. Der, and N. Ay. 2013. Information Driven Self-Organization of Complex Robotic Behaviors. *PLoS One* **8**:e63400. [2]

Massaro, D. W. 1997. Perceiving Talking Faces : From Speech Perception to a Behavioral Principle. Cambridge, MA: MIT Press. [14]

Massimini, M., G. Tononi, and R. Huber. 2009. Slow Waves, Synaptic Plasticity and Information Processing: Insights from Transcranial Magnetic Stimulation and High-Density EEG Experiments. *Eur. J. Neurosci.* **29**:1761–1770. [14]

Mathews, Z., R. Cetnarski, and P. F. M. J. Verschure. 2015. Visual Anticipation Biases Conscious Perception but Not Bottom-up Visual Processing. *Front. Psychol.* **5**:1443. [14, 15]

Mathews, Z., and P. F. M. J. Verschure. 2011. PASAR-DAC7: An Integrated Model of Prediction, Anticipation, Sensation, Attention and Response for Artificial Sensorimotor Systems. *Inf. Sci.* **186**:1–19. [14]

Matthews, P. B. 1991. The Human Stretch Reflex and the Motor Cortex. *Trends Neurosci.* **14**:87–91. [3]

Maturana, H. R., and F. J. Varela. 1980. Autopoiesis and Cognition: The Realization of Living. Dordrecht: Reidel Publ. [2, 4]

Maye, A., and A. K. Engel. 2011. A Computational Model of Sensorimotor Contingencies for Object Perception and Control of Behavior. In: IEEE Intl. Conf. on Robotics and Automation (ICRA), pp. 3810–3815. Shanghai: IEEE. [11]

———. 2012. Time Scales of Sensorimotor Contingencies. In: Advances in Brain Inspired Cognitive Systems, ed. H. Zhang et al., pp. 240–249. Heidelberg: Springer. [1, 11]

———. 2013. Extending Sensorimotor Contingency Theory: Prediction, Planning, and Action Generation. *Adapt. Behav.* **21**:423–436. [1, 11]

McBeath, M. K., D. M. Shaffer, and M. K. Kaiser. 1995. How Baseball Outfielders Determine Where to Run to Catch Fly Balls. *Science* **28**:569–573. [16]

McCarthy, J., and P. J. Hayes. 1969. Some Philosophical Problems from the Standpoint of Artificial Intelligence. *Mach. Intell.* **4**:463–502. [14]

McGann, M. 2007. Enactive Theorists Do It on Purpose: Toward an Enactive Account of Goals and Goal-Directedness. *Phenom. Cogn. Sci.* **6**:463–483. [20]

———. 2010. Perceptual Modalities: Modes of Presentation or Modes of Interaction? *J. Conscious. Stud.* **17**:72–94. [15]

McGeer, T. 1990. Passive Dynamic Walking. *Int. J. Rob. Res.* **9**:62–82. [20]

McGeer, V. 2007. The Regulative Dimension of Folk Psychology. In: Folk Psychology Re-Assessed, ed. D. Hutto and M. Ratcliffe, pp. 137–156. Dordrecht: Springer. [12]

Mead, G. H. 1934. Mind, Self, & Society. Chicago: Univ. of Chicago Press. [14]

———. 1938. The Philosophy of the Act. Chicago: Univ. of Chicago Press. [1, 9, 11]

Melloni, L. 2015. Consciousness as Inference in Time: A Commentary on Victor Lamme. In: Open Mind, ed. T. K. Metzinger and J. M. Windt. Frankfurt: MIND Group. [12]

Melloni, L., C. M. Schwiedrzik, N. Muller, E. Rodriguez, and W. Singer. 2011. Expectations Change the Signatures and Timing of Electrophysiological Correlates of Perceptual Awareness. *J. Neurosci.* **31**:1386–1396. [15]

Memelink, J., and B. Hommel. 2013. Intentional Weighting: A Basic Principle in Cognitive Control. *Psychol. Res.* **77**:249–259. [10]

Menary, R. 2007. Cognitive Integration: Mind and Cognition Unbounded. London: Palgrave Macmillan. [13, 20]

———. 2009. Intentionality, Cognitive Integration and the Continuity Thesis. *Topoi* **28**:31–43. [13]

———. 2010. The Extended Mind and Cognitive Integration. In: The Extended Mind, ed. R. Menary, pp. 227–244. Cambridge, MA: MIT Press. [20]

———. 2012. Cognitive Practices and Cognitive Character. *Philos. Explor.* **15** 147–164. [13]

———. 2013. The Enculturated Hand. In: The Hand, an Organ of the Mind—What the Manual Tells the Mental, ed. Z. Radman, pp. 561–593. Cambridge, MA: MIT Press. [13]

———. 2014. Neural Plasticity, Neuronal Recycling and Niche Construction. *Mind Lang.* **29**:286–303. [13]

———. 2015. What? Now. Predictive Coding and Enculturation: A Reply to Regina E. Fabry. In: Open MIND: 25(R), ed. T. Metzinger and J. M. Wind. Frankfurt am Main: MIND Group. [13]

Menary, R., and M. Kirchhoff. 2014. Cognitive Transformations and Extended Expertise. *Educ. Philos. Theory* **46**:610–623. [13]

Merker, B. 2005. The Liabilities of Mobility: A Selection Pressure for the Transition to Consciousness in Animal Evolution. *Conscious. Cogn.* **14**:89–114. [14]

———. 2013. The Efference Cascade, Consciousness, and Its Self: Naturalizing the First Person Pivot of Action Control. *Front. Psychol.* **4**:501. [15]

Merleau-Ponty, M. 1962. Structure of Behavior. Boston: Beacon Press. [11]

———. 2012. Phenomenology of Perception, transl. D. Landes. London: Routledge. [16]

Merleau-Ponty, M., and J. M. Edie. 1964. The Primacy of Perception: And Other Essays on Phenomenological Psychology, the Philosophy of Art, History and Politics. Evanston: Northwestern Univ. Press. [14]

Meteyard, L., S. R. Cuadrado, B. Bahrami, and G. Vigliocco. 2012. Coming of Age: A Review of Embodiment and the Neuroscience of Semantics. *Cortex* **48**:788–804. [9, 16]

Metzinger, T., ed. 1995a. Conscious Experience. Thorverton: Imprint Academic. [12]

———. 1995b. The Problem of Consciousness. In: Conscious Experience, ed. T. Metzinger, pp. 3–37. Paderborn: Schöningh. [12]

———. 2003. Being No One: The Self-Model Theory of Subjectivity. Cambridge, MA: MIT Press. [12, 14]

———. 2006. Conscious Volition and Mental Representation: Toward a More Fine-Grained Analysis. In: Disorders of Volition, ed. N. Sebanz and W. Prinz. Cambridge, MA: Bradford Books: MIT Press. [15]

———. 2009. The Ego Tunnel: The Science of the Mind and the Myth of the Self. New York: Basic Books. [12]

———. 2010. The No-Self-Alternative. In: The Oxford Handbook of the Self. Oxford, ed. S. Gallagher pp. 279–296. Oxford: Oxford Univ. Press. [12]

———. 2013. The Myth of Cognitive Agency: Subpersonal Thinking as a Cyclically Recurring Loss of Mental Autonomy. *Front. Psychol.* **4**:931. [12]

———. 2015. M-Autonomy. *J. Conscious. Stud.* **22**:270–302. [12]

———. 2016. Suffering. In: The Return of Consciousness, ed. K. Almqvist and A. Haag. Stockholm: Axel and Margaret Ax:son Johnson Foundation. [12]

Miall, C., and D. M. Wolpert. 1996. Forward Models for Physiological Motor Control. *Neural Netw.* **9**:1265–1279. [3]

Michael, J., and L. De Bruin. 2015. How Direct Is Social Perception? *Conscious. Cogn.* **36**:373–375. [6]

Michael, J., K. Sandberg, J. Skewes, et al. 2014. Continuous Theta Burst Demonstrates a Causal Role of Premotor Homunculus in Action Interpretation. *Psychol. Sci.* **25**:963–972. [6]

Michel, E., M. Roethlisberger, R. Neuenschwander, and C. M. Roebers. 2011. Development of Cognitive Skills in Children with Motor Coordination Impairments at 12-Month Follow-Up. *Child Neuropsychol.* **17**:151–172. [3]

Miele, D. B., T. D. Wager, J. P. Mitchell, and J. Metcalfe. 2011. Dissociating Neural Correlates of Action Monitoring and Metacognition of Agency. *J. Cogn. Neurosci.* **23**:3620–3636. [15]

Millikan, R. 1984. Language, Thought, and Other Biological Categories. Cambridge, MA: Bradford Books/MIT Press. [13]

———. 1993. White Queen Psychology and Other Essays for Alice. Cambridge, MA: Bradford Books/MIT Press. [13]

Milner, A. D., and M. A. Goodale. 1995. The Visual Brain in Action. Oxford: Oxford Univ. Press. [5, 14]

———. 2008. Two Visual Systems Re-Viewed. *Neuropsychologia* **46**:774–785. [4]

Mineka, S., and M. Cook. 1993. Mechanisms Involved in the Observational Conditioning of Fear. *J. Exp. Psychol. Gen.* **122**:23. [4]

Mirolli, M., and G. Baldassarre. 2013. Functions and Mechanisms of Intrinsic Motivations. In: Intrinsically Motivated Learning in Natural and Artificial Systems, ed. G. Baldassarre and M. Mirolli, pp. 49–72. Heidelberg: Springer. [4]

Mischel, W. 1968. Personality and Assessment. New York: Wiley. [5]

Montesano, L., M. Lopes, A. Bernardino, and J. Santos-Victor. 2008. Learning Object Affordances: From Sensory-Motor Coordination to Imitation. *IEEE Trans. Robot.* **24**:15–26. [18]

Moore, D. G., and R. George. 2011. ACORNS: A Tool for the Visualisation and Modelling of Atypical Development. *J. Intell. Disab. Res.* **55**:956–972. [3]

Moore, J., and P. Haggard. 2008. Awareness of Action: Inference and Prediction. *Conscious. Cogn.* **17**:136–144. [12]

Moore, T., and M. Fallah. 2001. Control of Eye Movements and Spatial Attention. *PNAS* **98**:1273–1276. [20]

Moran, J. M., L. L. Young, R. Saxe, et al. 2011. Impaired Theory of Mind for Moral Judgment in High-Functioning Autism. *PNAS* **108**:2688–2692. [4]

Moran, R. J., P. Campo, M. Symmonds, et al. 2013. Free Energy, Precision and Learning: The Role of Cholinergic Neuromodulation. *J. Neurosci.* **33**:8227–8236. [6]

Moravec, H. 1983. The Stanford Cart and the CMU Rover. *IEEE Proc.* **71**:872–884. [20]

———. 1988. Mind Children: The Future of Robot and Human Intelligence. Cambridge, MA: Harvard Univ. Press. [18]

Moren, J., A. Ude, A. Koene, and G. Cheng. 2008. Biologically-Based Top-Down Attention Modulation for Humanoid Interactions. *Int. J. Hum. Robotics* **5**:3–24. [18]

Morlino, G., C. Gianelli, A. M. Borghi, and S. Nolfi. 2015. Learning to Manipulate and Categorize in Human and Artificial Agents *Cogn. Sci.* **39**:39–64. [2]

Morsella, E. 2005. The Function of Phenomenal States: Supramodular Interaction Theory. *Psychol. Rev.* **112**:1000–1021. [8]

Morsella, E., J. A. Bargh, and P. M. Gollwitzer, eds. 2009. Oxford Handbook of Human Action. Oxford: Oxford Univ. Press. [17]

Morton, J. M. 2004. Understanding Developmental Disorders: A Causal Modelling Approach. Oxford: Blackwell. [3]

Moseley, R., F. Carota, O. Hauk, B. Mohr, and F. Pulvermüller. 2012. A Role for the Motor System in Binding Abstract Emotional Meaning. *Cereb. Cortex* **22**:1634–1647. [4, 9]

Moseley, R. L., B. Mohr, M. V. Lombardo, et al. 2013. Brain and Behavioural Correlates of Action Semantic Deficits in Autism. *Front. Hum. Neurosci.* **7**:725 [9]

Moseley, R. L., Y. Shtyrov, B. Mohr, et al. 2015. Lost for Emotion Words: What Motor and Limbic Brain Activity Reveals About Autism and Semantic Theory. *NeuroImage* **104**:413–422. [9]

Mostofsky, S. H., P. Dubey, V. K. Jerath, et al. 2006. Developmental Dyspraxia Is Not Limited to Imitation in Children with Autism Spectrum Disorders. *J. Int. Neuropsychol. Soc.* **12**:2–3. [3]

Möttönen, R., R. Dutton, and K. E. Watkins. 2013. Auditory-Motor Processing of Speech Sounds. *Cereb. Cortex* **23**:1190–1197. [9]

Moulin-Frier, C., S. M. Nguyen, and P.-Y. Oudeyer. 2014. Self-Organization of Early Vocal Development in Infants and Machines: The Role of Intrinsic Motivation. *Front. Psychol.* **4**:1006. [4]

Mumford, D. 1992. On the Computational Architecture of the Neocortex. II. *Biol. Cybern.* **66**:241–251. [6]

Mussa-Ivaldi, F. A., and E. Bizzi. 2000. Motor Learning through the Combination of Primitives. *Phil. Trans. R. Soc. B* **355**:1755–1769. [15]

Nagel, S. K. 2010. Too Much of a Good Thing? Enhancement and the Burden of Self-Determination. *Neuroethics* **3**:109–119. [4]

Nagel, S. K., C. Carl, T. Kringe, R. Märtin, and P. König. 2005. Beyond Sensory Substitution: Learning the Sixth Sense. *J. Neural Eng.* **2**:R13. [4, 19]

Nagel, T. 1974. What Is It Like to Be a Bat? *Philos. Rev.* 435–450. [12, 14]

———. 1986. The View from Nowhere. New York: Oxford Univ. Press. [12]

Nahmias, E., S. Morris, T. Nadelhoffer, and J. Turner. 2005. Surveying Freedom: Folk Intuitions About Free Will and Moral Responsibility. *Philos. Psychol.* **18**:561–584. [12]

Nair, V., and E. H. Geoffrey. 2006. Inferring Motor Programs from Images of Handwritten Digits. In: Proc. of the Neural Information Processing Systems (NIPS 2005), ed. Y. Weiss et al., pp. 515–522. Cambridge, MA: MIT Press. [10]

Nardini, M., R. Bedford, and D. Mareschal. 2010. Fusion of Visual Cues Is Not Mandatory in Children. *PNAS* **107**:17041–17046. [3]

Natale, L., F. Orabona, G. Metta, and G. Sandini. 2007. Sensorimotor Coordination in a "Baby" Robot: Learning About Objects through Grasping. *Progr. Brain. Res.* **164**:403–424. [11]

Neisser, U. 1967. Cognitive Psychology. New York: Appleton-Century-Crofts. [17]

Neumann, O., and W. Prinz, eds. 1990. Relationships between Perception and Action: Current Approaches. Berlin: Springer. [17]

Newell, A. 1990. Unified Theories of Cognition. Cambridge, MA: Harvard Univ. Press. [14]

Newell, A., and H. A. Simon. 1963. GPS, a Program That Simulates Human Thought. In: Computers and Thought, ed. J. F. E. A. Feigenbaum. New York: Mc Graw-Hill. [14]

———. 1972. Human Problem Solving. Englewood Cliffs, NJ: Prentice-Hall. [1]

Newen, A., and A. Bartels. 2007. Animal Minds and the Possession of Concepts. *Philos. Psychol.* **20**:283–308. [4]

Niell, C. M., and M. P. Stryker. 2010. Modulation of Visual Responses by Behavioral State in Mouse Visual Cortex. *Neuron* **65**:472–479. [11]

Nieuwenstein, M. R., T. Wierenga, R. D. Morey, et al. 2015. On Making the Right Choice: A Meta-Analysis and Large-Scale Replication Attempt of the Unconscious Thought Advantage. *Judgm. Decis. Mak.* **10**:1–17. [14]

Nisbett, R. E., and T. D. Wilson. 1977. Telling More Than We Can Know: Verbal Reports on Mental Processes. *Psychol. Rev.* **84**:231–259. [12]

Noë, A. 2004. Action in Perception. Cambridge, MA: MIT Press. [1, 7, 11, 13, 16, 17, 20]

———. 2006. Experience without the Head. In: Perceptual Experience, ed. T. Gendler and A. Hawthorne, pp. 411–434. New York: Clarendon/Oxford Univ. Press. [6]

———. 2009. Out of Our Heads. New York: Hill & Wang. [1]

Nolfi, S. 2009. Behavior and Cognition as a Complex Adaptive System: Insights from Robotic Experiments. In: Handbook of the Philosophy of Science, ed. C. Hooker, vol. 10. Amsterdam: Elsevier. [2]

Nolfi, S., and D. Floreano. 2001. Evolutionary Robotics. The Biology, Intelligence, and Technology of Self-Organizing Machines. Cambridge, MA: MIT Press. [2]

Nordmann, A. 2004. Converging Technologies: Shaping the Future of European Societies. Luxembourg: European Comm. [19]

Norman, D. A., and T. Shallice. 1986. Attention to Action: Willed and Automatic Control of Behavior. In: Consciousness and Self Regulation: Advances in Research, ed. R. J. Davidson et al., vol. 4, pp. 1–18. New York: Plenum Press. [12]

Northoff, G., A. Heinzel, M. de Greck, et al. 2006. Self-Referential Processing in Our Brain: A Meta-Analysis of Imaging Studies on the Self. *NeuroImage* **31**:440–457. [5]

Nowotny, H. 1999. The Place of People in Our Knowledge. *Eur. Rev.* **7**: 247–262. [19]

Nowotny, H., P. Scott, and M. Gibbons. 2001. Re-Thinking Science. Knowledge and the Public in an Age of Uncertainty. Cambridge: Polity Press. [19]

———. 2005. Re-Thinking Science: Mode 2 in Societal Context. In: Knowledge Creation, Diffusion, and Use in Innovation Networks and Knowledge Clusters, ed. E. Carayannis and D. Campbell, pp. 39–51. Westport: Praeger. [19]

Odling-Smee, F. J., K. N. Laland, and M. W. Feldman. 2003. Niche Construction: The Neglected Process in Evolution. Princeton: Princeton Univ. Press. [13]

Olshausen, B. A., and D. J. Field. 1996. Emergence of Simple-Cell Receptive Field Properties by Learning a Sparse Code for Natural Images. *Nature* **381**:607–609. [4]

O'Regan, J. K. 2011. Why Red Doesn't Sound Like a Bell: Understanding the Feel of Consciousness. New York: Oxford Univ. Press. [1, 13–15, 17]

O'Regan, J. K., and A. Noë. 2001. A Sensorimotor Account of Vision and Visual Consciousness. *Behav. Brain Sci.* **24**:939–973; discussion 973–1031. [1, 2, 4–6, 8, 9, 11, 13–16, 18–20]

Osei-Bryson, K., and O. Ngwenyama. 2011. Using Decision Tree Modelling to Support Peircian Abduction in Is Research: A Systematic Approach for Generating and Evaluating Hypotheses for Systematic Theory Development. *Inf. Sys. J.* **21**:407–440. [13]

Oudeyer, P.-Y. 2010. On the Impact of Robotics in Behavioral and Cognitive Sciences: From Insect Navigation to Human Cognitive Development. *IEEE Trans. Auton. Ment. Dev.* **2**:2–16. [4]

Oudeyer, P.-Y., A. Baranes, and F. Kaplan. 2013. Intrinsically Motivated Learning of Real-World Sensorimotor Skills with Developmental Constraints. In: Intrinsically Motivated Learning in Natural and Artificial Systems, ed. G. Baldassarre and M. Mirolli, pp. 303–365. New York: Springer. [4]

Oudeyer, P.-Y., F. Kaplan, V. V. Hafner, and A. Whyte. 2005. The Playground Experiment: Task-Independent Development of a Curious Robot. In: AAAI Spring Symp. on Developmental Robotics, ed. D. Bank and L. Meeden, pp. 42–47. Stanford: AAAI. [2]

Oudeyer, P.-Y., and L. Smith. 2016. How Evolution May Work through Curiosity-Driven Developmental Process. *Top. Cogn. Sci.*, in press. [4]

Ozçalışkan, S., and S. Goldin-Meadow. 2009 When Gesture-Speech Combinations Do and Do Not Index Linguistic Change. *Lang. Cogn. Process.* **24**:190–217. [10]

Özçalışkan, Ş., S. Goldin-Meadow, D. Gentner, and C. Mylander. 2009. Does Language About Similarity Foster Children's Similarity Comparisons? *Cognition* **112**:217–228. [20]

Oztop, E., D. Wolpert, and M. Kawato. 2005. Mental State Inference Using Visual Control Parameters. *Cogn. Brain Res.* **22**:129–151. [18]

Ozyurek, A. 2014. Hearing and Seeing Meaning in Speech and Gesture: Insights from Brain and Behaviour. *Phil. Trans. R. Soc. B* **369**:1651. [10]

Pacherie, E. 2008. The Phenomenology of Action: A Conceptual Framework. *Cognition* **107**:179–217. [15]

———. 2014. Can Conscious Agency Be Saved? *Topoi* **33**:33–45. [7]

Paivio, A. 2007. Mind and Its Evolution: A Dual Coding Theoretical Approach. Mahwah, NJ: Lawrence Erlbaum [4]

Palmer, C. J., B. Paton, J. Hohwy, and P. G. Enticott. 2013. Movement under Uncertainty: The Effects of the Rubber-Hand Illusion Vary Along the Nonclinical Autism Spectrum. *Neuropsychologia* **51**:1942–1951. [7]

Panksepp, J. 2005. Affective Consciousness: Core Emotional Feelings in Animals and Humans. *Conscious. Cogn.* **14**:30–80. [15]

Pascolo, P. B., and A. Cattarinussi. 2012. On the Relationship between Mouth Opening and "Broken Mirror Neurons" in Autistic Individuals. *J. Electromyogr. Kinesiol.* **22**:98–102. [3]

Pascual-Leone, A., and V. Walsh. 2001. Fast Backprojections from the Motion to the Primary Visual Area Necessary for Visual Awareness. *Science* **292**:510–512. [15]

Pastor, P., H. Hoffmann, T. Asfour, and S. Schaal. 2009. Learning and Generalization of Motor Skills by Learning from Demonstration. IEEE Intl. Conf. on Robotics and Automation (ICRA), pp. 763–768. Kobe: IEEE. [18]

Pastor, P., L. Righetti, M. Kalakrishnan, and S. Schaal. 2011. Online Movement Adaptation Based on Previous Sensor Experiences. Proc. IEEE/RSJ Intl. Conf. on Intelligent Robots and Systems (IROS), pp. 365–371. San Francisco: IEEE. [18]

Paton, B., J. Hohwy, and P. Enticott. 2011. The Rubber Hand Illusion Reveals Proprioceptive and Sensorimotor Differences in Autism Spectrum Disorders. *J. Autism Dev. Disord.* **42**:1870–1883. [7]

Pavlov, I. P. 1927. Conditioned Reflexes: An Investigation of the Physiological Activity of the Cerebral Cortex. London: Oxford Univ. Press. [14]

Pearl, J. 2000. Causality. Cambridge: Cambridge Univ. Press. [7]

———. 2009. Causal Inference in Statistics: An Overview. *Stat. Surv.* **3**:96–146. [18]

Peirce, C. S. 1887. Logical Machines. Modern Logic. *Am. J. Psychol.* **1**:165–170. [20]

———. 1931. Collected Papers of Charles Sanders Peirce: Vols. 1–6 (1931–1935), ed. Charles Hartshorne and Paul Weiss; Vols. 7–8 (1958), ed. Arthur W. Burks. Cambridge, MA: Harvard Univ. Press. [1, 9, 13]

Pellicano, E., and D. Burr. 2012. When the World Becomes Too Real: A Bayesian Explanation of Autistic Perception. *Trends Cogn. Sci.* **16**:504–510. [6]

Pezzulo, G. 2011. Grounding Procedural and Declarative Knowledge in Sensorimotor Anticipation. *Mind Lang.* **26**:78–114. [2, 4]

———. 2014. Why Do You Fear the Bogeyman? An Embodied Predictive Coding Model of Perceptual Inference. *Cogn. Affect. Behav. Neurosci.* **14**:902–911. [2]

Pezzulo, G., L. Barca, A. L. Bocconi, and A. M. Borghi. 2010. When Affordances Climb into Your Mind: Advantages of Motor Simulation in a Memory Task Performed by Novice and Expert Rock Climbers. *Brain Cogn.* **73**:68–73. [2]

Pezzulo, G., L. W. Barsalou, A. Cangelosi, et al. 2011. The Mechanics of Embodiment: A Dialogue on Embodiment and Computational Modeling. *Front. Psychol.* **2**:1–21. [2, 4, 16]

Pezzulo, G., M. Candidi, H. Dindo, and L. Barca. 2013. Action Simulation in the Human Brain: Twelve Questions. *New Ideas Psychol.* **31**:270–290. [2]

Pezzulo, G., and C. Castelfranchi. 2007. The Symbol Detachment Problem. *Cogn. Process.* **8**:115–131. [4]

———. 2009. Thinking as the Control of Imagination: A Conceptual Framework for Goal-Directed Systems. *Psychol. Res.* **73**:559–577. [2, 4]

Pezzulo, G., and H. Dindo. 2011. What Should I Do Next? Using Shared Representations to Solve Interaction Problems. *Exp. Brain Res.* **211**:613–630. [2, 4]

Pezzulo, G., F. Donnarumma, and H. Dindo. 2013. Human Sensorimotor Communication: A Theory of Signaling in Online Social Interactions. *PLoS One* **8**:e79876. [2]

Pezzulo, G., D. Rigoli, and K. Friston. 2015. Active Inference, Homeostatic Regulation and Adaptive Behavioural Control. *Prog. Neurobiol.* **134**:17–35. [2]

Pezzulo, G., M. A. van der Meer, C. S. Lansink, and C. Pennartz. 2014. Internally Generated Sequences in Learning and Executing Goal-Directed Behavior. *Trends Cogn. Sci.* **18**:647–657. [2]

Pezzulo, G., P. F. M. J. Verschure, C. Balkenius, and C. M. A. Pennartz. 2014. The Principles of Goal-Directed Decision-Making: From Neural Mechanisms to Computation and Robotics. *Phil. Trans. R. Soc. B* **369**:20130470. [2]

Pfeifer, R., and J. C. Bongard. 2006. How the Body Shapes the Way We Think: A New View of Intelligence. Cambridge, MA: MIT Press. [1, 11, 14, 19]

Pfeifer, R., M. Lungarella, and F. Iida. 2012. The Challenges Ahead for Bio-Inspired "Soft" Robotics. *Commun. ACM* **55**:76–87. [20]

Pfeifer, R., and C. Scheier. 1999. Understanding Intelligence. Cambridge, MA: MIT Press. [2, 14, 18, 20]

Piaget, J. 1952. The Origins of Intelligence in Children. New York: International Universities Press. [4]

———. 1954. The Construction of Reality in the Child. New York: Ballentine. [2, 17]

Pickering, M. J., and A. Clark. 2014. Getting Ahead: Forward Models and Their Place in Cognitive Architecture. *Trends Cogn. Sci.* **18**:451–456. [3]

Pickering, M. J., and S. Garrod. 2007. Do People Use Language Production to Make Predictions During Comprehension? *Trends Cogn. Sci.* **11**:105–110. [1]

———. 2013a. How Tightly Are Production and Comprehension Interwoven? *Front. Psychol.* **4**:238. [10]

———. 2013b. An Integrated Theory of Language Production and Comprehension. *Behav. Brain Sci.* **36**:329–347. [4, 9, 20]

Plenz, D., and T. C. Thiagarajan. 2007. The Organizing Principles of Neuronal Avalanches: Cell Assemblies in the Cortex? *Trends Neurosci.* **30**:101–110. [9]

Polanyi, M. 1958. Personal Knowledge: Towards a Post-Critical Philosophy. London: Routledge & Kegan Paul. [19]

Popper, K. 1935/2005. The Logic of Scientific Discovery. London: Taylor & Francis. [19]

Postle, N., K. L. McMahon, R. Ashton, M. Meredith, and G. I. de Zubicaray. 2008. Action Word Meaning Representations in Cytoarchitectonically Defined Primary and Premotor Cortices. *NeuroImage* **43**:634–644. [9]

Pothos, E. M., and A. J. Wills. 2011. Formal Approaches in Categorization. Cambridge: Cambridge Univ. Press. [5]

Power, M. 1997. The Audit Society: Rituals of Verification. Oxford: Oxford Univ. Press. [19]

Pratt, M. L., H. C. Leonard, H. Adeyinka, and E. L. Hill. 2014. The Effect of Motor Load on Planning and Inhibition in Developmental Coordination Disorder. *Res. Dev. Disabil.* **35**:1579–1587. [3]

Prescott, T. J. 2015. The Me in the Machine. *New Sci.* **3013**:36–39. [19]

Prescott, T. J., N. Lepora, and P. F. M. J. Verschure. 2014. A Future of Living Machines? International Trends and Prospects in Biomimetic and Biohybrid Systems. Proc. SPIE 9055, Bioinspiration, Biomimetics, and Bioreplication, http://dx.doi.org/10.1117/12.2046305. (accessed Oct. 12, 2015). [19]

Prescott, T. J., M. J. Pearson, B. Mitchinson, J. C. W. Sullivan, and A. G. Pipe. 2009. Whisking with Robots from Rat Vibrissae to Biomimetic Technology for Active Touch. *IEEE Robot. Autom. Mag.* **16**:42–50. [4]

Price, H. 2002. Boltzmann's Time Bomb. *Br. J. Philos. Sci.* **53**:83–119. [14]

Prinz, J. J., and L. W. Barsalou. 2000. Steering a Course for Embodied Representation. In: Cognitive Dynamics: Conceptual and Representational Change in Humans and Machines, ed. E. Dietrich and A. B. Markman, pp. 51–77. Cambridge, MA: Cambridge Univ. Press. [5]

Prinz, W. 1997. Perception and Action Planning. *Eur. J. Cogn. Psychol.* **9**:129–154. [5]

———. 2012. Open Minds: The Social Making of Agency and Intentionality. Cambridge, MA: MIT Press. [17, 20]

Prinz, W., G. Aschersleben, and I. Koch. 2009. Cognition and Action. In: Oxford Handbook of Human Action, ed. E. Morsella et al., pp. 35–71. Oxford: Oxford Univ. Press. [17]

Prinz, W., M. Beisert, and A. Herwig, eds. 2013. Action Science: Foundations of an Emerging Discipline. Cambridge, MA: MIT Press. [17]

Prinz, W., and B. Hommel, eds. 2002. Common Mechanisms in Perception and Action: Attention and Performance XIX. Oxford: Oxford Univ. Press. [17]

Psillos, S. 2000. Abduction: Between Conceptual Richness and Computational Complexity. In: Abduction and Induction, ed. P. A. Flach and A. C. Kakas, pp. 50–74. Dordrecht: Springer Science+Business Media. [13]

Pulvermüller, F. 1999. Words in the Brain's Language. *Behav. Brain Sci.* **22**:253–336. [9]

———. 2001. Brain Reflections of Words and Their Meaning. *Trends Cogn. Sci.* **5**:517–524. [9]

———. 2002a. A Brain Perspective on Language Mechanisms: From Discrete Neuronal Ensembles to Serial Order. *Prog. Neurobiol.* **67**:85–111. [9]

———. 2002b. The Neuroscience of Language. Cambridge: Cambridge Univ. Press. [9]

———. 2005. Brain Mechanisms Linking Language and Action. *Nat. Rev. Neurosci.* **6**:576–582. [2, 4, 9, 15]

———. 2010. Brain Embodiment of Syntax and Grammar: Discrete Combinatorial Mechanisms Spelt out in Neuronal Circuits. *Brain Lang.* **112**:167–179. [9]

———. 2013. How Neurons Make Meaning: Brain Mechanisms for Embodied and Abstract-Symbolic Semantics. *Trends Cogn. Sci.* **17**:458–470. [9]

Pulvermüller, F., C. Cook, and O. Hauk. 2012. Inflection in Action: Semantic Motor System Activation to Noun- and Verb-Containing Phrases Is Modulated by the Presence of Overt Grammatical Markers. *NeuroImage* **60**:1367–1379. [9]

Pulvermüller, F., and L. Fadiga. 2010. Active Perception: Sensorimotor Circuits as a Cortical Basis for Language. *Nat. Rev. Neurosci.* **11**:351–360. [1, 4, 9, 10, 14]

Pulvermüller, F., and M. Garagnani. 2014. From Sensorimotor Learning to Memory Cells in Prefrontal and Temporal Association Cortex: A Neurocomputational Study of Disembodiment *Cortex* **57**:1–21. [9, 10]

Pulvermüller, F., O. Hauk, V. V. Nikulin, and R. J. Ilmoniemi. 2005. Functional Links between Motor and Language Systems. *Eur. J. Neurosci.* **21**:793–797. [9, 15]

Pulvermüller, F., F. Hummel, and M. Härle. 2001. Walking or Talking? Behavioral and Neurophysiological Correlates of Action Verb Processing. *Brain Lang.* **78**:143–168. [9]

Pulvermüller, F., M. Huss, F. Kherif, et al. 2006. Motor Cortex Maps Articulatory Features of Speech Sounds. *PNAS* **103**:7865–7870. [9]

Pulvermüller, F., J. Kiff, and Y. Shtyrov. 2012. Can Language-Action Links Explain Language Laterality? An ERP Study of Perceptual and Articulatory Learning of Novel Pseudowords. *Cortex* **48**:471–481. [9]

Pulvermüller, F., and A. Knoblauch. 2009. Discrete Combinatorial Circuits Emerging in Neural Networks: A Mechanism for Rules of Grammar in the Human Brain? *Neural Netw.* **22**:161–172. [9]

Pulvermüller, F., R. L. Moseley, N. Egorova, Z. Shebani, and V. Boulenger. 2014. Motor Cognition–Motor Semantics: Action Perception Theory of Cognition and Communication. *Neuropsychologia* **55**:71–84. [2, 9, 17]

Pulvermüller, F., Y. Shtyrov, and R. J. Ilmoniemi. 2003. Spatio-Temporal Patterns of Neural Language Processing: An MEG Study Using Minimum-Norm Current Estimates. *NeuroImage* **20**:1020–1025. [9]

———. 2005. Brain Signatures of Meaning Access in Action Word Recognition. *J. Cogn. Neurosci.* **17**:884–892. [9]

Putnam, H. 1960. Minds and Machines. In: Dimensions of Mind, ed. S. Hook, pp. 148–179. New York: New York Univ. Press. [14]

Pylyshyn, Z. W. 1973. What the Mind's Eye Tells the Mind's Brain: A Critique of Mental Imagery. *Psychol. Bull.* **80**:1–24. [5]

———. 2001. Seeing, Acting, and Knowing. *Behav. Brain Sci.* **24**:999. [5]

Quinton, J. C., N. Catenacci Volpi, L. Barca, and G. Pezzulo. 2013. The Cat Is on the Mat. Or Is It a Dog? Dynamic Competition in Perceptual Decision Making. *IEEE Trans. Syst. Man Cybern. A Syst. Hum.* **44**:539–551. [2]

Qureshi, A. W., I. A. Apperly, and D. Samson. 2010. Executive Function Is Necessary for Perspective Selection, Not Level-1 Visual Perspective Calculation: Evidence from a Dual-Task Study of Adults. *Cognition* **117**:230–236. [12]

Ramsey, W. M. 2007. Representation Reconsidered. Cambridge: Cambridge Univ. Press. [5]

Rao, R. P., and D. H. Ballard. 1999. Predictive Coding in the Visual Cortex: A Functional Interpretation of Some Extra-Classical Receptive-Field Effects. *Nat. Neurosci.* **2**:79–87. [1, 6, 14]

Rao, R. P. N., A. Shon, and A. Meltzoff. 2007. A Bayesian Model of Imitation in Infants and Robots. In: Imitation and Social Learning in Robots, Humans, and Animals, ed. C. L. Nehaniv and K. Dautenhahn, pp. 217–247. New York: Cambridge Univ. Press. [18]

Raposo, A., H. E. Moss, E. A. Stamatakis, and L. K. Tyler. 2009. Modulation of Motor and Premotor Cortices by Actions, Action Words and Action Sentences. *Neuropsychologia* **47**:388–396. [9]

Rasolzadeh, B., M. Björkman, K. Huebner, and D. Kragic. 2010. An Active Vision System for Detecting, Fixating and Manipulating Objects in the Real World. *Int. J. Rob. Res.* **29**:133–154. [18]

Reb, J., and T. Connolly. 2009. Myopic Regret Avoidance: Feedback Avoidance and Learning in Repeated Decision Making. *Organ. Behav. Hum. Decis. Process.* **109**:182–189. [12, 15]

Reddy, V. 2008. How Infants Know Minds. Cambridge, MA: Harvard Univ. Press. [3]

Rescorla, R. A., and A. R. Wagner. 1972. A Theory of Pavlovian Conditioning: Variations in the Effectiveness of Reinforcement and Nonreinforcement. In: Classical Conditioning II, Current Theory and Research, ed. A. H. Black and W. F. Prokasy, pp. 64–99. New York: Appleton-Century-Crofts. [14]

Revonsuo, A. 1995. Consciousness, Dreams and Virtual Realities. *Philos. Psychol.* **8**:35–58. [14]

———. 2006. Inner Presence: Consciousness as a Biological Phenomenon. Cambridge, MA: MIT Press. [14]

Reynolds, R. F., and A. M. Bronstein. 2003. The Broken Escalator Phenomenon. Aftereffect of Walking onto a Moving Platform. *Exp. Brain Res.* **151**:301–308. [15]

Richardson, M. J., K. Shockley, B. R. Fajen, M. A. Riley, and M. T. Turvey. 2008. Ecological Psychology: Six Principles for an Embodied-Embedded Approach to Behavior. In: Handbook of Cognitive Science: An Embodied Approach, ed. P. Calvo and T. Gomila, pp. 161–188. Amsterdam: Elsevier. [2]

Riečanský, I., N. Paul, S. Kölble, S. Stieger, and C. Lamm. 2014. Beta Oscillations Reveal Ethnicity Ingroup Bias in Sensorimotor Resonance to Pain of Others. *Soc. Cogn. Affect. Neurosci.* **10**:893–901. [4]

Rigato, S., J. Begum Ali, J. L. J. van Velzen, and A. J. Bremner. 2014. The Neural Basis of Somatosensory Remapping Develops in Human Infancy. *Curr. Biol.* **24**:1222–1226. [3]

Righetti, L., M. Kalakrishnan, P. Pastor, et al. 2014. An Autonomous Manipulation System Based on Force Control and Optimization. *Auton. Robots* **36**:11–30. [20]

Rigoli, D., J. P. Piek, R. Kane, and J. Oosterlaan. 2012a. An Examination of the Relationship between Motor Coordination and Executive Functions in Adolescents. *Dev. Med. Child Neurol.* **54**:1025–1031. [3]

———. 2012b. Motor Coordination, Working Memory, and Academic Achievement in a Normative Adolescent Sample: Testing a Mediation Model. *Arch. Clin. Neuropsychol.* **27**:766–780. [3]

Rigoni, D., S. Kühn, G. Sartori, and M. Brass. 2011. Inducing Disbelief in Free Will Alters Brain Correlates of Preconscious Motor Preparation. *Psychol. Sci.* [12]

Rigoni, D., H. Wilquin, M. Brass, and B. Burle. 2013. When Errors Do Not Matter: Weakening Belief in Intentional Control Impairs Cognitive Reaction to Errors. *Cognition* **127**:264–269. [12]

Riley, M. A., M. J. Richardson, K. Shockley, and V. C. Ramenzoni. 2011. Interpersonal Synergies. *Front. Psychol.* **2**:38. [20]

Rilling, J. K., M. F. Glasser, S. Jbabdi, J. Andersson, and T. M. Preuss. 2011. Continuity, Divergence, and the Evolution of Brain Language Pathways. *Front. Evol. Neurosci.* **3**:11. [9]

Rilling, J. K., M. F. Glasser, T. M. Preuss, et al. 2008. The Evolution of the Arcuate Fasciculus Revealed with Comparative DTI. *Nat. Neurosci.* **11**:426–428. [9]

Riva, G. 2000. From Telehealth to E-Health: Internet and Distributed Virtual Reality in Health Care. *Cyberpsychol. Behav.* **3**:989–998. [19]

———. 2011. The Key to Unlocking the Virtual Body: Virtual Reality in the Treatment of Obesity and Eating Disorders. *J. Diabetes Sci. Technol.* **5**:283–292. [19]

Rivlin, E., S. J. Dickinson, and A. Rosenfeld. 1995. Recognition by Functional Parts. *Comput. Vis. Image Underst.* **62**:164–176. [18]

Rizzolatti, G., and L. Craighero. 2004. The Mirror-Neuron System. *Annu. Rev. Neurosci.* **27**:169–192. [1, 2, 4, 9, 11, 20]

Rizzolatti, G., M. Fabbri-Destro, and L. Cattaneo. 2009. Mirror Neurons and Their Clinical Relevance. *Nat. Clin. Pract. Neurol.* **5**:24–34. [9]

Rizzolatti, G., L. Fadiga, V. Gallese, and L. Fogassi. 1996. Premotor Cortex and the Recognition of Motor Actions. *Cogn. Brain Res.* **3**:131–141. [9]

Rizzolatti, G., L. Fogassi, and V. Gallese. 2002. Motor and Cognitive Functions of the Ventral Premotor Cortex. *Curr. Opin. Neurobiol.* **12**:149–154. [1]

Rizzolatti, G., L. Riggio, I. Dascola, and C. Umiltá. 1987. Reorienting Attention across the Horizontal and Vertical Meridians: Evidence in Favor of a Premotor Theory of Attention. *Neuropsychologia* **25**:31–40. [20]

Rizzolatti, G., and C. Sinigaglia. 2008. Mirrors in the Brain: How Our Minds Share Actions and Emotions, transl. F. Anderson. Oxford: Oxford Univ. Press. [17]

———. 2010. The Functional Role of the Parieto-Frontal Mirror Circuit: Interpretations and Misinterpretations. *Nat. Rev. Neurosci.* **11**:264–274. [3, 9, 15]

Robalino, N., and A. Robson. 2012. The Economic Approach to "Theory of Mind." *Phil. Trans. R. Soc. B* **367**:2224–2233. [15]

Robinson, W. S. 2010. Epiphenomenalism. *Wiley Interdiscip. Rev. Cogn. Sci.* **1**:539–547. [14]

Rochat, P. 2010. The Innate Sense of the Body Develops to Become a Public Affair by 2–3 Years. *Neuropsychologia* **48**:738–745. [15]

Roco, M. C., and W. S. Bainbridge. 2003. Converging Technologies for Improving Human Performance: Nanotechnology, Biotechnology, Information Technology and Cognitive Science. Berlin: Springer. [19]

Roebers, C. M., and M. Kauer. 2009. Motor and Cognitive Control in a Normative Sample of 7-Year-Olds. *Dev. Sci.* **12**:175–181. [3]

Roepstorff, A., and C. Frith. 2004. What's at the Top in the Top-Down Control of Action? Script-Sharing and "Top-Top" Control of Action in Cognitive Experiments. *Psychol. Res.* **68**:189–198. [12]

Roese, N. J., K. Epstude, F. Fessel, et al. 2009. Repetitive Regret, Depression, and Anxiety: Findings from a Nationally Representative Survey. *J. Soc. Clin. Psychol.* **28**:671–688. [12]

Rosch, E., and C. B. Mervis. 1975. Family Resemblances: Studies in the Internal Structure of Categories. *Cogn. Psychol.* **7**:573–605. [9]

Rosenbaum, D. A. 2009. Human Motor Control (2nd edition). London: Academic Press. [17]

———. 2013. Cognitive Foundations of Action Planning and Control. In: Action Science: Foundations of an Emerging Discipline, ed. W. Prinz et al., pp. 89–112. Cambridge, MA: MIT Press. [17]

Rosenbaum, D. A., R. A. Carlson, and R. O. Gilmore. 2001. Acquisition of Intellectual and Perceptual-Motor Skills. *Annu. Rev. Psychol.* **52**:453–470. [2]

Rosenbaum, D. A., F. Marchak, H. Barnes, et al. 1990. Constraints for Action Selection: Overhand versus Underhand Grips. In: Attention and Performance, ed. M. Jeannerod, pp. 321–342. Hillsdale, NJ: Lawrence Erlbaum. [3]

Rosenbleuth, A., and N. Wiener. 1945. The Role of Models in Science. *Phil. Sci.* **12**:316–321. [19]

Rosenthal, D. M. 1986. Two Concepts of Consciousness. *Philos. Stud.* **49**:329–359. [12]

———. 2005. Consciousness and Mind. Oxford: Clarendon. [15]

———. 2008. Consciousness and Its Function. *Neuropsychologia* **46**:829–840. [14]

Rothenberg, M., ed. 1998. Presidential Address to the American Association for the Advancement of Science, [August 22, 1850]. In: The Papers of Joseph Henry, vol. 8, pp. 101–102. Washington, D.C.: Smithsonian Institute Press. [19]

Rowe, M. L., and S. Goldin-Meadow. 2009. Early Gesture Selectively Predicts Later Language Learning. *Dev. Sci.* **12**:182–187. [20]

Rowland, H. A. 1883. A Plea for Pure Science. *Science* **2**:242–250. [19]

Rowlands, M. 1999. The Body in Mind: Understanding Cognitive Processes. Cambridge: Cambridge Univ. Press. [13]

———. 2006. Body Language. Cambridge, MA: MIT Press. [16]

Roy, D. 2005. Semiotic Schemas: A Framework for Grounding Language in Action and Perception. *Artif. Intell.* **167**:170–205. [2]

Rupert, R. D. 1998. On the Relationship between Naturalistic Semantics and Individuation Criteria for Terms in a Language of Thought. *Synthese* **117**:95–131. [4]

———. 2001. Coining Terms in the Language of Thought: Innateness, Emergence, and the Lot of Cummins's Argument against the Causal Theory of Mental Content. *J. Philos.* **98**:499–530. [4]

———. 2009. Cognitive Systems and the Extended Mind. Oxford: Oxford Univ. Press. [4]

———. 2011. Embodiment, Consciousness, and the Massively Representational Mind. *Phil. Top.* **39**:99–120. [5]

Russell, S. J., and P. Norvig. 2003. Artificial Intelligence: A Modern Approach. Upper Saddle River, NJ: Prentice-Hall. [18]

Ryle, G. 1949. The Concept of Mind. New York: Routledge. [14]

Salinas, E., and T. J. Sejnowski. 2001. Gain Modulation in the Central Nervous System: Where Behavior, Neurophysiology, and Computation Meet. *Neuroscientist* 7:430–440. [11]

Salomon, R., M. Lim, B. Herbelin, G. Hesselmann, and O. Blanke. 2013. Posing for Awareness: Proprioception Modulates Access to Visual Consciousness in a Continuous Flash Suppression Task. *J. Vis.* **13**:2. [15]

Samson, D., I. A. Apperly, J. J. Braithwaite, B. J. Andrews, and S. E. Bodley Scott. 2010. Seeing It Their Way: Evidence for Rapid and Involuntary Computation of What Other People See. *J. Exp. Psychol. Hum. Percept. Perform.* **36**:1255–1266. [12]

Sánchez-Fibla, M., U. Bernardet, E. Wasserman, et al. 2010. Allostatic Control for Robot Behavior Regulation: A Comparative Rodent-Robot Study. *Adv. Complex Syst.* **13**:377–403. [14, 15]

Sánchez-Fibla, M., A. Duff, and P. F. M. J. Verschure. 2011. The Acquisition of Intentionally Indexed and Object Centered Affordance Gradients: A Biomimetic Controller and Mobile Robotics Benchmark. In: IEEE/RSJ Intl. Conf. Intelligent Robots and Systems (IROS), pp. 1115–1121. San Francisco: IEEE. [11]

Sanchez-Vives, M. V., and M. Slater. 2005. From Presence to Consciousness through Virtual Reality. *Nat. Rev. Neurosci.* **6**:332–339. [19]

Sarewitz, D. 2012. Blue-Sky Bias Should Be Brought Down to Earth. *Nature* **481**:7. [19]

Sastry, S. 1999. Nonlinear Systems: Analysis, Stability and Control. New York: Springer. [20]

Saxe, R. R., S. Whitfield-Gabrieli, J. Scholz, and K. A. Pelphrey. 2009. Brain Regions for Perceiving and Reasoning About Other People in School-Aged Children. *Child Dev.* **80**:1197–1209. [16]

Saxena, A., J. Driemeyer, and A. Y. Ng. 2008. Robotic Grasping of Novel Objects Using Vision. *Int. J. Rob. Res.* **27**:157–173. [18]

Schaal, S. 1999. Is Imitation Learning the Route to Humanoid Robots? *Trends Cogn. Sci.* **3**:233–242. [18]

Schaal, S., P. Mohajerian, and A. Ijspeert. 2007. Dynamics Systems vs. Optimal Control: A Unifying View. *Prog. Brain Res.* **165**:425–445. [18]

Schacter, S., and J. E. Singer. 1962. Cognitive, Social and Physiological Determinants of Emotional State. *Psychol. Rev.* **69**:379–399. [7]

Scheffler, I. 1974. Four Pragmatists: A Critical Introduction to Peirce, James, Mead, and Dewey. London: Routledge & Kegan Paul. [13]

Scherer, K. R. 2001. Appraisal Considered as a Process of Multilevel Sequential Checking. In: Appraisal Processes in Emotion: Theory, Methods, Research, ed. K. R. Scherer et al., pp. 92–120. Oxford: Oxford Univ. Press. [5]

Schilbach, L., B. Timmermans, V. Reddy, et al. 2013. Toward a Second-Person Neuroscience. *Behav. Brain Sci.* **36**:393–414. [1, 3, 20]

Schmidhuber, J. 1991a. Adaptive Confidence and Adaptive Curiosity. Technical Report FKI-149-91. Munich: Institut für Informatik, Technische Universität München. [2]

———. 1991b. Curious Model-Building Control Systems. In: IEEE International Joint Conference on Neural Networks, pp. 1458–1463. Singapore: IEEE. [10]

Schmidt, R. E., and M. Van der Linden. 2013. Feeling Too Regretful to Fall Asleep: Experimental Activation of Regret Delays Sleep Onset. *Cogn. Ther. Res.* **37**:872–880. [12]

Schmitz, C., N. Martin, and C. Assaiante. 2002. Building Anticipatory Postural Adjustment During Childhood: A Kinematic and Electromyographic Analysis of Unloading in Children from 4 to 8 Years of Age. *Exp. Brain Res.* **142**:354–364. [3]

Schneider, D. M., A. Nelson, and R. Mooney. 2014. A Synaptic and Circuit Basis for Corollary Discharge in the Auditory Cortex. *Nature* **513**:189–194. [15]

Scholl, B. J. 2001. Objects and Attention: The State of the Art. *Cognition* **80**:1–46. [14]

Scholl, B. J., and D. J. Simons. 2001. Change Blindness, Gibson, and the Sensorimotor Theory of Vision. *Behav. Brain Sci.* **24**:1004–1006. [5]

Schomers, M., E. Kirilina, A. Weigand, M. Bajbouj, and F. Pulvermüller. 2015. Causal Influence of Articulatory Motor Cortex on Comprehending Single Spoken Words: TMS Evidence. *Cereb. Cortex* **25**:3894–3902. [9]

Schubotz, R. 2007. Prediction of External Events without Motor System: Towards a New Framework. *Trends Cogn. Sci.* **11**:211–218. [1, 11]

Schultz, W., and A. Dickinson. 2000. Neuronal Coding of Prediction Errors. *Annu. Rev. Neurosci.* **23**:473–500. [12]

Schurz, G. 2008. Patterns of Abduction. *Synthese* **164**:201–234. [13]

Schütz-Bosbach, S., and W. Prinz. 2007. Perceptual Resonance: Action-Induced Modulation of Perception. *Trends Cogn. Sci.* **11**:349–355. [1]

———. 2015. Mirrors Match Minds. In: New Frontiers in Mirror Neuron Research, ed. P. F. Ferrari and G. Rizzolatti, pp. 198–221. Oxford: Oxford Univ. Press. [17]

Schwartz, A. B. 2004. Cortical Neural Prosthetics. *Annu. Rev. Neurosci.* **27**:487–507. [20]

Searle, J. R. 1969. Speech Acts: An Essay in the Philosophy of Language. Cambridge: Cambridge Univ. Press. [9, 10]

———. 1975. Indirect Speech Acts. *Syntax and Semantics* **3**:59–82. [9]

———. 1979. Expression and Meaning. Cambridge: Cambridge Univ. Press. [9]

———. 1998. How to Study Consciousness Scientifically. *Phil. Trans. R. Soc. B* **353**:1935–1942. [14]

Sebanz, N., H. Bekkering, and G. Knoblich. 2006. Joint Action: Bodies and Minds Moving Together. *Trends Cogn. Sci.* **10**:70–76. [1, 20]

Sebanz, N., and G. Knoblich. 2009. Prediction in Joint Action: What, When, and Where. *Top. Cogn. Sci.* **1**:353–367. [4]

Sebanz, N., G. Knoblich, and W. Prinz. 2003. Representing Others' Actions: Just Like One's Own? *Cognition* **88**:B11–21. [12]

Semini, C., V. Barasuol, T. Boaventura, M. Frigerio, and J. Buchli. 2013. Is Active Impedance the Key to a Breakthrough for Legged Robots? In: Proc. Intl. Symp. on Robotics Research (ISRR), Springer Star Series. Zürich: ETH-Zürich. [20]

Senju, A., V. Southgate, S. White, and U. Frith. 2009. Mindblind Eyes: An Absence of Spontaneous Theory of Mind in Asperger Syndrome. *Science* **325**:883–885. [4]

Seth, A. K. 2009. Explanatory Correlates of Consciousness: Theoretical and Computational Challenges. *Cogn. Comput.* **1**:50–63. [14]

———. 2013. Interoceptive Inference, Emotion, and the Embodied Self. *Trends Cogn Sci.* **17**:565–573. [6, 7, 15]

———. 2014. A Predictive Processing Theory of Sensorimotor Contingencies: Explaining the Puzzle of Perceptual Presence and Its Absence in Synesthesia. *Cogn. Neurosci.* **5**:97–118. [2, 7, 12, 15]

———. 2015. The Cybernetic Bayesian Brain: From Interoceptive Inference to Sensorimotor Contingencies. In: Open Mind, ed. T. Metzinger and J. Windt. Frankfurt: MIND Group. [2, 6, 12, 15]

Seth, A. K., Z. Dienes, A. Cleeremans, M. Overgaard, and L. Pessoa. 2008. Measuring Consciousness: Relating Behavioural and Neurophysiological Approaches. *Trends Cogn. Sci.* **12**:314–321. [15]

Seth, A. K., K. Suzuki, and H. D. Critchley. 2011. An Interoceptive Predictive Coding Model of Conscious Presence. *Front. Psychol.* **2**:395. [6]

Shadmehr, R., M. A. Smith, and J. W. Krakauer. 2010. Error Correction, Sensory Prediction, and Adaptation in Motor Control. *Annu. Rev. Neurosci.* **33**:89–108. [4]

Shallice, T., and R. P. Cooper. 2013. Is There a Semantic System for Abstract Words? *Front. Hum. Neurosci.* **7**:175. [9]

Shapiro, L. 2011. Embodied Cognition. Oxford: Routledge. [7]

Shea, N. J., A. Boldt, D. Bang, et al. 2014. Supra-Personal Cognitive Control and Metacognition. *Trends Cogn. Sci.* **18**:186–193. [12]

Shebani, Z., and F. Pulvermüller. 2013. Moving the Hands and Feet Specifically Impairs Working Memory for Arm- and Leg-Related Action Words. *Cortex* **49**:222–231. [9]

Shepherd, J. 2012. Free Will and Consciousness: Experimental Studies. *Conscious. Cogn.* **21**:915–927. [12]

Sherman, J. W., B. Gawronski, and Y. Trope. 2014. Dual-Process Theories of the Social Mind. New York: Guilford Press. [5]

Shipp, S., R. A. Adams, and K. J. Friston. 2013. Reflections on Agranular Architecture: Predictive Coding in the Motor Cortex. *Trends Neurosci.* **36**:706–716. [6]

Shtyrov, Y., A. Butorina, A. Nikolaeva, and T. Stroganova. 2014. Automatic Ultrarapid Activation and Inhibition of Cortical Motor Systems in Spoken Word Comprehension. *PNAS* **111**:E1918–1923. [9]

Shtyrov, Y., V. V. Nikulin, and F. Pulvermüller. 2010. Rapid Cortical Plasticity Underlying Novel Word Learning. *J. Neurosci.* **30**:16864–16867. [9]

Shubina, K., and J. K. Tsotsos. 2010. Visual Search for an Object in a 3D Environment Using a Mobile Robot. *Comput. Vis. Image Underst.* **114**:535–547. [10]

Siegel, M., T. H. Donner, R. Oostenveld, P. Fries, and A. K. Engel. 2008. Neuronal Synchronization Along the Dorsal Visual Pathway Reflects the Focus of Spatial Attention. *Neuron* **60**:709–719. [20]

Singer, T., B. Seymour, J. P. O'Doherty, et al. 2006. Empathic Neural Responses Are Modulated by the Perceived Fairness of Others. *Nature* **439**:466–469. [12]

Singer, W. 2001. Consciousness and the Binding Problem. *Ann. NY Acad. Sci.* **929**:123–146. [8]

Singer, W., and C. Gray. 1995. Visual Feature Integration and the Temporal Correlation Hypothesis. *Annu. Rev. Neurosci.* **18**:555–586. [8, 9]

Skerry, A. E., S. Carey, and E. S. Spelke. 2013. First-Person Action Experience Reveals Sensitivity to Action Efficiency in Prereaching Infants. *PNAS* **110**:18728–18733. [3]

Skinner, B. F. 1938. The Behavior of Organisms: An Experimental Analysis. New York: Appleton-Century-Crofts. [4]

———. 1953. Science and Human Behavior. New York: Macmillan. [17]

Sloman, A. 2001. Evolvable Biologically Plausible Visual Architectures. In: Proc. British Machine Vision Conf., pp. 313–322. Manchester: BMVC. [18]

Smallwood, J., and J. W. Schooler. 2015. The Science of Mind Wandering: Empirically Navigating the Stream of Consciousness. *Annu. Rev. Psychol.* **66**:487–518. [5]

Smith, C. H., D. A. Oakley, and J. Morton. 2013. Increased Response Time of Primed Associates Following an "Episodic" Hypnotic Amnesia Suggestion: A Case of Unconscious Volition. *Conscious. Cogn.* **22**:1305–1317. [12]

Smolensky, P. 1990. Tensor Product Variable Binding and the Representation of Symbolic Structures in Connectionist Systems. *Artif. Intell.* **46**:159–216. [5]

Sommerfeld, R. D., H. J. Krambeck, D. Semmann, and M. Milinski. 2007. Gossip as an Alternative for Direct Observation in Games of Indirect Reciprocity. *PNAS* **104**:17435–17440. [12]

Sommerville, J. A., A. L. Woodward, and A. Needham. 2005. Action Experience Alters 3-Month-Old Infants' Perception of Others' Actions. *Cognition* **96**:B1–11. [3]

Song, D., K. Huebner, V. Kyrki, and D. Kragic. 2010. Learning Task Constraints for Robot Grasping Using Graphical Models. In: IEEE/RSJ International Conference on Intelligent Robots and Systems (IROS), pp. 1579–1585. Taipei: IEEE. [18]

Soon, C. S., M. Brass, H. J. Heinze, and J. D. Haynes. 2008. Unconscious Determinants of Free Decisions in the Human Brain. *Nat. Neurosci.* **11**:543–545. [12, 14, 16]

Southgate, V., and K. Begus. 2013. Motor Activation During the Prediction of Nonexecutable Actions in Infants. *Psychol. Sci.* **24**:828–835. [3]

Southgate, V., K. Begus, S. Lloyd-Fox, V. di Gangi, and A. F. Hamilton. 2014. Goal Representation in the Infant Brain. *NeuroImage* **85**:294–301. [3]

Southgate, V., G. Csibra, and M. H. Johnson. 2008. Infants Attribute Goals Even to Biomechanically Impossible Actions. *Cognition* **107**:1059–1069. [3]

Southgate, V., M. H. Johnson, I. El Karoui, and G. Csibra. 2010. Motor System Activation Reveals Infants' on-Line Prediction of Others' Goals. *Psychol. Sci.* **21**:355–359. [3]

Southgate, V., M. H. Johnson, T. Osborne, and G. Csibra. 2009. Predictive Motor Activation During Action Observation in Human Infants. *Biol. Lett.* **5**:769–772. [3]

Speelman, C., and K. Kirsner. 2005. Beyond the Learning Curve: The Construction of Mind. Oxford: Oxford Univ. Press. [15]

Spelke, E. S., and S. A. Lee. 2012. Core Systems of Geometry in Animal Minds. *Phil. Trans. R. Soc. B* **367**:2784–2793. [5]

Sperber, D. 1975. Rethinking Symbolism. Cambridge: Cambridge Univ. Press. [4]

———. 1996. Explaining Culture: A Naturalistic Approach. Oxford: Wiley-Blackwell. [12]

Sperry, R. W. 1952. Neurology and the Mind-Brain Problem. *Am. Sci.* **40**:291–312. [15]

Spunt, R. P., E. B. Falk, and M. D. Lieberman. 2010. Dissociable Neural Systems Support Retrieval of How and Why Action Knowledge. *Psychol. Sci.* **21**:1593–1598. [3]

Squire, L. R., C. E. L. Stark, and R. E. Clark. 2004. The Medial Temporal Lobe. *Annu. Rev. Neurosci.* **27**:279–306. [5]

Srinivasan, M. V., S. B. Laughlin, and A. Dubs. 1982. Predictive Coding: A Fresh View of Inhibition in the Retina. *Proc. R. Soc. Lond. B* **216**:427–459. [6]

Stalnaker, R. C. 2002. Common Ground. *Linguist. Philos.* **25**:701–721. [9]

Stark, L., and K. Bowyer. 1996. Generic Object Recognition Using Form and Function. New York: World Scientific. [18]

Stark, M., P. Lies, M. Zillich, J. Wyatt, and B. Schiele. 2008. Functional Object Class Detection Based on Learned Affordance Cues. In: Computer Vision Systems, pp. 435–444. Heidelberg: Springer. [18]

Steels, L., and T. Belpaeme. 2005. Coordinating Perceptually Grounded Categories through Language: A Case Study for Colour. *Behav. Brain Sci.* **28**:469–489; discussion 489–529. [4]

Steiner, A. P., and A. D. Redish. 2014. Behavioral and Neurophysiological Correlates of Regret in Rat Decision-Making on a Neuroeconomic Task. *Nat. Neurosci.* **17**:995–1002. [15]

Stern, D. N. 1985. The Interpersonal World of the Infant: A View from Psychoanalysis and Developmental Psychology. New York: Basic Books. [13]

Stöckel, T., C. M. L. Hughes, and T. Schack. 2012. Representation of Grasp Postures and Anticipatory Motor Planning in Children. *Psychol. Res.* **76**:768–776. [3]

Stoffregen, T. A., B. G. Bardy, and B. Mantel. 2006. Affordances in the Design of Enactive Systems. *Virtual Real.* **10**:4–10. [19]

Stoianov, I., A. Genovesio, and G. Pezzulo. 2016. Prefrontal Goal-Codes Emerge as Latent States in Probabilistic Value Learning. *J. Cogn. Neurosci.* **28**:140–157. [2]

Stokes, D. E. 1992. Basic Science and Technological Innovation. Washington, D.C.: Brookings Institution Press. [19]

Suddendorf, T., D. R. Addis, and M. C. Corballis. 2009. Mental Time Travel and the Shaping of the Human Mind. *Phil. Trans. R. Soc. B* **364**:1317–1324. [7]

Suzuki, K., S. N. Garfinkel, H. D. Critchley, and A. K. Seth. 2013. Multisensory Integration across Exteroceptive and Interoceptive Domains Modulates Self-Experience in the Rubber-Hand Illusion. *Neuropsychologia* **51**:2909–2917. [7, 15]

Tatler, B. W. 2001. Re-Presenting the Case for Representation. Commentary to O'Regan & Noë: A Sensorimotor Account of Vision and Visual Consciousness. *Behav. Brain Sci.* **24**:1006–1007. [5]

Tenenbaum, J. B., T. L. Griffiths, and C. Kemp. 2006. Theory-Based Bayesian Models of Inductive Learning and Reasoning. *Trends Cogn. Sci.* **10**:309–318. [4]

Tenenbaum, J. B., C. Kemp, T. L. Griffiths, and N. D. Goodman. 2011. How to Grow a Mind: Statistics, Structure, and Abstraction. *Science* **331**:1279–1285. [2, 4]

Tettamanti, M., G. Buccino, M. C. Saccuman, et al. 2005. Listening to Action-Related Sentences Activates Fronto-Parietal Motor Circuits. *J. Cogn. Neurosci.* **17**:273–281. [9]

Teufel, C., P. C. Fletcher, and G. Davis. 2010. Seeing Other Minds: Attributed Mental States Influence Perception. *Trends Cogn. Sci.* **14**:376–382. [6]

Thakkar, K. N., J. S. Peterman, and S. Park. 2014. Altered Brain Activation During Action Imitation and Observation in Schizophrenia: A Translational Approach to Investigating Social Dysfunction in Schizophrenia. *Am. J. Psychiatry* **171**:539–548. [15]

Thelen, E., G. Schöner, C. Scheier, and L. Smith. 2001. The Dynamics of Embodiment: A Field Theory of Infant Perseverative Reaching. *Behav. Brain Sci.* **24**:1–33. [2, 4]

Thelen, E., and L. B. Smith. 1996. A Dynamic Systems Approach to the Development of Cognition and Action. Cambridge, MA: MIT Press. [3, 4, 17]

Thompson, E. 2007. Mind in Life: Biology, Phenomenology, and the Sciences of Mind. Boston: Harvard Univ. Press. [7, 11, 15, 16]

———. 2014. The Embodied Mind: An Interview with Evan Thompson. *Tricycle*, Fall 2014. [16]

Thompson, E., and F. J. Varela. 2001. Radical Embodiment: Neural Dynamics and Consciousness. *Trends Cogn Sci.* **5**:418–425. [6, 15, 16]

Thorndike, E. L. 1911. Animal Intelligence: Experimental Studies. New York: Macmillan. [17]

———. 1932. The Fundamentals of Learning. New York: Teachers College Bureau of Publications. [4]

Thrun, S., M. Montemerlo, H. Dahlkamp, et al. 2006. The Robot That Won the DARPA Grand Challenge. *J. Field Robotics* **23**:661–692. [20]

Tkach, D., J. Reimer, and N. G. Hatsopoulos. 2008. Observation-Based Learning for Brain-Machine Interfaces. *Curr. Opin. Neurobiol.* **18**:589–594. [20]

Tobler, P., J. P. O'Doherty, R. Dolan, and W. Schultz. 2006. Human Neural Learning Depends on Reward Prediction Errors in the Blocking Paradigm. *J. Neurophysiol.* **95**:301–310. [12]

Todorov, E., and M. I. Jordan. 2002. Optimal Feedback Control as a Theory of Motor Coordination. *Nat. Neurosci.* **5**:1226–1235. [3]

Tolman, E. C. 1932. Purposive Behavior in Animals and Man. New York: Century Co. [14]

———. 1948. Cognitive Maps in Rats and Men. *Psychol. Rev.* **55**:189–208. [14]

———. 1959. Principles of Purposive Behavior. In: Psychology: The Study of a Science, ed. S. Koch, vol. 2, pp. 92–157. New York: McGraw-Hill. [17]

Tomasello, M. 2009. The Cultural Origins of Human Cognition. Cambridge, MA: Harvard Univ. Press. [5]

———. 2014. A Natural History of Human Thinking. Cambridge, MA: Harvard Univ. Press. [4, 15]

Tononi, G. 2008. Consciousness as Integrated Information: A Provisional Manifesto. *Biol. Bull.* **215**:216–242. [12, 14]

———. 2012. Phi: A Voyage from the Brain to the Soul: Pantheon Books. [14]

Tononi, G., and G. M. Edelman. 1998. Neuroscience: Consciousness and Complexity. *Science* **282**:1846–1851. [12, 14]

Tononi, G., and C. Koch. 2014. Consciousness: Here, There but Not Everywhere. *arXiv* **1405**:7089. [14]

Torres, E. B. 2013. Atypical Signatures of Motor Variability Found in an Individual with ASD. *Neurocase* **19**:150–165. [20]

Torres, E. B., M. Brincker, R. W. Isenhower, et al. 2013. Autism: The Micro-Movement Perspective. *Front. Integr. Neurosci.* **7**:32. [20]

Trevarthen, C. 1998. The Concept and Foundations of Infant Intersubjectivity. In: Intersubjective Communication and Emotion in Early Ontogeny, ed. S. Bråten, pp. 15–46. Cambridge: Cambridge Univ. Press. [17]

Trevarthen, C., and J. T. Delafield-Butt. 2013. Autism as a Developmental Disorder in Intentional Movement and Affective Engagement. *Front. Integr. Neurosci.* **7**:49. [20]

Trimmer, B. 2013. A Journal of Soft Robotics: Why Now? *Soft Robotics* **1**:1–4. [20]

Trivedi, D., C. D. Rahn, W. M. Kier, and I. D. Walker. 2008. Soft Robotics: Biological Inspiration, State of the Art, and Future Research. *Appl. Bionics Biomech.* **5**:99–117. [19]

Trivers, R. 2011. The Folly of Fools: The Logic of Deceit and Self-Deception in Human Life. New York: Basic Books. [12]

Trungpa, C. 2002. Cutting through Spiritual Materialism. London: Shambhala Publications. [20]

Tsakiris, M. 2010. My Body in the Brain: A Neurocognitive Model of Body-Ownership. *Neuropsychologia* **48**:703–712. [15]

Tsakiris, M., and P. Haggard. 2005. The Rubber Hand Illusion Revisited: Visuotactile Integration and Self-Attribution. *J. Exp. Psychol. Hum. Percept. Perform.* **31**:80. [7]

Tsakiris, M., S. Schütz-Bosbach, and S. Gallagher. 2007. On Agency and Body-Ownership: Phenomenological and Neurocognitive Reflections. *Conscious. Cogn.* **16**:645–660. [7]

Tsakiris, M., A. Tajadura-Jimenez, and M. Costantini. 2011. Just a Heartbeat Away from One's Body: Interoceptive Sensitivity Predicts Malleability of Body-Representations. *Proc. R. Soc. Lond. B* **278**:2470–2476. [15]

Tsotsos, J. K. 1992. On the Relative Complexity of Active vs. Passive Visual Search. *Int. J. Comput. Vis.* **7**:127–141. [10]

———. 1995. Behaviorist Intelligence and the Scaling Problem. *Artif. Intell.* **75**:135–160. [10]

————. 2011. A Computational Perspective on Visual Attention. Cambridge, MA: MIT Press. [10]

Tsuchiya, N., and R. Adolphs. 2007. Emotion and Consciousness. *Trends Cogn. Sci.* **11**:158–167. [14]

Turing, A. M. 1950. Computing Machinery and Intelligence. In: The New Media Reader, ed. N. Wardrip-Fruin and N. Montfort, pp. 50–64. Cambridge, MA: MIT Press. [18]

Uhlhaas, P. J., G. Pipa, B. Lima, et al. 2009. Neural Synchrony in Cortical Networks: History, Concept and Current Status. *Front. Integr. Neurosci.* **3**:17. [14]

van Ackeren, M. J., D. Casasanto, H. Bekkering, P. Hagoort, and S. A. Rueschemeyer. 2012. Pragmatics in Action: Indirect Requests Engage Theory of Mind Areas and the Cortical Motor Network. *J. Cogn. Neurosci.* **24**:2237–2247. [9]

Van de Cruys, S., K. Evers, R. Van der Hallen, et al. 2014. Precise Minds in Uncertain Worlds: Predictive Coding in Autism. *Psychol. Rev.* **121**:649–675. [6]

van der Hoort, B., A. Guterstam, and H. H. Ehrsson. 2011. Being Barbie: The Size of One's Own Body Determines the Perceived Size of the World. *PLoS One* **6**:e20195. [7]

van der Meer, A. L. 1997. Keeping the Arm in the Limelight: Advanced Visual Control of Arm Movements in Neonates. *Eur. J. Paediatr. Neurol.* **1**:103–108. [3]

Van Doorn, G., B. Paton, J. Howell, and J. Hohwy. 2015. Attenuated Self-Tickle Sensation Even under Trajectory Perturbation. *Conscious. Cogn.* **36**:147–153. [7]

Van Gulick, R. 2001. Still Room for Representations. Commentary to O'Regan & Noë: A Sensorimotor Account of Vision and Visual Consciousness. *Behav. Brain Sci.* **24**:1007–1008. [5]

Van Hemmen, J. L., and A. B. Schwartz. 2008. Population Vector Code: A Geometric Universal as Actuator. *Biol. Cybern.* **98**:509–518. [20]

van Swieten, L. M., E. van Bergen, J. H. G. Williams, et al. 2010. A Test of Motor (Not Executive) Planning in Developmental Coordination Disorder and Autism. *J. Exp. Psychol. Hum. Percept. Perform.* **36**:493–499. [3]

Vapnik, V. N. 2000. The Nature of Statistical Learning Theory. New York: Springer. [8]

Varela, F. J. 1996. Neurophenomenology: A Methodological Remedy for the Hard Problem. *J. Conscious. Stud.* **3**:330–349. [20]

————. 1999. The Specious Present: A Neurophenomenology of Time Consciousness. In: Naturalizing Phenomenology: Issues in Contemporary Phenomenology and Cognitive Science, ed. J. Petitot et al., pp. 226–314. Stanford: Stanford Univ. Press. [16]

Varela, F. J., E. Thompson, and E. Rosch. 1992. The Embodied Mind: Cognitive Science and Human Experience. Cambridge, MA: MIT Press. [1, 9, 11, 15, 16, 18–20]

Varraine, E., M. Bonnard, and J. Pailhous. 2002. The Top Down and Bottom up Mechanisms Involved in the Sudden Awareness of Low Level Sensorimotor Behavior. *Brain Res. Cogn. Brain Res.* **13**:357–361. [12]

Velliste, M., S. Perel, M. C. Spalding, A. S. Whitford, and A. B. Schwartz. 2008. Cortical Control of a Prosthetic Arm for Self-Feeding. *Nature* **453**:1098–1101. [20]

Venezia, J. H., K. Saberi, C. Chubb, and G. Hickok. 2012. Response Bias Modulates the Speech Motor System During Syllable Discrimination. *Front. Psychol.* **3**:157. [9]

Verschure, P. F. M. J. 1992. Taking Connectionism Seriously: The Vague Promise of Subsymbolism and an Alternative. In: Proc. 14th Annual Conf. of the Cognitive Science Society, pp. 653–658. Hillsdale, NJ: Erlbaum. [14]

Verschure, P. F. M. J. 1997. Connectionist Explanation: Taking Positions in the Mind-Brain Dilemma. In: Neural Networks and a New Artificial Intelligence, ed. G. Dorffner, pp. 133–188. London: Thompson. [14]

———. 1998. Synthetic Epistemology: The Acquisition, Retention, and Expression of Knowledge in Natural and Synthetic Systems. In: Proc. World Conf. on Computational Intelligence, pp. 147–153. Anchorage: IEEE. [14]

———. 2012a. Consciousness Solves Pervasive Intentionality. 16th Annual Meeting of the Association for the Scientific Study of Consciousness, p. 167. Brighton: ASSC. [14]

———. 2012b. The Distributed Adaptive Control Architecture of the Mind, Brain, Body Nexus. *Biol. Inspired Cogn. Arch.* **1**:55–72. [14, 15]

———. 2013. Formal Minds and Biological Brains II: From the Mirage of Intelligence to a Science and Engineering of Consciousness. *IEEE Expert* **28**:33–36. [14, 19]

———. 2016. From Big Data Back to Big Ideas: The Risks of a Theory Free Data Rich Science of Mind and Brain and a Solution. *Connect. Sci.*, in press. [14, 19]

Verschure, P. F. M. J., B. Krose, and R. Pfeifer. 1992. Distributed Adaptive Control: The Self-Organization of Structured Behavior. *Rob. Auton. Syst.* **9**:181–196. [14]

Verschure, P. F. M. J., C. M. A. Pennartz, and G. Pezzulo. 2014. The Why, What, Where, When and How of Goal-Directed Choice: Neuronal and Computational Principles. *Phil. Trans. R. Soc. B* **369**:20130483. [2, 4, 14]

Verschure, P. F. M. J., and R. Pfeifer. 1993. Environment Interaction: A Case Study in Autonomous Systems. In: From Animals to Animats 2, p. 210. Cambridge, MA: MIT Press. [2]

Verschure, P. F. M. J., T. Voegtlin, and R. J. Douglas. 2003. Environmentally Mediated Synergy between Perception and Behaviour in Mobile Robots. *Nature* **425**:620–624. [2, 4, 14, 15]

Vico, G. 1730. Scienza Nuova Seconda: The New Science of Giambattista Vico, rev. transl. 3rd edition by Thomas Goddard Bergin and Max Harold Fisch. Ithaca: Cornell Univ. Press, 1948; Cornell Paperbacks, 1976. [14]

Vigliocco, G., S. T. Kousta, P. A. Della Rosa, et al. 2014. The Neural Representation of Abstract Words: The Role of Emotion. *Cereb. Cortex* **24**:1767–1777. [4, 9]

Vigliocco, G., P. Perniss, and D. Vinson. 2014. Language as a Multimodal Phenomenon: Implications for Language Learning, Processing and Evolution. *Phil. Trans. R. Soc. B* **369**:20130292. [10]

Vishton, P. M., N. J. Stephens, L. A. Nelson, et al. 2007. Planning to Reach for an Object Changes How the Reacher Perceives It. *Psychol. Sci.* **18**:713–719. [15]

Viviani, P. 2002. Motor Competence in the Perception of Dynamic Events: A Tutorial. In: Common Mechanisms in Perception and Action: Attention and Performance XIX, ed. W. Prinz and B. Hommel, pp. 406–442. Oxford: Oxford Univ. Press. [17]

Vohs, K. D., and J. W. Schooler. 2008. The Value of Believing in Free Will: Encouraging a Belief in Determinism Increases Cheating. *Psychol. Sci.* **19**:49–54. [12]

von der Malsburg, C. 1973. Self-Organization of Orientation-Sensitive Cells in the Striate Cortex. *Kybernetik* **14**:85–100 [8]

von der Malsburg, C., W. Phillips, and W. Singer, eds. 2010. Dynamic Coordination in the Brain. Strüngmann Forum Reports, vol. 5. J. Lupp, series ed. Cambridge, MA: MIT Press. [8]

von Hippel, W., and R. Trivers. 2011. The Evolution and Psychology of Self-Deception. *Behav. Brain Sci.* **34**:1–16; discussion 16–56. [12]

von Hofsten, C. 2004. An Action Perspective on Motor Development. *Trends Cogn. Sci.* **8**:266–272. [2, 4]

von Hofsten, C., and L. Ronnqvist. 1988. Preparation for Grasping an Object: A Developmental Study. *J. Exp. Psychol. Hum. Percept. Perform.* **14**:610–621. [3]

von Holst, E., and H. Mittelstaedt. 1950. Das Reafferenzprinzip. *Naturwissenschaften* **37**:464–476. [1, 4]

Vosgerau, G. 2010. Memory and Content. *Conscious. Cogn.* **19**:838–846. [5]

Vygotsky, L. S. 1978. Mind in Society: The Development of Higher Psychological Processes. Cambridge, MA: Harvard Univ. Press. [4, 13]

———. 1979. The Genesis of Higher Mental Functions. In: The Concept of Activity in Soviet Psychology, ed. J. V. Wertsch, pp. 144–188. Armonk, NY: Sharpe. [17]

Walker, E., and R. J. Lewine. 1990. Prediction of Adult-Onset Schizophrenia from Childhood Home Movies of the Patients. *Am. J. Psychiatry* **147**:1052–1056. [15, 20]

Wang, S. S. H., A. D. Kloth, and A. Badura. 2014. The Cerebellum, Sensitive Periods, and Autism. *Neuron* **83**:518–532. [3]

Wang, Y., and A. F. Hamilton. 2012. Social Top-Down Response Modulation (STORM): A Model of the Control of Mimicry in Social Interaction. *Front. Hum. Neurosci.* **6**:153–153. [3]

Ward, L. M. 2011. The Thalamic Dynamic Core Theory of Conscious Experience. *Conscious. Cogn.* **20**:464–486. [14]

Warlaumont, A. S., J. A. Richards, J. Gilkerson, and D. K. Oller. 2014. A Social Feedback Loop for Speech Development and Its Reduction in Autism. *Psychol. Sci.* **25**:1314–1324. [3]

Watkins, S., L. Shams, O. Josephs, and G. Rees. 2007. Activity in Human V1 Follows Multisensory Perception. *NeuroImage* **37**:572–578. [12]

Webb, B. 1994. Robotic Experiments in Cricket Phonotaxis. In: From Animals to Animats 3, ed. D. Cliff et al., pp. 45–54. Cambridge, MA: MIT Press. [13]

Weber, A. M., and G. Vosgerau. 2012. Grounding Action Representations. *Rev. Phil. Psych.* **3**:53–69. [4]

Wegner, D. M. 2003. The Illusion of Conscious Will. Cambridge, MA: MIT Press. [12, 14]

Weigelt, M., and T. Schack. 2010. The Development of End-State Comfort Planning in Preschool Children. *Exp. Psychol.* **57**:476–782. [3]

Weisberg, J., M. van Turennout, and A. Martin. 2007. A Neural System for Learning About Object Function. *Cereb. Cortex* **17**:513–521. [1]

Wellman, H. M., D. Cross, and J. Watson. 2001. Meta-Analysis of Theory-of-Mind Development: The Truth About False Belief. *Child Dev.* **72**:655–684. [5]

Wellman, H. M., P. L. Harris, M. Banerjee, and A. Sinclair. 1995. Early Understanding of Emotion: Evidence from Natural Language. *Cogn. Emot.* **9**:117–149. [4]

Wennekers, T., M. Garagnani, and F. Pulvermüller. 2006. Language Models Based on Hebbian Cell Assemblies. *J. Physiol. Paris* **100**:16–30. [9]

Wennekers, T., and G. Palm. 2007. Modelling Generic Cognitive Functions with Operational Hebbian Cell Assemblies. In: Neural Network Research Horizons, ed. M. L. Weiss, pp. 225–294. New York: Nova Science Publishers. [9]

Westermann, G., and E. R. Miranda. 2004. A New Model of Sensorimotor Coupling in the Development of Speech. *Brain Lang.* **89**:393–400. [11]

Wheeler, M. 2005. Reconstructing the Cognitive World: The Next Step. Cambridge, MA: MIT Press. [16]

Whyatt, C., and C. Craig. 2013. Sensory-Motor Problems in Autism. *Front. Integr. Neurosci.* **7**:51. [20]

Wilkes, D., and J. K. Tsotsos. 1992. Active Object Recognition. In: IEEE Computer Society Conf. on Computer Vision and Pattern Recognition, pp. 136–141. Urbana: IEEE. [10]

Wilson, M. 2002. Six Views of Embodied Cognition. *Psychon. Bull. Rev.* **9**:625–636. [15, 20]

Wilson-Mendenhall, C. D., and L. W. Barsalou. 2016. A Fundamental Role for the Human Conceptual System in Emotion. In: Handbook of Emotion (4th edition), ed. L. F. Barrett et al. New York: Guilford Press, in press. [5]

Winkielman, P., K. C. Berridge, and J. L. Wilbarger. 2005. Unconscious Affective Reactions to Masked Happy versus Angry Faces Influence Consumption Behavior and Judgments of Value. *Pers. Soc. Psychol. Bull.* **31**:121–135. [12]

Winograd, T., and F. Flores. 1986. Understanding Computers and Cognition: A New Foundation for Design. Norwood, NJ: Ablex Publishing Corp. [1]

Witt, J. K., D. R. Proffitt, and W. Epstein. 2005. Tool Use Affects Perceived Distance, but Only When You Intend to Use It. *J. Exp. Psychol. Hum. Percept. Perform.* **31**:880–888. [15]

Wittgenstein, L. 1953. Philosophical Investigations. Oxford: Blackwell. [9]

Wodlinger, B., J. E. Downey, E. C. Tyler-Kabara, et al. 2015. Ten-Dimensional Anthropomorphic Arm Control in a Human Brain–Machine Interface: Difficulties, Solutions, and Limitations. *J. Neural Eng.* **12**:016011. [20]

Wohlschlager, A. 2000. Visual Motion Priming by Invisible Actions. *Vision Res.* **40**:925–930. [15]

Wolpert, D. M., J. A. Diedrichsen, and J. R. Flanagan. 2011. Principles of Sensorimotor Learning. *Nat. Rev. Neurosci.* **12**:739–751. [1, 18]

Wolpert, D. M., K. Doya, and M. Kawato. 2003. A Unifying Computational Framework for Motor Control and Social Interaction. *Phil. Trans. R. Soc. B* **358**:593–602. [2, 3]

Wolpert, D. M., and J. R. Flanagan. 2001. Motor Prediction. *Curr. Biol.* **11**:729–732. [1]

Wolpert, D. M., and Z. Ghahramani. 2000. Computational Principles of Movement Neuroscience. *Nat. Neurosci. Suppl.* **3**:1212–1217. [7]

Wolpert, D. M., and M. Kawato. 1998. Multiple Paired Forward and Inverse Models for Motor Control. *Neural Netw.* **11**:1317–1329. [18]

Woodward, J. 2003. Making Things Happen. New York: Oxford Univ. Press. [7]

Wurtz, R. H., K. McAlonan, J. Cavanaugh, and R. A. Berman. 2011. Thalamic Pathways for Active Vision. *Trends Cogn. Sci.* **5**:177–184. [6]

Wynne, B. 2007. Taking European Knowledge Society Seriously. Report of the Expert Group on Science and Governance to the Science, Economy and Society Directorate. http://ec.europa.eu/research/science-society/document_library/pdf_06/european-knowledge-society_en.pdf (accessed Oct. 12, 2015). [19]

Wyss, R., P. König, and P. F. M. J. Verschure. 2006. A Model of the Ventral Visual System Based on Temporal Stability and Local Memory. *PLoS Biol.* **4**:e120. [4]

Yarbus, A. L. 1967. Eye Movements and Vision. New York: Plenum Press. [20]

Young, L., A. Bechara, D. Tranel, H. Damasio, M. Hauser, and A. Damasio, A. 2010. Damage to Ventromedial Prefrontal Cortex Impairs Judgment of Harmful Intent. *Neuron* **65**:845–851. [4]

Yu, C., and L. B. Smith. 2013. Joint Attention without Gaze Following: Human Infants and Their Parents Coordinate Visual Attention to Objects through Eye-Hand Coordination. *PLoS One* **8**:e79659. [10, 20]

Yurovsky, D., L. B. Smith, and C. Yu. 2013. Statistical Word Learning at Scale: The Baby's View Is Better. *Dev. Sci.* **166**:959–966. [4]

Zalla, T., N. Labruyère, A. Clément, and N. Georgieff. 2010. Predicting Ensuing Actions in Children and Adolescents with Autism Spectrum Disorders. *Exp. Brain Res.* **201**:809–819. [3]

Zeki, S., and S. Shipp. 1988. The Functional Logic of Cortical Connections. *Nature* **335**:311–317. [6]

Ziemke, T. 2003. What's That Thing Called Embodiment? In: Proc. 25th Annual Conference of the Cognitive Science Society, ed. R. Alterman and D. Kirsh, pp. 1134–1139. Mahwah, NJ: Lawrence Erlbaum. [18]

Ziman, J. 1978. Reliable Knowledge: An Exploration of the Grounds for Belief in Science. Cambridge: Cambridge Univ. Press. [19]

Zmyj, N., J. Jank, S. Schütz-Bosbach, and M. M. Daum. 2011. Detection of Visual-Tactile Contingency in the First Year after Birth. *Cognition* **120**:82–89. [3]

Subject Index

simulation (*continued*)
  internal simulation theory 311, 312
skill learning 24, 41, 44, 52, 53, 58, 75
social cognition 10, 28, 36, 38, 42–47,
    58, 90, 91, 190, 285, 288, 291, 333,
    340, 355
  in artificial agents 9
  spectator theory 10, 339
social mirroring 10, 288, 306, 307, 339
societal implications of a paradigm shift
    201, 321, 329–331, 352–356
soft robotics 343, 345
sparsity assumption 137, 165
spatial navigation 27, 68, 71
speech 43, 53, 143–145, 154–156, 168,
    169, 232, 279, 316
  child-directed 25, 276
  perception 6, 143–145, 244
structuralism 235, 237, 238
structural priors 7, 137
symbolic grounding 148, 237, 239
synaptic gain control 100, 107
synesthesia 71, 72
systems engineering 312, 313, 346–348,
    355

temporality 12, 56, 179, 200, 249, 285,
    286, 347
theory of event coding 162, 163
theory of mind 3, 9, 45, 63, 103, 104,
    241, 248, 288, 339, 349
thinking 3, 19, 22, 29, 90, 194, 215, 216,
    220–223, 230, 232, 318
  enactive 354
tool use 27, 70, 73, 175, 182, 183, 188,
    334, 352

top-down control 44, 107, 197, 208, 209,
    297, 299, 301, 302, 308
transient memory system 237, 254, 258
trust 208, 209, 276
Turing test 309, 310, 318–320
type-token binding 87

unconscious inference 98, 103, 104, 170,
    212
unconscious processing 245, 246, 247,
    254, 255, 265, 268
unobservability assumption 97, 102, 103

values 203, 204, 236, 247, 249, 268, 294
Vico, Giambattista 256, 326, 329, 331
virtualization memory 9, 253
virtual reality 11, 13, 172, 278, 328, 342
  brain-based 253
  clinical applications 236, 350
  self-generated 244
vision 71, 171, 176, 182, 184, 186, 194,
    249, 270, 275, 294, 314, 338, 345
  active 21, 170, 171, 176, 345
  computer 171, 312, 313, 316
visuomotor mapping 36, 38, 39
voluntary action 9, 161, 205, 246, 264,
    267, 272, 302, 304, 349

wakefulness 194, 263
word form circuits 142–147, 151
working memory 42, 43, 85, 86, 91, 144,
    162, 250, 270

Further Titles in the Strüngmann Forum Report Series[1]

Better Than Conscious? Decision Making, the Human Mind, and Implications For Institutions
edited by Christoph Engel and Wolf Singer, ISBN 978-0-262-19580-5

Clouds in the Perturbed Climate System: Their Relationship to Energy Balance, Atmospheric Dynamics, and Precipitation
edited by Jost Heintzenberg and Robert J. Charlson, ISBN 978-0-262-01287-4

Biological Foundations and Origin of Syntax
edited by Derek Bickerton and Eörs Szathmáry, ISBN 978-0-262-01356-7

Linkages of Sustainability
edited by Thomas E. Graedel and Ester van der Voet, ISBN 978-0-262-01358-1

Dynamic Coordination in the Brain: From Neurons to Mind
edited by Christoph von der Malsburg, William A. Phillips and Wolf Singer, ISBN 978-0-262-01471-7

Disease Eradication in the 21st Century: Implications for Global Health
edited by Stephen L. Cochi and Walter R. Dowdle, ISBN 978-0-262-01673-5

Animal Thinking: Contemporary Issues in Comparative Cognition
edited by Randolf Menzel and Julia Fischer, ISBN 978-0-262-01663-6

Cognitive Search: Evolution, Algorithms, and the Brain
edited by Peter M. Todd, Thomas T. Hills and Trevor W. Robbins, ISBN 978-0-262-01809-8

Evolution and the Mechanisms of Decision Making
edited by Peter Hammerstein and Jeffrey R. Stevens, ISBN 978-0-262-01808-1

Language, Music, and the Brain: A Mysterious Relationship
edited by Michael A. Arbib, ISBN 978-0-262-01962-0

Cultural Evolution: Society, Technology, Language, and Religion
edited by Peter J. Richerson and Morten H. Christiansen, ISBN 978-0-262-01975-0

Schizophrenia: Evolution and Synthesis
edited by Steven M. Silverstein, Bita Moghaddam and Til Wykes, ISBN 978-0-262-01962-0

Rethinking Global Land Use in an Urban Era
edited by Karen C. Seto and Anette Reenberg, ISBN 978-0-262-02690-1

Trace Metals and Infectious Diseases
edited by Jerome O. Nriagu and Eric P. Skaar, ISBN 978-0-262-02919-3

Translational Neuroscience: Toward New Therapies
edited by Karoly Nikolich and Steven E. Hyman, ISBN: 9780262029865

---

[1] available at https://mitpress.mit.edu/books/series/str%C3%BCngmann-forum-reports-0

Printed in the United States
by Baker & Taylor Publisher Services